ASTRONOMY AND
ASTROPHYSICS LIBRARY

M. Ollivier · T. Encrenaz · F. Roques · F. Selsis ·
F. Casoli

# Planetary Systems

Detection, Formation and Habitability
of Extrasolar Planets

 Springer

Marc Ollivier
Université Paris-Sud 11
Inst. Astrophysique Spatiale
Bâtiment 121
91405 Orsay CX
France
marc.ollivier@ias.u-psud.fr

Thérèse Encrenaz
Observatoire de Paris
Section de Meudon
5 place Jules Janssen
92195 Meudon CX
France
therese.encrenaz@obspm.fr

Francoise Roques
Observatoire de Paris
Section de Meudon
5 place Jules Janssen
92195 Meudon CX
France
francoise.roques@obspm.fr

Franck Selsis
Laboratoire d'Astrophysique de Bordeaux (LAB)
Université de Bordeaux–CNRS
BP 89
33271 Floirac CX
France
franck.selsis@obs.u-bordeaux1.fr

Fabienne Casoli
CNES
2 Place Maurice Quentin
75039 Paris CX01
France
fabienne.casoli@cnes.fr

*Cover image:* ESO PR Photo 14a/05. The first planet outside of our solar system to be imaged orbits a brown dwarf (centre-right) at a distance that is nearly twice as far as Neptune is from the sun. The photo is based on three near-infrared exposures (in the H, K and L' wavebands) with the NACO adaptive-optics facility at the 8.2-m VLT Yepun telescope at the ESO Paranal Observatory. © ESO. Credit: ESO and Gael Chauvin, Ben Zuckerman and Anne-Marie Lagrange.

ISBN: 978-3-540-75747-4        e-ISBN: 978-3-540-75748-1

Library of Congress Control Number: 2008933296

© Springer-Verlag Berlin Heidelberg 2009

*Cover design:* eStudio Calamar S.L.

Printed on acid-free paper

9 8 7 6 5 4 3 2 1

springer.com

# Preface

Are there inhabitable worlds elsewhere in the Universe? Or better, inhabited worlds? Or is the third planet of the solar system really special? More than two thousand years ago, some Greek philosophers were already speculating about the existence of Earth-like planets. For the atomists like Epicure, it was a matter of principle: there should be planets around every star. But despite these high expectations and several research programs, only nine planets orbiting a main sequence star were known at the beginning of the nineties: the nine planets of the solar system.

Since then, the situation has changed drastically. While the nine planets of the solar system are now only eight (Pluto having lost this appellation), more and more extrasolar planets are now being detected, observed and catalogued by the astronomers.

Indeed, extrasolar planet, or exoplanet, is the name given by astrophysicists to these new worlds—a name not yet in all dictionaries—as well as exoEarth, exoJupiter, exoUranus, depending on the mass of the exoplanet. In mid-2008, there were more than three hundreds of these extrasolar planets, many of them belonging to planetary systems with at least two planets orbiting the same star. The solar system is thus far from being the only one of its kind: a revolution that both astronomy and planetary science have awaited for a long time.

And, indeed, it is a revolution. Exoplanets are not at all what astronomers expected. Actually, they do not look like if they were twins, or even cousins of the Earth, Jupiter and solar system planets. Most of them are giant, likely gaseous planets. This is not completely surprising since our detection methods, at least in the first years of this story, were not sensitive enough to detect less massive planets. But what was a real shock was the discovery of dozens of such exoJupiters 50 or 100 times closer to their star than Jupiter is from the Sun—a really hot place to live. How were these hot Jupiters formed? Another mystery is the rather elliptical orbits of many extrasolar planets, while planets in our solar system are on rather circular orbits. In the end, could it be that our solar system has exceptional features? The discovery of extrasolar planets has then led researchers to reconsider all theories of planetary formation and evolution. For example, the role of a phenomenon called *migration* is now widely recognized: most planets do not stay in the region of the planetary system where they were formed. With the discovery of more and more "superEarths," with masses between several Earth masses and the Uranus mass, it also seems that the distinction between telluric and giant planets is fading away.

Exoplanets were a challenge not only for the theoreticians, but also for observers. These tiny dots are not yet directly observable except in very exceptional cases. It is an indirect method (velocimetry) that has yielded the majority of the discoveries, but astrometry, microlensing, transits, and pulsar timing are among the methods that observers have invented to circumvent this problem. The ultimate goal for the next 20 years, however, is to detect exoEarths—planets with a mass close to that of Earth, and located at the right distance from their star for life to be able to develop. Thanks to innovative methods, detecting exoEarths should be feasible in the next 10 years, while detecting signs of life (biosignatures) in their atmospheres is still an immense challenge, and the goal of extremely ambitious space projects.

This book attempts another challenge, which is to draw a picture as complete as possible of this field while it is still quickly evolving. The first chapters describe what is currently known of exoplanets, from a description of the detection methods and of the observed properties of the known objects to the dynamics of planetary systems and the structure and evolution of planets in general. It appears that the solar system planets are still the reference for all models. The last two chapters deal with current and future detection projects, and the final goal—the search for life on exoplanets.

One could hope that the field of exoplanet research has reached a mature state and the major results that one can get with present-day techniques are known. However, the "other worlds" are still capable of amazing us. In such a dynamic field, the foreword is the best place for the latest news. Indeed, three months after this book was completed, the space mission COROT discovered an enigmatic object between a star and a planet with a density twice that of platinum. The team using the HARPS spectrograph on the 3.6 m telescope at the European Southern Observatory announced that a system of three superEarths orbits the star HD40307. When will be the first announcement of the discovery of a true Earth twin? Two years, five years, ten years from now? Let us guess that exoplanets will surprise us again.

# Acknowledgements

We are grateful to J. Lequeux for his very helpful comments.

We also wish to thank Mr. Storm Dunlop for his collaboration and his excellent translation of the manuscript.

# Contents

# Chapter 1
# Planetary Systems

## 1.1 Introduction

The question of the existence of inhabited worlds outside the Solar System goes
back to antiquity. Even at the time of the Greeks, who, following the tradition of
Aristotle, placed the Earth at the centre of the Universe, voices were raised, suggest-
ing a heliocentric system. History has preserved the name of one celebrated pioneer,
Aristarchos of Samos. However, his works remained forgotten for more than a mil-
lennium, until the Copernican revolution at the end of the 15th century. 'An infinite
number of suns exist, an infinite number of earths turn around those suns just as the
seven planets turn around our sun': These visionary views were held, four centuries
ago, by Giordano Bruno. This intuitive view continued to grow among scientists,
along with the astronomical discoveries of recent centuries: with the discovery of
the nature of stars; of the place of the Sun in the Galaxy; and of the extragalactic
Universe. Because the Sun is just one unremarkable star among the 100,000 million
that populate our galaxy, why should it be the only star to have a train of planets?
And why should life be confined to our own Earth?

## 1.2 The Plurality of Worlds: A Question as Old as the Hills

### 1.2.1 From Antiquity to the Copernican Revolution

We can trace the concept of the existence of other worlds as far back as ancient
Greece. It is described by the philosopher Epicurus (341–270 BC), in particular,
in the following words, written to Herodotus: 'There are infinite worlds similar
to and different from our own.' In his work *De Caelo*, the philosopher Aristotle
(384–322 BC) also wondered about the existence of other worlds. Two centuries
later, in his work *De natura rerum*, Lucretius, the Roman philosopher, also sub-
scribed to the concept of the plurality of worlds.

In the Middle Ages, the question of the plurality of worlds was the subject of
debate, with a succession of confrontations between theologians in the 13th and 14th

M. Ollivier et al., *Planetary Systems*. Astronomy and Astrophysics Library,
DOI 978-3-540-75748-1_1, © Springer-Verlag Berlin Heidelberg 2009

centuries. According to Albert Magnus (1193–1280), 'the concept of the plurality of
worlds represents "one of the most marvellous and noble questions in Nature" '. His
disciple Thomas Aquinas (1224–1274) also favoured the concept of the existence of
other worlds. Jean Buridan (*c*. 1295–1358), Rector of Paris University, and William
of Okham (*c*. 1280–1347) also both maintained that other worlds could exist.

In 1543, the year of the death of its author, the publication of *De revolutionibus*,
by Nicholas Copernicus, signalled, with the advent of the heliocentric system, the
end of the Aristotelian system which located the Earth at the centre of the Universe.
A few decades later, Giordano Bruno (1548–1600) became its ardent champion, as
well as the passionate advocate for an infinite number of possible worlds, which he
described in certain of his works: *De l'infinito universo et mondi*, which appeared
in 1584, and then *De immenso e innumerabilibus*, published in 1591. In 1600, he
paid for his writings with his life, condemned as a heretic by the Inquisition. The
heliocentric view did, however, come to convince the world of scholars and philosophers (Fig. 1.1), thanks to the work of observers such as Tycho Brahe (1541–1601),
Johannes Kepler (1571–1630) and Galileo Galilei (1564–1642). With the discoveries of Galileo, published in 1610 in his famous work *Sidereus Nuncius*, on the
presence of mountains on the surface of the Moon and the existence of satellites

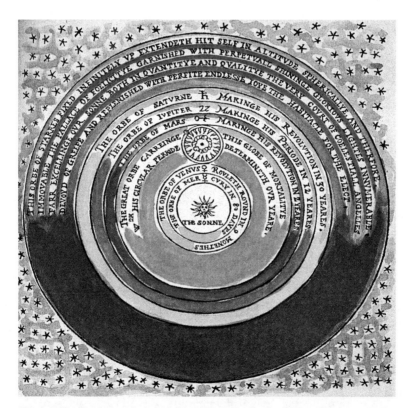

**Fig. 1.1** The *Perfit description of the caelestiall orbes* by Thomas Digges (1576). A supporter of
the Copernican system, the author described, beyond the orbit of Saturn 'an orb of the fixed stars
whose dimensions are infinite in altitude' (After Verdet, 2002)

of Jupiter, the question of the existence of other inhabited worlds arose yet again. Kepler, in the *Dissertatio*, wondered about the habitability of the Moon and Jupiter's satellites; but there were still many opponents, notably among the Church.

### 1.2.2  The First Theories on the Formation of the World

The first attempt to account for the formation of the Earth and the Universe came with René Descartes (1596–1650). In *Principia philosophiae*, which appeared in 1644, Descartes described the latter as a set of whirls, or vortices, the centres of which were occupied by the Sun and other stars. To Descartes, the stars played exactly the same role as the Sun. Descartes' concept therefore opened the door to the possibility of an infinite number of possible worlds.

In 1686, Bovier de Fontenelle (1657–1757), published his *Entretiens sur la pluralité des mondes* (Fig. 1.2). In this work, Fontenelle supports the idea of the habitability of the planets and the satellites in our Solar System, but he also puts forward the idea of an infinite number of planets. 'Our Sun has planets that he illuminates,

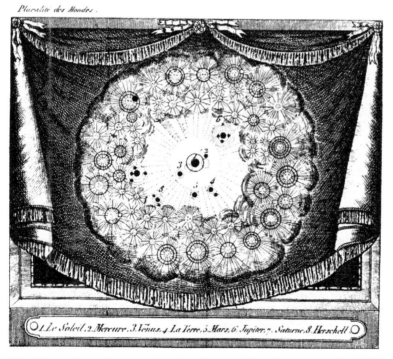

**Fig. 1.2** An illustration from the first edition of *Entretiens sur la pluralité des mondes*, by Fontenelle (1686), showing planets outside the Solar System, orbiting other fixed stars (After Dick, 1982)

why should each fixed Star not also have some that it illuminates?' A few years later, Christiaan Huygens (1629–1695), in his work *Cosmotheoros* published in 1698, took up Fontenelle's suggestions, but this time based them, not on the theory of vortices, but on the physical analogy between the Earth and the other planets in the Solar System.

The year 1687 saw the start of a new era. The publication by Isaac Newton (1642–1727) of the *Philosophiae naturalis principia mathematica* laid down the fundamental basis of the laws of universal gravitation, and provided an irrefutable demonstration of the Copernican theory. It opened the way for the nebular theory, which was proposed by Immanuel Kant (1724–1804, Fig. 1.3) in his *Allgemeine Naturgeschichte und Theorie des Himmels* [Universal Natural History and Theories of the Heavens], which appeared in 1755. In this work, Kant explains the origin of the Solar System by the collapse of a rotating cloud, which flattens into a disk, within which planets subsequently form. Subsequently finalized by the work of Pierre-Simon de Laplace (1749–1804, Fig. 1.4), which was published in 1796 in his *Exposition du système du monde*, this model is the basis for the view of planetary formation that is generally accepted today. Like Huygens, Kant saw the stars of the Milky Way as objects comparable with the Sun. He put forward the idea that these

**Fig. 1.3** Immanuel Kant (1724–1804)

**Fig. 1.4** Pierre-Simon de Laplace (1749–1827). Together with Immanuel Kant, he proposed the nebular theory, explaining the origin of the Solar System, which is generally accepted today

stars might be endowed with planets, and that these could be habitable. The idea was taken up by 19th-century astronomers, in particular by Camille Flammarion (1842–1925) in his *Astronomie Populaire*.

## 1.3 First Searches for Other Worlds

The idea of the probable existence of extrasolar planets (or 'exoplanets') was therefore definitely in the minds of astronomers, but until the end of the 20th century, they did not have the requisite observational methods to detect them. The sought-after exoplanets do, after all, give off very weak radiation, but this is drowned by that from their parent stars. Before the advent of large telescopes in the 8–10-m class, exoplanets could not be detected by direct imaging because of their intrinsically weak flux, and their angular proximity to their parent stars. For an observer located outside the Solar System, the ratio of the flux emitted by Jupiter and the Sun, in visible light, is of the order of $10^{-9}$. Note that in the mid-infrared ($10\,\mu$m) it is of the order of $10^{-6}$, which is more favourable, but still beyond the range of the instruments available at the end of the 20th century. As for the angular distance of Jupiter from the Sun, it is 0.5 as for an observer at 10 pc[1] from the Sun. The light from Jupiter is thus utterly negligible compared with the light emitted by the Sun.

### 1.3.1 The First Astrometric Searches

At the beginning of the 20th century, astronomers wondered about the presence of possible companions around nearby stars. Because it is accepted that the Sun is a perfectly ordinary star in the Galaxy, nothing ruled out the existence of numerous stellar systems, similar to our own. The whole problem, however, lay in detecting these systems. We have seen that direct detection was impossible. Another method had to be used, namely measurement of the periodic motion of the central star caused by the companion, relative to the centre of mass of the overall system.

In the middle of the 19th century, the German astronomer Friedrich Bessel (1784–1846) had become famous by detecting, using this indirect method, the presence of a low-mass companion orbiting Sirius. It was still not a planet; in 1930 Chandrasekhar would show that the mysterious object is a white dwarf, the final stage in the evolution of low-mass stars.

Bessel's discovery opened the way to the search for low-mass stellar companions by using the astrometric technique. During the course of the 20th century, with more powerful telescopes, and more sensitive detectors, several astronomers embarked on a search for exoplanets. Piet Van de Kamp (1901–1995, Fig. 1.5), from the astrometric curve of Barnard's Star, announced the discovery of one or more companions. Announced in 1944, this result was finally disproved by other investigations which showed the presence of systematic errors that were linked to the instrumentation used for Van de Kamp's measurements. He announced another discovery in 1974, this time orbiting the star Epsilon Eridani but, once again, this result was challenged by other measurements. At the beginning of the 1980s, the prevailing view within the astronomical community was that astrometric techniques were not sufficiently accurate to allow the detection of exoplanets around nearby stars, bearing in mind the performance of the instruments that were then available.

### 1.3.2 The Velocimetry Method

The second indirect method to detect exoplanets, known as velocimetry, consists of measuring, by means of the Doppler Effect, the periodic fluctuations in the velocity of a star that has a companion, relative to the centre of gravity of the system

---

[1] 1 parsec (pc) is the distance at which one AU (the average distance between the Sun and Earth, 149.6 million kilometres) subtends an angle of one second of arc. Because of the Earth's revolution around the Sun, a star that lies at a distance of 1 pc describes a small ellipse, whose semi-major axis is one second of arc (1 pc $= 2.05 \times 10^5$ AU $= 3.26$ light-years).

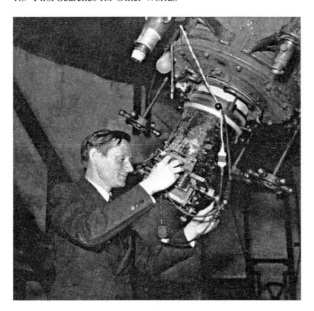

**Fig. 1.5** The astronomer Piet Van de Kamp (1901–1995) believed that he had discovered the first extrasolar planet orbiting Barnard's Star. The observations were later explained by a systematic error linked to the telescope that was used

(Fig. 1.6). To detect the equivalent of a Jupiter-sized planet orbiting the Sun, it is necessary to be able to measure velocity differences of around 10 m/s. To obtain such precision, astronomers construct spectrographs, operating in the visible spectral range with very high spectral resolution ($>10^5$ in resolving power), ca-

**Fig. 1.6** The principles on which the velocimetric method is based. (**a**) The star and its companion orbit their common centre of gravity of the system. (**b**) The Doppler Effect allows the observer to detect the motion of the star (provided that the orbit of the planet is not in a plane perpendicular to the line of sight) (After Casoli & Encrenaz, 2005)

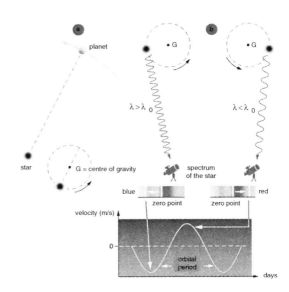

pable of measuring several hundred spectral lines simultaneously. Such a method started to be employed at the beginning of the 1980s, with the principal targets being solar-type stars, and was eventually crowned with success. In 1977, the Swiss astronomer Michel Mayor started a systematic search for stellar companions with the 1.93-m telescope at the Observatoire de Haute Provence (Fig. 1.7).

At the same time, other teams were carrying out similar programmes, notable ones being those of G. Marcy and P. Butler in the United States, and of G. Walker and B. Campbell in Canada. The first few years of searching failed to produce any results. Finally, in 1995, Michel Mayor and Didier Queloz announced the discovery of the first exoplanet orbiting a solar-type star, 51 Peg. With a mass that is at least half that of Jupiter, it orbits its star in just 4 days! (Fig. 1.8). The reason why 51

**Fig. 1.7** The 193-cm telescope at the Observatoire de Haute Provence. Equipped with the high-resolution Elodie spectrograph, it enabled the first exoplanet orbiting a solar-type star, 51 Peg, to be discovered in 1995

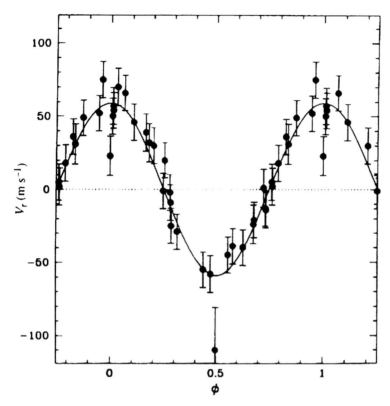

**Fig. 1.8** The velocity curve of the star 51 Pegasi, measured by the team led by M. Mayor at the Haute Provence Observatory (After Mayor & Queloz, 1995)

Peg b had not been discovered sooner it that the various teams carrying out these searches did not envisage the existence of giant exoplanets so close to their parent stars. This was the start of a long series of discoveries. In January 1996, G. Marcy announced the discovery of two new exoplanets, one orbiting 47 UMa, the other 70 Vir. Twelve years later, the number of exoplanets exceeds 250; most of which have been discovered by velocimetry.

### 1.3.3 The First Results and the Problems Raised

Given the instrumentation available at the beginning of the 1990s, the velocimetry method was only able to detect giant exoplanets, with masses comparable to that of Jupiter. Less massive exoplanets did not sufficiently perturb the motion of the star for that perturbation to be detectable. Astronomers therefore searched for 'exo-Jupiters' and, by analogy with the giant planets in the Solar System, for objects with orbital periods of a few years, or even more. This is why the discovery of 51 Peg b caused such surprise. This exoplanet, whose mass is about half that of Jupiter, lies at a distance of just 0.05 AU from its star and its orbital period is just 4 days! These surprising characteristics were to be found in many exoplanets, discovered in the months and years that followed (Fig. 1.9).

Astronomers thus found themselves faced with an unexpected situation. There were certainly exoplanets around solar-type stars, but for a considerable number of them, their properties did not resemble in any way those of the planets of the Solar System.

### 1.3.4 Planets Around Pulsars

51 Peg b was the first exoplanet to be discovered around a main-sequence star, i.e., a star in its hydrogen-burning phase (like the Sun). However, the first true exoplanet, which is orbiting a neutron star, was discovered in 1992 by a technique known as pulsar timing.

In 1967, systematic observation of radio sources led J. Bell and A. Hewish to discover a new form of stars at the end of their lifetimes: neutron stars. Endowed with an intense magnetic field and an extremely high rate of rotation, these objects periodically emit – every 1.3 s for the first to be discovered – the intense radio signals that led to their detection, whence the name pulsar (pulsating star, Fig. 1.10). In 1969, a similar object was discovered in the Crab Nebula. This established the link between pulsars and the explosions that mark the final stages of the evolution of the most massive stars.

The periodic signal emitted by pulsars has another advantage: if the pulsar is orbited by one or more companions, the periodic signal is perturbed by them, and its analysis enables us to work back to the characteristics of the one or more companions. Several tentative detections were announced in the 1970s, but were soon discounted: some perturbations in the emission curves of pulsars are caused by pulsations in the neutron stars. Ten years later, in 1985, the team led by A. Lyne announced the discovery of a planet orbiting the pulsar PSR 1829–10. But, once again, appearances were deceptive: the observed effect is an artefact related to the period of the Earth's revolution around the Sun.

Finally, in 1992, the first discovery of an exoplanet was announced. The Polish astronomer, Alexander Wolszczan discovered two planets orbiting the pulsar PSR 1257+12, which is notable for its extremely fast rotation (1.5 ms). A third planet, the size of the Moon, appears to complete the system (Fig. 1.11). After this first discovery, other millisecond pulsars with planets appear to have been detected. Although these pulsars do not have the slightest resemblance to the Sun, the exoplanets discovered around them are certainly the first planetary systems ever detected.

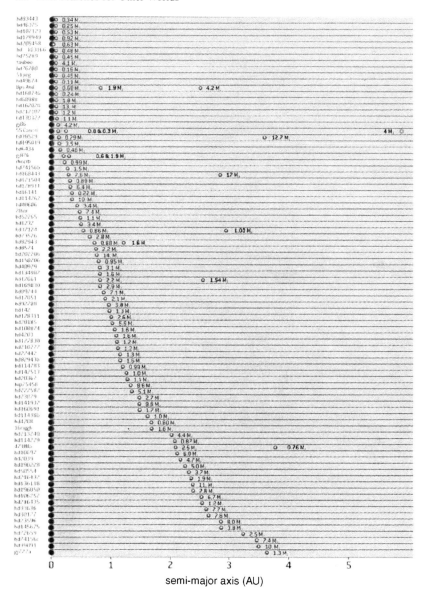

**Fig. 1.9** Exoplanets that have been observed (up to 2005), shown as a function of their semi-major axes (After Casoli & Encrenaz, 2005)

**Fig. 1.10** The radio emission from a pulsar is caused by the intense magnetic field that produces it, coupled with an extremely high rate of rotation. Observers periodically receive the radio emission radiated along the axis of the magnetic field if the latter, as it sweeps round, passes close to the direction of the Earth (After Mayor & Frei, 2001)

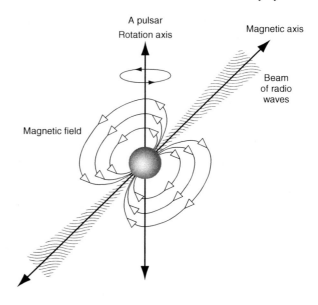

**Fig. 1.11** The planetary system around PSR 1257+32, compared with the inner Solar System. (The size of the bodies is not to scale.) (After M. Mayor & Frei, 2001)

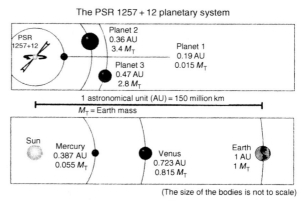

(The size of the bodies is not to scale)

### 1.3.5  The Search for Protoplanetary Disks

In parallel with the first discoveries of exoplanets, the study of circumstellar disks around young stars allowed us to gain a better understanding of the first stages in the life of a star, and the formation of the Solar System. According to the scenario generally accepted today, the Solar System formed from a disk, resulting from the collapse of a rapidly rotating nebula of dust and gas. Once the planets formed, the disk subsequently dispersed, probably during the phase of intense activity (known as the 'T-Tauri' phase) that the young Sun experienced at the very beginning of its lifetime. Such disks could exist around young stars neighbouring the Sun. Cooler than the star, they would emit radiation in the near infrared, distinct from that of the star, whose maximum is at shorter wavelengths. Their detection from Earth would therefore be far easier than that of possible exoplanets.

The discovery of protoplanetary disks has a rich history, which extends back to the 1970s, with, in particular, optical and infrared observations of T-Tauri stars (Herbig, 1977) and other young objects (Cohen and Kuhi, 1979), as well as theoretical work (Lynden-Bell and Pringle, 1974).

In 1983, the first disk surrounding a young star was actually discovered around Vega by the IRAS satellite. This Earth-orbiting satellite, designed to carry out a deep survey of the sky in the mid and far infrared, was able to measure the temperature and the dimensions of the disk. However, it was not, strictly speaking, a protoplanetary disk of the sort that existed in the Solar System, given Vega's spectral type, and its age (some 100 million years). IRAS discovered that about 20 per cent of stars of type A are surrounded by a disk.

A year later, the first image of another disk was obtained by coronagraph observations from the ground. This disk lies around the star ß Pictoris, a type-A star less than 100 million years old. These observations were subsequently repeated using the Hubble Space Telescope. The images showed a highly flattened disk, with a radius of several hundred Astronomical Units (AU). Spectroscopic observations in the UV and visible regions then revealed the presence of atomic absorption lines shifted by the Doppler effect towards the red, that were variable over time, which were attributed to episodes, in which comets, captured by the star, fell into it. ß Pictoris was therefore the first system that revealed certain analogies with the early history of the Solar System.

Since the discovery of ß Pictoris, numerous protoplanetary disks have been discovered, most notably by the Hubble Space Telescope. In many cases, observations undertaken in parallel from the ground, in the visible, infrared, and at millimetre wavelengths, have revealed bipolar flows, emitted symmetrically, perpendicular to the plane of the disk. This is the case with what are known as Herbig-Haro objects (observable in the visible) and YSO (young stellar objects), which are hidden within a cocoon of dust and are detectable only in the infrared. Such objects are found, in particular, in the Orion Nebula. The typical age of these objects, deduced from observations, is no more than a few million years, which corresponds to the phase of planetary formation in the Solar System. In the case of older stars (such as Vega and ß Pictoris), the circumstellar disks are called debris disks, and may consist of the remnants following a phase of planetary formation. The discovery of a gap in the disk around ß Pic may be the signature of a planet having accreted the material missing from the gap (see Chap. 6).

## 1.4 The Solar System: A Typical Planetary System?

Let us see why the discovery of planets as massive as Jupiter orbiting very close to their stars was such a surprise to astronomers, by studying the case of the Solar System.

### 1.4.1 The Sun as an Average Star

The Sun's characteristics make it an 'average' star. It was born 4.6 billion years ago from a cloud that was rich in gas (primarily hydrogen) and dust, which collapsed into a disk. The star is now in the main-sequence phase (several thousand million years) during which it transforms the hydrogen that it contains into helium (Fig. 1.12). In a second phase, the helium is itself transformed into carbon, nitrogen, and oxygen. A star with the mass of the Sun is destined to become, in about 5 billion years, a red giant, and its outer envelope will finally be ejected to form a planetary nebula, while the core of the star will collapse to form a white dwarf. Stars that are far more massive than the Sun undergo a different evolutionary path: they become red supergiants, and then explode as supernovae. During this final phase, the heavy elements (magnesium, silicon, metals, etc.), which exist in the Universe and particularly on Earth, are formed from C, N, and O.

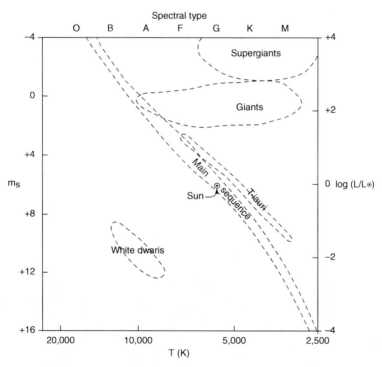

**Fig. 1.12** The Hertzsprung-Russell diagram shows the relation between the temperature and the luminosity of stars. Most of the stars are located along the Main Sequence, where they convert their hydrogen into helium. Later they evolve into giants and white dwarfs, or into supergiants and supernovae, depending upon their initial mass

## 1.4.2 Brown Dwarfs: Between Stars and Planets

What is the boundary between stars and planets? A star is a body in which thermonuclear reactions are taking place, the primary one being the conversion of hydrogen into helium. Calculations of stellar internal structure show that a mass of 74 Jupiter masses (74 $M_J$) is required to initiate the process.

A planet, in contrast, is an object orbiting the Sun (or a star), which does not emit any visible radiation in its own right; but only reflects the light from the star that it accompanies. Historically, the term 'planet' described the nine largest bodies in the Solar System, recognized in the 20th century as orbiting the Sun. This definition excluded the asteroids (also known as 'minor planets') because of their smaller size, as well as the trans-Neptunian bodies, other than Pluto, recently discovered in the Kuiper Belt. Since 2006, Pluto is officially no longer a planet; it is classed as a 'dwarf planet' and is a particularly massive trans-Neptunian object (TNO).

So a planet has insufficient mass to start thermonuclear reactions in its core. Nevertheless, there are two possible scenarios. If the mass of the object is less than 13 $M_J$, no thermonuclear reaction is possible. For a mass lying between 13 and 74 $M_J$, the internal temperature is sufficient to allow the onset of the first cycle of reactions, which involve deuterium only, and which last about 10 million years. The object is then known as a brown dwarf. During its short active phase, it appears like a cool star; subsequently, it becomes as difficult to detect as a planet. Like stars, brown dwarfs may, in theory, form as the result of gravitational collapse of a molecular cloud, either as an isolated object or as companion to a star.

## 1.4.3 A Specific Planetary System: The Solar System

The model for star formation beginning with the collapse of an interstellar cloud became established initially through our knowledge of the Solar System itself. This has one primary characteristic: all the planets orbit the Sun in the same direction (direct revolution, as seen from the Sun's north pole) on almost circular orbits, and all very close to the plane of the Earth's orbit (Table 1.1). The latter (the ecliptic) is chosen as the reference plane (Fig. 1.13). Most of the planets also rotate in the same direction, with the axis of rotation almost perpendicular to the ecliptic. This property strongly supports formation from a protoplanetary disk, as was suggested by Kant and Laplace in the 18th century.

## 1.4.4 The Formation of the Planets by Nucleation

According to the standard model of formation that is generally accepted nowadays, the Sun and the Solar System formed $4.55 \pm 0.10 \times 10^9$ years ago (Fig. 1.14). We have information about the ages of the planetary material from measurements of

**Table 1.1** Orbital properties of the planets in the solar system (After Encrenaz et al., 2004)

| Name | Semi-major axis (AU) | Eccentricity | Inclination/ ecliptic | Sidereal period (years) |
|------|----------------------|--------------|-----------------------|-------------------------|
| Mercury | 0.38710 | 0.205631 | 7.0048 | 0.2408 |
| Venus | 0.72333 | 0.006773 | 3.39947 | 0.6152 |
| Earth | 1.00000 | 0.01671 | 0.0000 | 1.0000 |
| Mars | 1.52366 | 0.093412 | 1.8506 | 1.8807 |
| Jupiter | 5.20336 | 0.048393 | 1.3053 | 11.856 |
| Saturn | 9.53707 | 0.054151 | 2.4845 | 29.424 |
| Uranus | 19.1913 | 0.047168 | 0.7699 | 83.747 |
| Neptune | 30.0690 | 0.008586 | 1.7692 | 163.723 |

the abundance of radioactive atoms with very long half-lives, which have been applied to terrestrial rocks, lunar samples, and meteorites. A fragment of an interstellar cloud, which was turbulent and unstable, and in rapid rotation, collapsed into a disk. This disk consisted primarily of gas (most of which was hydrogen and helium), as well as dust particles consisting of heavier elements.

The material in the centre, under the influence of its own gravity, collected together to form a star, the future Sun. At greater distances from the centre, the solid particles clumped together, at first under the influence of local instabilities, and then as the result of multiple collisions. The phenomenon subsequently tended to accelerate. The planetoids that arose attracted the surrounding particles through the effects of their gravity. Computer simulations of the dynamical evolution of such a system of particles show that these multiple collisions act to cause the orbits of the fragments that are thus formed to become circular and co-planar.

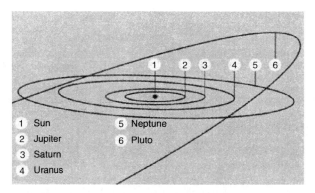

**Fig. 1.13** The orbits of the giant planets and Pluto. Like those of the terrestrial planets, they are very close to the plane of the ecliptic (defined by Earth's orbit), whereas Pluto's orbit lies away from it (After Encrenaz, 2003)

**Fig. 1.14** A schematic representation of the formation of the Solar System. (**a–b**) Collapse into a disk; (**c–d**) formation of planetoids; (**e–f**) accretion by collisions; (**g**) growth by the gravitational attraction of the larger bodies; (**h**) because of the effects of multiple collisions, the orbits slowly become circular and co-planar (After Encrenaz, 2003)

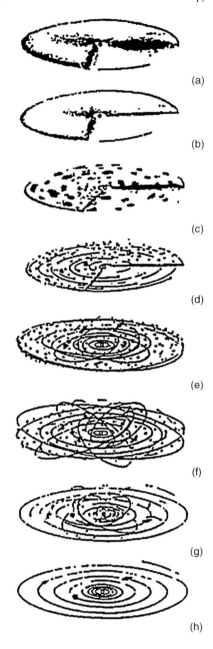

(a)

(b)

(c)

(d)

(e)

(f)

(g)

(h)

## 1.4.5 Terrestrial and Giant Planets

From a physical point of view, the planets in the Solar System exhibit one particular characteristic. They may be classified into two distinct categories (see Table 1.2). At a heliocentric distance $R_h$ less than 2 AU, the four terrestrial planets (Mercury, Venus, Earth, and Mars) are relatively small and dense; with the exception of Mercury, which lacks one, they have atmospheres whose masses are negligible relative to their overall mass; and the number of satellites varies between 0 and 2. Beyond 5 AU, the four giant planets (Jupiter, Saturn, Uranus, and Neptune) have large volumes and are low in density. Their atmospheres are dense, and they do not have a 'surface' as we understand it; they all have systems of rings and numerous satellites.

How may the classification of the planets into two such distinct classes be explained? According to the standard model described below, this is purely the natural consequences of the method by which the planets were formed.

Actually, the chemical composition of the protoplanetary disk should reflect (to a first approximation) the cosmic abundances. Hydrogen is the most abundant element, then helium, followed by oxygen and nitrogen, and then the heavier elements (Table 1.3). Close to the Sun ($R_h < 2$ AU), the temperature is sufficiently high for the elements C, N, and O, which are bound to hydrogen in the form of $CH_4$, $NH_3$, $H_2O$, $CO_2$, etc., to be in gaseous form. The only solid materials available to form planetary nuclei were the refractory materials, in particular silicates and metals. Because these elements are far less abundant than the lighter elements (see Table 1.3), the mass available to form planetary nuclei was limited. In contrast, beyond 4 AU, the temperature was sufficiently low for the elements C, N, and O to be in solid

Table 1.2 Physical characteristics of the planets in the solar system (after Encrenaz et al., 2004). The obliquity is the angle between the rotation axis of the planet and the perpendicular to the ecliptic (where the North Pole of the ecliptic is $0°$)

| Name | Diameter (Earth = 1) | Mass (Earth = 1) | Density (g/cm$^3$) | Surface gravity (m/s$^2$) | Sidereal rotation | Obliquity | Atmosph. Comp. |
|---|---|---|---|---|---|---|---|
| Mercury | 0.382 | 0.055 | 5.44 | 3.78 | 58.646 d | 0 | – |
| Venus | 0.949 | 0.815 | 5.25 | 8.60 | 243 d (retr.) | $2°07'$ | $CO_2$ (97%) |
| Earth | 1 | 1 | 5.52 | 9.78 | 23 h 56 min 04 s | $23°26'$ | $N_2$ (78%) $O_2$ (21%) |
| Mars | 0.533 | 0.107 | 3.94 | 3.72 | 24 h 37 min 23 s | $23°59'$ | $CO_2$ (95%) |
| Jupiter | 11.19 | 317.80 | 1.24 | 24.8 | 9 h 50 min to 9 h 56 min | $3°08'$ | $H_2$, He, $CH_4$, $NH_3$ |
| Saturn | 9.41 | 95.1 | 0.63 | 10.5 | 10 h 14 min to 10 h 39 min | $26°44'$ | $H_2$, He, $CH_4$, $NH_3$ |
| Uranus | 3.98 | 14.6 | 1.21 | 8.5 | 17 h 06 min | $98°$ | $H_2$, He, $CH_4$ |
| Neptune | 3.81 | 17.2 | 1.67 | 10.8 | 15 h 48 min | $29°$ | $H_2$, He, $CH_4$ |

**Table 1.3** Cosmic abundances of elements in the solar photosphere today, and in meteorites (C1 chondrites). Indirect estimates are shown within square brackets (After Grevesse et al., 2005)

| Element | Photosphere | Meteorites | Element | Photosphere | Meteorites |
|---|---|---|---|---|---|
| 1 H | 12.00 | $8.25 \pm 0.05$ | 44 Ru | $1.84 \pm 0.07$ | $1.77 \pm 0.08$ |
| 2 He | $[10.93 \pm 0.01]$ | 1.29 | 45 Rh | $1.12 \pm 0.12$ | $1.07 \pm 0.02$ |
| 3 Li | $1.05 \pm 0.10$ | $3.25 \pm 0.06$ | 46 Pd | $1.69 \pm 0.04$ | $1.67 \pm 0.02$ |
| 4 Be | $1.38 \pm 0.09$ | $1.38 \pm 0.08$ | 47 Ag | $0.94 \pm 0.25$ | $1.20 \pm 0.06$ |
| 5 B | $2.70 \pm 0.20$ | $2.75 \pm 0.04$ | 48 Cd | $1.77 \pm 0.11$ | $1.71 \pm 0.03$ |
| 6 C | $8.39 \pm 0.05$ | $7.40 \pm 0.06$ | 49 In | $1.60 \pm 0.20$ | $0.80 \pm 0.03$ |
| 7 N | $7.78 \pm 0.06$ | $6.25 \pm 0.07$ | 50 Sn | $2.00 \pm 0.30$ | $2.08 \pm 0.04$ |
| 8 O | $8.66 \pm 0.05$ | $8.39 \pm 0.02$ | 51 Sb | $1.00 \pm 0.30$ | $1.03 \pm 0.07$ |
| 9 F | $4.56 \pm 0.30$ | $4.43 \pm 0.06$ | 52 Te | $2.19 \pm 0.04$ | |
| 10 Ne | $[7.84 \pm 0.06]$ | $-1.06$ | 53 I | $1.51 \pm 0.12$ | |
| 11 Na | $6.17 \pm 0.04$ | $6.27 \pm 0.03$ | 54 Xe | $[2.27 \pm 0.02]$ | $-1.97$ |
| 12 Mg | $7.53 \pm 0.09$ | $7.53 \pm 0.03$ | 55 Cs | $1.07 \pm 0.03$ | |
| 13 Al | $6.37 \pm 0.06$ | $6.43 \pm 0.02$ | 56 Ba | $2.17 \pm 0.07$ | $2.16 \pm 0.03$ |
| 14 Si | $7.51 \pm 0.04$ | $7.51 \pm 0.02$ | 57 La | $1.13 \pm 0.05$ | $1.15 \pm 0.06$ |
| 15 P | $5.36 \pm 0.04$ | $5.40 \pm 0.04$ | 58 Ce | $1.70 \pm 0.10$ | $1.58 \pm 0.02$ |
| 16 S | $7.14 \pm 0.05$ | $7.16 \pm 0.04$ | 59 Pr | $0.58 \pm 0.10$ | $0.75 \pm 0.03$ |
| 17 Cl | $5.50 \pm 0.30$ | $5.23 \pm 0.06$ | 60 Nd | $1.45 \pm 0.05$ | $1.43 \pm 0.03$ |
| 18 Ar | $[6.18 \pm 0.08]$ | $-0.45$ | 62 Sm | $1.01 \pm 0.06$ | $0.92 \pm 0.04$ |
| 19 K | $5.08 \pm 0.07$ | $5.06 \pm 0.05$ | 63 Eu | $0.52 \pm 0.06$ | $0.49 \pm 0.04$ |
| 20 Ca | $6.31 \pm 0.04$ | $6.29 \pm 0.03$ | 64 Gd | $1.12 \pm 0.04$ | $1.03 \pm 0.02$ |
| 21 Sc | $3.17 \pm 0.10$ | $3.04 \pm 0.04$ | 65 Tb | $0.28 \pm 0.30$ | $0.28 \pm 0.03$ |
| 22 Ti | $4.90 \pm 0.06$ | $4.89 \pm 0.03$ | 66 Dy | $1.14 \pm 0.08$ | $1.10 \pm 0.04$ |
| 23 V | $4.00 \pm 0.02$ | $3.97 \pm 0.03$ | 67 Ho | $0.51 \pm 0.10$ | $0.46 \pm 0.02$ |
| 24 Cr | $5.64 \pm 0.10$ | $5.63 \pm 0.05$ | 68 Er | $0.93 \pm 0.06$ | $0.92 \pm 0.03$ |
| 25 Mn | $5.39 \pm 0.03$ | $5.47 \pm 0.03$ | 69 Tm | $0.00 \pm 0.15$ | $0.08 \pm 0.06$ |
| 26 Fe | $7.45 \pm 0.05$ | $7.45 \pm 0.03$ | 70 Yb | $1.08 \pm 0.15$ | $0.91 \pm 0.03$ |
| 27 Co | $4.92 \pm 0.08$ | $4.86 \pm 0.03$ | 71 Lu | $0.06 \pm 0.10$ | $0.06 \pm 0.06$ |
| 28 Ni | $6.23 \pm 0.04$ | $6.19 \pm 0.03$ | 72 Hf | $0.88 \pm 0.08$ | $0.74 \pm 0.04$ |
| 29 Cu | $4.21 \pm 0.04$ | $4.23 \pm 0.06$ | 73 Ta | $-0.17 \pm 0.03$ | |
| 30 Zn | $4.60 \pm 0.03$ | $4.61 \pm 0.04$ | 74 W | $1.11 \pm 0.15$ | $0.62 \pm 0.03$ |
| 31 Ga | $2.88 \pm 0.10$ | $3.07 \pm 0.06$ | 75 Re | $0.23 \pm 0.04$ | |
| 32 Ge | $3.58 \pm 0.05$ | $3.59 \pm 0.05$ | 76 Os | $1.45 \pm 0.10$ | $1.34 \pm 0.03$ |
| 33 As | $2.29 \pm 0.05$ | | 77 Ir | $1.38 \pm 0.05$ | $1.32 \pm 0.03$ |
| 34 Se | $3.33 \pm 0.04$ | | 78 Pt | $1.64 \pm 0.03$ | |
| 35 Br | $2.56 \pm 0.09$ | | 79 Au | $1.01 \pm 0.15$ | $0.80 \pm 0.06$ |
| 36 Kr | $[3.28 \pm 0.08]$ | $-2.27$ | 80 Hg | $1.13 \pm 0.18$ | |
| 37 Rb | $2.60 \pm 0.15$ | $2.33 \pm 0.06$ | 81 Tl | $0.90 \pm 0.20$ | $0.78 \pm 0.04$ |
| 38 Sr | $2.92 \pm 0.05$ | $2.88 \pm 0.04$ | 82 Pb | $2.00 \pm 0.06$ | $2.02 \pm 0.04$ |
| 39 Y | $2.21 \pm 0.02$ | $2.17 \pm 0.04$ | 83 Bi | $0.65 \pm 0.03$ | |
| 40 Zr | $2.59 \pm 0.04$ | $2.57 \pm 0.02$ | 90 Th | $0.06 \pm 0.04$ | |
| 41 Nb | $1.42 \pm 0.06$ | $1.39 \pm 0.03$ | 92 U | $< -0.47$ | $-0.52 \pm 0.04$ |

form, as ices of $H_2O$, $CH_4$, $NH_3$, etc. The large amount of material available to form planetesimals allowed the formation of nuclei that were far larger than the terrestrial planets. Planetary formation models predict that when the mass of a nucleus reaches some ten Earth masses, the gravitational field becomes sufficient to cause the surrounding protosolar material to collapse into a disk, one which largely consists of hydrogen and helium. Within this disk, which lies in the planet's equatorial plane, a series of satellites will be formed in its turn. The planets that have formed in this manner are very large and low in density because of the contribution from protosolar gas: they are the giant planets. The heliocentric distance beyond which the gases become frozen is known as the ice line. At the time the planets were formed, it was probably located around 4 AU.

In summary, the current scenario of Solar-System formation is able to account for the overall properties of the Solar System. In this model, based on accretion around solid particles, the small, dense planets are expected to form close to the Sun, whereas the giant planets form at greater distances.

The challenge is therefore the following: to create a model for star formation that would take account of the properties of the newly discovered planetary systems, but also to ensure that the Solar System could be incorporated as a specific case within the overall model.

# Bibliography

Casoli, F. and Encrenaz, T., *Planètes extrasolaires, les nouveaux mondes,* Belin, Paris (2005)

Cohen, M. and Kuhi, L.V., 'Observational studies of star formation – conclusions', *Astrophys. J.,* **227**, L105–L106 (1979)

Dick, S.J., *Plurality of Worlds: The Origins of the Extraterrestrial Life Debate from Democritus to Kant,* Cambridge University Press, Cambridge (1982)

Encrenaz, T., *Le système solaire,* Flammarion, Paris (2003)

Encrenaz, T., Bibring, J.-P. and Blanc, M., et al., *The Solar System,* 3rd edn, Springer-Verlag, Heidelberg (2004)

Grevesse, N., Asplund, M. and Sauval, A.J., 'The new chemical composition', in *Element Stratification in Stars: 40 Years of Atomic Diffusion,* (eds.) Alecian, G., Richard, O. and Vauclair, S., EAS Publication Series, **17**, 21–30 (2005)

Herbig, G.H., 'Radial velocities and spectral types of T-Tauri stars', *Astrophys. J.,* **214**, 747–758 (1977)

Lynden-Bell, D. and Pringle, J.E., 'The evolution of viscous disks and the origin of the nebular variables', *Mon. Not. R. Astron. Soc.,* **168**, 603–637 (1974)

Mayor, M. and Frei, P.-Y., *Les nouveaux mondes du cosmos,* Editions du Seuil, Paris (2001)

Mayor, M. and Queloz, D., 'A Jupiter-mass companion to a solar-type star', *Nature,* **378**, 355–359 (1995)

Smith, B.A. and Terrile, R.J., 'A circumstellar disk around Beta Pictoris', *Science,* **226**, 1421–1424 (1984)

Taylor, S.R., *Solar System Evolution,* Cambridge University Press, Cambridge (1992)

Verdet, J.-P., *Voir et rêver le monde,* Larousse, Paris (2002)

# Chapter 2
# Detection Methods

Surprising though it may seem, practically all of the exoplanets that have been currently detected have never been 'seen' directly, in the sense that no images obtained
with a telescope exist for these objects. The reason is very simple, and lies in the
extreme difficulty of detecting these objects. This also explains why it was necessary to wait for the beginning of the 1990s for the first systems to be discovered,
when far more exotic astrophysical objects (quasars, pulsars, etc.) had been imaged
for several decades. In this chapter, we discuss the different techniques employed to
detect and determine the properties of exoplanets and their environment.

## 2.1 The Extent of the Problem

The direct observation of an exoplanet – in the sense of being able to separate physically the photons from the planet from those of the central star, sufficient to obtain
an image of the two objects – is a problem that is as simple to state as it is difficult
to achieve in practice. It may be largely summarized by three critical points which
strongly influence the detection methods that may be envisaged:

- the contrast between the star and the planet,
- the angular distance between the two objects,
- the environment of the Earth and of the exoplanet (and more generally, of the
  exosystem as a whole).

### 2.1.1 Contrast Between Star and Planet

The contrast between an exoplanet and its parent star depends on several factors:

   **(a) the spectral region in which observations are made:** The spectrum of an
exoplanet primarily consists of two components (see Chap. 7): one component arising from the reflection of the star's light by the planet, and the other being its own
emission component. In the case of a planet like the Earth, orbiting a solar-type

M. Ollivier et al., *Planetary Systems*. Astronomy and Astrophysics Library,
DOI 978-3-540-75748-1_2, © Springer-Verlag Berlin Heidelberg 2009

**Fig. 2.1** Comparative spectra
of the Sun and the Earth as
they would be seen from a
distance of 10 parsecs. The
Earth's spectrum clearly
shows the component
consisting of reflected
sunlight (0.1–4 μm) and the
component linked to the
planet's own emission
(beyond 4 μm)

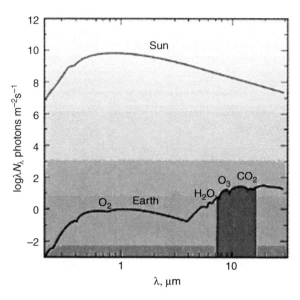

star (Fig. 2.1), the reflected component is dominant in the visible and near-infrared regions, while the planet's own emission component dominates in the thermal infrared (beyond 4 μm).

   **(b) the nature of the stellar and planetary objects:** The two spectral components just mentioned obviously depend on the nature of the various objects forming the planetary system. The spectral distribution and the amplitude of the reflected component depend on the star's effective temperature (and thus on the type of star), but also on the albedo, the size, and the distance of the planet from its star. The planetary emission itself depends exclusively on the nature of the planet – whether terrestrial or giant – and primarily on its effective temperature, but also on the atmospheric composition, which is indicated by the presence of absorption lines in the emission spectrum (Fig. 2.1).

   **(c) the distance between the objects:** The distance between the star and the planet is one of the essential factors involved in the calculation of the radiative equilibrium temperature of the planet. For example, an object of the size of Jupiter has an effective temperature of about 1500 K at 0.05 AU (i.e., as is the case with a hot Jupiter), whereas it is only about 110 K at 5.2 AU (Jupiter's location in the Solar System). The contrast is directly affected: at a wavelength of 10 μm it is about $10^4$ for a hot Jupiter, but it rises to $10^9$ for a true Jupiter-like planet.

## 2.1.2 Angular Separation Between the Objects

Because of the significant distance between the Sun and the nearest stars – the closest stars are several parsecs from us – exoplanetary systems subtend very small angular distances. A Sun-Earth system, for example, at a distance of 10 parsecs would subtend an angle of one tenth of a second of arc. (One second of arc, abbreviated 'arcsec', is 1/3600th of a degree, one milliarcsec, abbreviated 'mas', is 1/1000th of an arcsec, one microsecond of arc abbreviated '$\mu$as', is 1/1000th of a mas.) This tiny angular separation means that methods with high angular resolution must be used to carry out direct observation (see Sect. 2.3). An illustration of these problems of contrast and angular separation may be framed as follows: Trying to observe a planet like the Earth orbiting a star like the Sun, the whole system being at a distance of 10 parsecs from us, resembles trying to observe a glow-worm 30 cm from a lighthouse in Marseille, when the observer is in Paris (at a distance of about 700 km).

## 2.1.3 Environment of the Earth and Exoplanets

Apart from its suite of planets, the Sun is surrounded by a disk of dust resulting from collisions between asteroids as well as cometary dust. This disk of dust lies in the plane of the ecliptic. The dust, both illuminated and heated by the Sun, has its own emission, which (in the visible region) is known as the zodiacal light. An observer in the Solar System who is examining the sky in the infrared therefore records a signal from this zodiacal emission. This emission is not negligible because, at a wavelength of 10 $\mu$m, the spatially integrated emission from the whole disk of dust is 300 times the Earth's emission.

In the same manner, it is not unreasonable to assume that exoplanetary systems also contain a disk of debris (see Chap. 5), whose emission is certainly not negligible, and may even be greater than that of the Solar System. The presence of this disk therefore is a source of a parasitic signal in any radiation that is detected. To illustrate this last point, we may return to our earlier analogy by imagining that this time our glow-worm and our observer are both bathed in the light from Marseille and Paris, which restricts the visibility of faint objects. If you need to be persuaded of this, all you need do is compare the number of stars visible from the centre of Paris with those visible on a clear night in the wide-open country.

**Table 2.1** Properties of certain typical objects which are assumed to be orbiting a solar-type star, lying at a distance of 10 parsecs

| Object | Radius ($R_{\oplus}$) | Mass ($M_{\oplus}$) | Dist. (AU) | Angular separation (mas) | Contrast (Vis) | Contrast (IR) |
|---|---|---|---|---|---|---|
| Jupiter | 11.18 | 317.83 | 5.2 | 520 | $5 \times 10^8$ | $5 \times 10^7$ |
| Saturn | 9.42 | 95.15 | 9.54 | 954 | $2 \times 10^9$ | $5 \times 10^8$ |
| Neptune | 3.94 | 17.23 | 30.06 | 3006 | $10^{11}$ | $10^{10}$ |
| Earth | 1 | 1 | 1 | 100 | $5 \times 10^9$ | $7 \times 10^6$ |
| Hot Jupiter | 10 | 300 | 0.05 | 5 | $4 \times 10^4$ | $10^3$ |

Table 2.1 summarizes the main factors for different planetary objects, assuming that they are in orbit around a star like the Sun, lying at a distance of 10 parsecs from Earth.

## 2.2 The Indirect Detection of Exoplanets

We have seen that the direct detection of an exoplanet orbiting a star, even one close to the Solar System, is extremely difficult. It is not, however, absolutely essential to be able to obtain an image of the planet to reveal its presence. In certain cases it is possible to detect the planet by observing the effect it has on its parent star, thus by observing the star itself. The techniques that are based on this principle are known as 'indirect' methods, in contrast to direct imaging. In this section we discuss the principal indirect methods of detection. Here, we should explain that these methods are by no means theoretical, but that, with a few exceptions, all the exoplanets currently detected have been found by indirect methods.

### 2.2.1 The Effect of a Planet on the Motion of Its Star

It is commonly said that a planet orbits a star. This assertion is true if one neglects the mass of the planet relative to that of the star, or if the motion of the planet is considered in very rough terms. In reality, the star and its planet (or planets) are bound by gravitation, and each of the bodies in the system (star and planet or planets) has a motion about the centre of mass (the centre of gravity) of the system. The position of the centre of mass $G$, in a system with N bodies, each of mass $m_i$, the centre of which is at $O_i$, is defined by the following vector relationship:

$$\sum_i m_i . \overrightarrow{GO_i} = \overrightarrow{0} \tag{2.1}$$

In particular, for a two-body system (the star being denoted by '*' and the single planet by 'p'), the centre of mass lies between the star and the planet, and the vectorial expression just given becomes an algebraic one:

$$\overline{GO_p} = \frac{m_*}{m_* + m_p} \overline{O_* O_p} \tag{2.2}$$

where $\overline{GO_p}$ denotes the distance between $G$ and $O_p$ and $\overline{O_* O_p}$ the distance between $O_*$ and $O_p$.

Each of the bodies thus follows an elliptical orbit, with the centre of mass of the system at one of the foci. If $a$ is the semi-major axis of the planet's orbit around the star, the semi-major axes of the stellar and planetary orbits relative to the barycentre of the system at one of the foci may be written:

$$a_* = \frac{m_p}{m_* + m_p} a \ and \ a_p = \frac{m_*}{m_* + m_p} a \qquad (2.3)$$

In the general case, the 'central' star also describes an orbit that is more or less complex depending on the number of planets in the system, certain properties of which may be described by observing:

- stellar motion projected on the plane of the sky. Here, we observe a variation in the position of the star relative to a fixed reference frame (consisting of very distant bodies, such as quasars). This method is known as 'astrometry'.
- stellar motion along the line of sight (radial motion). Here, we measure the star's velocity of approach or recession as a result of the motion around the centre of mass. This measurement method is known as 'radial velocimetry' or 'Doppler velocimetry' (after the experimental method used to measure the radial velocity).

In certain instances, when the star emits a periodic signal (as with a pulsar, for example), the motion of the source may be deduced from the changes in the pulsar's period as it is measured here on Earth. We shall return to this technique at the end of this section.

### 2.2.1.1 Astrometry

As the planet or planets move around the centre of mass, the central star also describes an orbit, the complexity of which depends on the number of planets in the system. In the case of our Solar System, the motion of the Sun is primarily caused by the presence of Jupiter, but also reveals the presence of other, less-massive planets (Fig. 2.2).

In the case of a system with a single planet (or where one of the objects dominates the others by a wide margin to such an extent that its effect masks those of all the other planets), the motion of the star is an ellipse. We shall be particularly concerned with such orbits, the general equation for which may be expressed, using polar coordinates having their origin at the focus, and in the orbital plane, by:

$$r(v) = \frac{a(1 - e^2)}{1 + e.\cos(v)} \qquad (2.4)$$

where $a$ is the semi-major axis, $e$ the eccentricity, and $v$ the angular position of the object relative to an origin given by the orbit's periastron.

The motion, projected onto the plane of the sky is deduced from the preceding expression by a change of reference frame that takes into account the orientation of the orbital plane relative to that of the sky, and also the conventions regarding the notation of the associated angles.

To describe the orientation of the orbit relative to the plane of the sky, it is normal to use the astronomical equivalent of Euler angles. The following are thus defined successively:

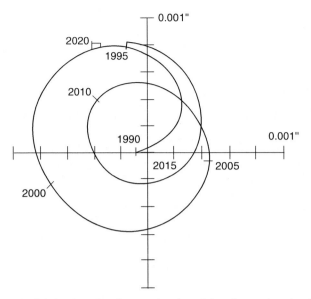

**Fig. 2.2** Motion of the Sun as a function of time (in years) on the plane of the sky as it would appear from a location 10 parsecs away and perpendicular to the plane of the ecliptic. This movement is mainly dominated by the giant planets (Jupiter, Saturn, Uranus, and Neptune). Note the scale: the variations in positions are approximately 1 mas (0.001 arcsec noted 0.001″)

- *the line of nodes*: this is the straight line given by the intersection of the orbital plane with the plane of the sky
- *the ascending and descending nodes*: these are the intersections of the line of nodes and the object's orbit. By convention, the ascending node is crossed when the object recedes from the observer, and the descending node when the object approaches the same observer
- *the position angle of the ascending node*, denoted $\Omega$: it describes the orientation of the line of nodes on the plane of the sky relative to celestial north. This angle lies between 0 and 180°
- *the inclination*, denoted *i*: this is the angle between the object's orbit and the plane of the sky. It is equal to 0 (or 180°) when the orbital plane and the plane of the sky coincide; *i* lies between 0 and 90° when the apparent motion of the object on the plane of the sky takes place in the trigonometrically direct sense, and lies between 90 and 180° in the opposite case (Fig. 2.3).

It can be shown (Bordé, 2003) that the position of the star in the reference frame based on the plane of the sky may be deduced from that in the plane of the orbit by 3 successive rotations:

- a rotation by the angle $-\Omega$ around the line of sight
- a rotation by angle *i* around the line of sight
- a rotation by angle $\omega$ around the normal to the plane of the orbit

**Fig. 2.3** Definition of the geometrical parameters describing the orientation of the orbit of a star relative to the plane of the sky

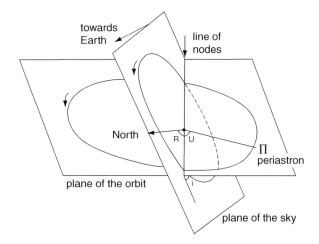

The expression for the path on the plane of the sky is thus obtained in polar coordinates, whose origin is the centre of mass of the system:

$$\rho = r\sqrt{1 - \sin^2 i \sin^2(v + \omega)}$$
$$\theta = \arctan\left[\cos i \tan(v + \omega)\right] + \Omega \tag{2.5}$$

which is still an elliptical path, and where the parameters defining it are the parameters of the initial orbit (in the reference frame of the orbit) and the observational geometry.

To detect an exoplanet by the astrometric method, it is thus necessary to be able to reconstruct the star's apparent path on the plane of the sky. By using Eq. (2.4), and by making the approximation that the star's true path is circular, and in the plane of the sky, it is possible to calculate the maximum value of the amplitude of the motion, which may be written as:

$$\delta\theta = \frac{a_*}{D} = \frac{m_p}{m_*} \cdot \frac{a_p}{D} \tag{2.6}$$

where $m_p$ and $m_*$ are the masses of the planet and the star, respectively; $a_p$ is the semi-major axis of the planet's orbit, and $D$ is the distance of the exoplanetary system from Earth. The value of $\delta\theta$ may be calculated for different cases (Table 2.2).

This table and Eq. (2.6) show clearly that the astrometric method is more sensitive (i.e., the amplitude of the motion of the star is the greater), if the planet is:

• massive
• located at a great distance from its star.

However, the greater the distance of the planet from the star, the longer the planet's orbital period – and consequently that of the star as well – (in accordance with Kepler's Third Law). This means that the time required to detect the planet

**Table 2.2** The amplitude of the apparent motion produced by the planets in the Solar System acting on the Sun, as if the latter were being observed from 5, 10, or 15 parsecs

| Object | $a_*(r_\odot)$ | $\delta\theta$ at 5 pc (mas) | $\delta\theta$ at 10 pc (mas) | $\delta\theta$ at 15 pc (mas) |
|---|---|---|---|---|
| Jupiter | 1.07 | 1.00 | 0.50 | 0.33 |
| Saturn | 0.59 | 0.55 | 0.27 | 0.18 |
| Neptune | 0.33 | 0.31 | 0.15 | 0.10 |
| Uranus | 0.18 | 0.17 | 0.08 | 0.06 |
| Earth | $6.5 \times 10^{-4}$ | $6.0 \times 10^{-4}$ | $3.0 \times 10^{-4}$ | $2.0 \times 10^{-4}$ |

(a significant fraction of the planet's orbital period around the star) also increases drastically with the planet's distance from the star. Nevertheless, the astrometric method remains one of the rare methods sensitive to low-mass objects, orbiting far from their star. In the case where the system is multiple (in practice, as soon as there are two planets with comparable effects), the path of the star rapidly becomes complex (see Fig. 2.2), and the accuracy of the reconstruction becomes limited by the resolution of the measurements.

Several techniques are possible to determine accurately the position of stars in the sky, and consequently their proper motion. The common factor with all these techniques is the necessity of being able to use a fixed reference frame, or one that is considered as such, on the sky. The principal difficulty is obviously that of finding the position of a fixed object in the sky to a greater degree of accuracy than the astrometric accuracy of the measurements (typically, to 1 μarcsec for the most accurate instruments). The positions of stars are generally referred to distant point objects, if possible, extragalactic ones. Nowadays, the ultimate reference frames consist of those based on quasars (quasi-stellar objects), which are actually the active nuclei of extremely distant galaxies. In the past, the position of nearby stars (their distances being determined by measurements of their parallax) was measured relative to more distant stars lying within the same field of view.

The first astrometric measurements aimed at detecting the presence of low-mass companions coincided with the systematic use of photographic plates for astrophysical observations. Peter Van de Kamp, who, in 1963, announced the detection of two planets orbiting Barnard's Star (the closest star to the Sun after the Alpha Centauri triple system, one component of which, Proxima Centauri, is the closest star to the Sun), based his measurements on the analysis of more than 2400 photographic plates taken between 1916 and 1963 by the various successive astronomers who used the 80-cm telescope at Sproul Observatory. In this case, a measurement consisted of locating on the photographic plate the coordinates of the photocentre of the reference objects (stars considered to be 'fixed'), and of the star for which one wanted to determine the proper motion. All that remained was to reconstruct the star's path, subtract the proper motion caused by parallax (the motion of the Earth in the Solar System), and the motion of the star in the Galaxy, to obtain the motion of the star that reflected the possible presence of a planet. The principal difficulty with this method is its low precision. The size of objects on a plate is, at best, 1 arcsec (that is the size of the physical spot forming the image with low atmospheric turbulence), and it is difficult to obtain a measurement on a photographic plate with a non-linear response, to better than one tenth of the size of the spot (i.e., 0.1 arcsec), to compare with the values in Table 2.2. Even with systematic measurements over several years, and using electronic cameras (CCD detectors), it is difficult, working from the ground and with classical telescopes, to obtain an astrometric accuracy better than 10 mas, which is too great for us to hope to detect planets – even giant ones.

The planets announced by Peter Van de Kamp remain, to this day, unconfirmed, despite the arrival of radial-velocity techniques that are far more accurate than astrometry with photographic plates. It is highly likely that Van de Kamp did not know how to overcome all the bias involved with the method.

Apart from the systematic use of electronic cameras (CCD devices), which have appreciably increased the accuracy and reproducibility of astrometric methods, two techniques have completely revolutionized this discipline and these are:

- observation from space
- interferometry.

Observation from space allows us to avoid all the problems arising from the atmosphere. In particular, turbulence, with its accompanying differential refraction, produces fluctuations in the position of an object against the plane of the sky. The degree of these fluctuations depends on the meteorological conditions under which the observations are made, and thus on the site's astronomical quality. The European whole-sky astrometric satellite, Hipparcos, was therefore able to obtain, with an accuracy of a few mas, the position and the proper motion of more than 120 000 stars in the Galaxy. This precision is at the upper limit for the detection of giant planets (see Table 2.3). The European GAIA project should, in 2012, allow us to attain an accuracy of a few $\mu$as. It will then be possible to search for giant planets using astrometry.

Interferometry is a technique which uses two or more telescopes, the light from which is recombined in pairs to obtain interference fringes. From measurements of these fringes, one can deduce information about the spatial structure of the object with an extremely high angular resolution (as would be obtained with a telescope having a diameter equivalent to the distance between the interferometric telescopes). In its astrometric mode, interferometry allows an extremely precise measurement of the position of an object against the sky. A diagram showing the principles behind the method is given in Fig. 2.4.

The light coming from the star of which one wants to measure the position, arrives at telescope 2 with a delay relative to the light arriving at telescope 1. This delay, caused by the angle $\theta$ between the direction of the star and the baseline, $B$, between the telescopes, creates an external path difference $B.\cos(\theta)$ between the two rays. This path difference may be compensated by the delay line which enables the optical path followed by the light from telescope 1 to be increased or decreased. Zero

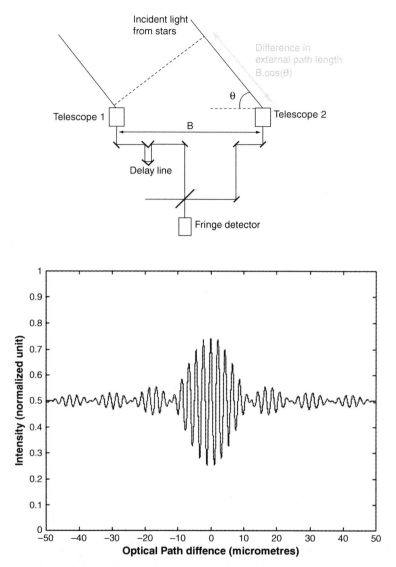

**Fig. 2.4** The principle behind interferometry (described fully in the text)

path difference (the optical paths followed by the light gathered by both telescopes 1 and 2 being equal) is determined by observation of the interference pattern (on the bottom of Fig. 2.4). It corresponds to the point at which the envelope over the interference fringes is a maximum. The precise determination of $B$, by optical metrology (with an accuracy of a few nanometres), and measurement of the difference in the external path length ($B.\cos\theta$) by measurement of the displacement of the delay line relative to zero path difference (again, with an accuracy of some few nanometres), provides an accurate determination of the angle $\theta$ between the telescope's baseline and the star on the sky.

In practice, interferometry, like the other methods, functions in a differential mode to avoid the effects of diurnal rotation and of the variation in the baseline during the observation. Here, the astrometric movement of the star is measured relative to a reference frame that is assumed to be fixed. The interferometer should therefore examine at least two objects simultaneously. This standard of performance should be obtained with the planned PRIMA (Phase Reference Imaging and Micro-arcsecond Astrometry) instrument on the European Very Large Telescope Interferometer (VLTI), under commissioning at the time this book was written.

The best astrometric measurements are currently obtained by radio interferometry with continental baselines (VLBI), where astrometric accuracy close to $100\,\mu$as is being attained. The optical interferometers that will shortly enter service should enable us to obtain astrometric accuracies that are at least 10 times better.

An additional stage in the search for the ultimate astrometric accuracy consists of combining interferometry with observations from space. The American SIM mission is a space-borne optical interferometer, which, around 2015, should be able to obtain an astrometric accuracy of $1\,\mu$as. To obtain such accuracy, it is necessary to know the length of the baseline to an accuracy of a few tens of picometres (in other words, the thickness on one atomic layer). Research and development efforts are currently under way to reach this objective, and allow the mission to be developed.

### 2.2.1.2 Radial Velocimetry

The astrometric method that we have just discussed, should enable us to measure and reconstruct the motion of the star in the plane of the sky. It omits, however, measurement of the star's motion along the line of sight, recession or approach relative to the observer. It is, however, impossible to observe, from Earth, the radial location of the star along the line of sight, because, by definition, the motion projected on the sky is non-existent. It is, however, possible to measure the radial velocity of the star by spectroscopy (see later). The expression for the radial velocity of the star may be derived from Eq. (2.4) and from the change of reference frame introduced in the preceding paragraph, which enables a conversion from the orbital reference frame to that of the sky, namely by rotations by the angles $-\Omega$, $i$, and $\omega$. The equation for the position of the star on the line of sight is then written as:

$$z(t) = r\sin(v(t)+\omega).\sin(i) \qquad (2.7)$$

By using Kepler's Second Law (the law of areas), which is expressed, for an elliptical orbit of period $P$, semi-major axis $a$, and eccentricity $i$, by the equation:

$$r^2\frac{dv}{dt} = 2\pi P^{-1}a^2\sqrt{1-e^2} \qquad (2.8)$$

and deriving the expression of the position of the star on the line of sight with respect to time, we obtain the equation for the star's radial velocity:

$$V_r = \frac{2\pi}{P}\frac{a_* \sin(i)}{\sqrt{1-e^2}}.[\cos(v+\omega)+e.\cos(\omega)] \qquad (2.9)$$

In what follows, we will set:

$$K_* = \frac{2\pi}{P}\frac{a_* \sin(i)}{\sqrt{1-e^2}} \qquad (2.10)$$

In the two-body case that we are considering, from the preceding equation, it is thus possible to derive an equation that is a function of the mass of the planet and the mass of the star. This equation is obtained using Kepler's Third Law, applied to the system consisting of the star and the planet:

$$a^3(m_* + m_p)^{-1} = P^2 G(4\pi^2)^{-1} \qquad (2.11)$$

where $a = a_* + a_p$ and $m_*.a_* = m_p.a_p$. As a result we have:

$$a_*^3 = P^2 G(4\pi^2)^{-1}.m_p^3 (m_* + m_p)^{-2} \qquad (2.12)$$

and we then obtain, using Eqs. (2.9) and (2.12) the equation:

$$\frac{(m_p \sin(i))^3}{(m_* + m_p)^2} = \frac{P}{2\pi G}K_*^3(1-e^2)^{3/2} \qquad (2.13)$$

The above equation shows that measurement of the radial velocity (Eq. 2.9) does not allow us to simultaneously determine $m_*$ and $m_p$. However, if we assume that the mass of the planet is negligible relative to the mass of the star, and also that the mass of the star may be estimated by some other means (for example from the position of the star on the Hertzsprung-Russel diagram, see Fig 1.12), then we may obtain an estimate of the product $m_p.\sin(i)$:

$$m_p \sin(i) \approx \left(\frac{P}{2\pi G}\right)^{1/3} K_*.m_*^{2/3}\sqrt{1-e^2} \qquad (2.14)$$

It will be noted that, whatever we do, we cannot derive equations that omit the system's angle of inclination ($i$). So measurement of radial velocities only allows us to derive a minimum mass for the planet. Only in the specific case where the system is seen edge on, and where the planet may also be detected by its transit of the star, is it possible to determine the individual masses of the system's planets. However, by assuming that planetary systems have a random orientation relative to the plane of the sky, it is possible to ignore the orientation statistically, and obtain a mass-distribution for the planets as a whole.

In the case of a circular orbit ($e = 0$), Eq. (2.14) may be expressed numerically as:

$$m_p.\sin(i) \approx 3.5\,10^{-2}K_*.P^{1/3} \ and \ K_* = \frac{2\pi}{P}\frac{m_p}{m_*}.\sin(i).a_p \qquad (2.15)$$

**Table 2.3** Properties and radial-velocity changes induced on the star by various planetary objects

|          | $m_p(m_\oplus)$ | $a_p$(AU) | $a_*(r_\odot)$ | $P$ (days) | $K_*(\text{m.s}^{-1})$ |
|----------|-----------------|-----------|----------------|------------|------------------------|
| Jupiter  | 317.83          | 5.2       | 1.07           | 4332.6     | 12.5                   |
| Saturn   | 95.15           | 9.54      | 0.59           | 10759.2    | 2.8                    |
| Uranus   | 14.54           | 19.18     | 0.18           | 30685.4    | 0.3                    |
| Neptune  | 17.23           | 30.06     | 0.33           | 60189      | 0.28                   |
| Earth    | 1               | 1         | $6.5 \times 10^{-4}$ | 365.25 | 0.09                  |
| 51 Peg b | 130             | 0.05      | 0.004          | 4.23       | 50.2                   |

where $K_*$ is in m.s$^{-1}$, $P$ in Earth years, and $m_p$ in Jupiter masses. Table 2.3 gives the value of $K_*$ for various specific instances.

If we continue to neglect the mass of the planet relative to the mass of the star, we may also deduce, from Kepler's Third Law, the semi-major axis of the planet's orbit around the centre of gravity (and thus around the star, because in this approximation, they are assumed to coincide) as a function of the mass of the star and of the orbital period:

$$a_p = \left( \frac{G}{4\pi^2} \right)^{1/3} P^{2/3} m_*^{1/3} \tag{2.16}$$

As Table 2.3 clearly shows, measurement of the radial velocity of a star to determine the presence of possible planets is a method that is biased towards:

- massive objects ($K_*$ is proportional to the mass of the planet),
- objects close to their parent star ($K$ is proportional to $P^{-1/3}$, and therefore greater, the smaller the value of $P$).

For massive objects, the radial-velocity method is therefore complementary to the astrometric method, which is more sensitive to distant objects. As with astrometry, to detect long-period objects the method requires regular observations and an instrumental stability that is monitored over time. We shall return to this point in the discussion of the equipment that enables these measurements to be carried out (Chap. 8).

Unlike the astrometric method, and as indicated earlier, measurement of the radial velocity does not allow us to determine the mass to better than the sine of the angle of inclination. So it does not allow us to observe systems that are viewed face-on. Finally, we should mention that in the case of multiple systems, the radial velocity, taken overall, consists of the sum of the different contributions of the planets, and of their periods of revolution. The detection of multiple systems (such as 55 Cancri, Upsilon Andromedae, etc.), is therefore carried out by analyzing the components of the radial velocity by subtracting the principal component, followed by analysis of the residuals until the measurement noise is reached.

Measurement of the radial velocity of a star is based on the Doppler-Fizeau effect:[1] any observer who receives a wave (of whatever nature) emitted at the frequency $v$ by a source in motion, detects it at the frequency $v + \delta v$, where $\delta v$ is positive (greater frequency and thus shorter wavelength) if the object is approaching, and negative (lower frequency and greater wavelength) if the object is receding.

A star is a source of electromagnetic waves: every object at a temperature $T$, has a thermal emission spectrum described, in the black-body model, by the Planck function. The star, as a source of electromagnetic waves may therefore equally be the source of a Doppler-Fizeau effect. The analogies with sound waves ceases here, however, because electromagnetic waves are transverse waves which propagate in a vacuum, whereas sound waves are longitudinal compression waves in a medium (air) that they require to be propagated (sound does not propagate in a vacuum).

In the relativistic expression of the Doppler-Fizeau effect, the wavelength observed is given as a function of the wavelength of the source by the equation:

$$\lambda_{\text{obs}} = \lambda_{\text{source}} \sqrt{\frac{1 + V_r/c}{1 - V_r/c}} \qquad (2.17)$$

where $V_r$ is the radial velocity of the source, positive when the object is receding, and $c$ is the velocity of light in vacuum. From this expression we can derive the relative shift in wavelength (or frequency) as:

$$\frac{\Delta\lambda}{\lambda} = \frac{V_r}{c} = -\frac{\Delta v}{v} \qquad (2.18)$$

So, for a hot Jupiter ($V_r \approx 50\,\text{ms}^{-1}$), the relative shift in wavelength is about $1.5 \times 10^{-7}$.

Measurement of the radial velocity of a star is therefore carried out by high-resolution spectroscopy. The spectrograph resolves the radiation from the star into its different components (corresponding to the 'notes' in our analogy with sound). Measurement of the shift in wavelength is made by observing the overall shift in the heavy-element absorption lines in the spectrum. Given the low amplitude of the shift ($\Delta\lambda/\lambda = 5 \times 10^{-7}$–$10^{-8}$, depending on the object), two conditions are required to obtain sufficient accuracy:

- using numerous lines to analyze the shift
- having a wavelength reference to calibrate the spectrum and obtain an absolute measurement of the radial velocity, allowing comparison of results obtained over several years (which is the time required to detect long-period planets).

The first point results in constraints on the spectral resolution of the spectrograph (the value $\lambda/\Delta\lambda$ where $\Delta\lambda$ is the spectral range covered by one element of the spectrum). The spectral resolution should be several tens of thousands for the best

---

[1] Fizeau's name should not be dissociated from that of Doppler. In fact, although we owe the observation of the effect to the latter, the true understanding of the phenomena in terms of the actual physics is the work of Fizeau.

modern instruments. There is also a constraint on the choice of targets. Stars that are too young, or are too hot, are difficult to observe (or have a reduced accuracy of measurement), because their spectra do not exhibit sufficient narrow lines for the amount of shift to be measured. The second point requires the use of a reference source, generally a spectral lamp (thorium–argon), or a cell containing a gas, whose absorption spectrum is known extremely accurately (generally a cell containing iodine vapour is used). In either of the solutions, the reference standard should be essentially stable, and thus requires a thermal environment, which itself needs to be stabilized. Measurements are made by exposing the spectrograph's detector to the stellar spectrum and the reference spectrum simultaneously. The best current instruments (such as HARPS on the 3.6-m telescope at La Silla) allow us to obtain a radial velocity to an accuracy better than $1~\text{ms}^{-1}$, and stable over a period of several years. The principal instruments used for the measurement of radial velocities and their method of operation are described later (Chap. 8).

### 2.2.1.3 The Timing of Pulsars

The detection of radial motion (along the line of sight for an observer on Earth) in a celestial body that is orbited by one or more planets may prove to be simplified if the body periodically emits a signal – electromagnetic waves, for example.

This is particularly the case with pulsars, which are neutron stars that have resulted from the explosion of a supernova, and which have the specific feature of emitting electromagnetic waves in a cone, which sweeps the sky in time with the rapid rotation of the star (with periods between a few milliseconds to several seconds), rather like the beam from a lighthouse. Some of these pulsars are visible from Earth, because the emission cone passes across the Earth at each rotation. They are then detectable as a periodic signal, which is easy to time. If the pulsar exists on its own, the Earth–pulsar distance does not vary and the period of the signal is absolutely constant. If the pulsar has planets, just like the other stars we have been discussing, then it will revolve around the centre of mass of the system, so that during the course of its revolution, the Earth–pulsar distance will increase and decrease, thus increasing or decreasing the travel time of the pulsar's signal. This variation in the travel distance, and thus in the travel time, is indicated by a variation in the period of the pulsar over the course of time.

Assuming, as previously, that the pulsar describes an orbit with a semi-major axis $a_*$ about the centre of mass, inclined at angle $i$ relative to the plane of the sky (where $i = 0$ when the orbit lies in the plane of the sky), the variation in the pulsar's period is given by:

$$\delta T_* = \frac{a_* \sin(i)}{c} \tag{2.19}$$

where $c$ is the propagation velocity of the wave from the pulsar in a vacuum (i.e., the velocity of light).

**Table 2.4** Amplitude of the period variations of a pulsar of one solar mass as a function of the planets in orbit

|         | $a_*(r_\odot)$      | $\delta T_*$ (ms) |
|---------|---------------------|-------------------|
| Jupiter | 1.07                | 250               |
| Saturn  | 0.59                | 137               |
| Uranus  | 0.18                | 42                |
| Neptune | 0.33                | 76                |
| Earth   | $6.5 \times 10^{-4}$ | 0.15             |

Just as in the case of the variations in the radial velocity, these variations of the pulsar's period have a period of their own, which is set by the orbital characteristics of the body orbiting the pulsar.

Table 2.4 gives the value of $\delta T_*$ for a pulsar of one solar mass accompanied by the bodies in our Solar System at their respective distances.

The accuracy of measurements of the variations in a pulsar's period caused by the presence of one or more companions increases with the intrinsic stability of the pulsar. This is the case with the old millisecond pulsars with a stable internal structure,[2] but which are revitalized by the accretion of material from a neighbouring red giant (Davis et al., 1985). So this applies to a small number of objects. Currently, the lowest-mass companions that are detectable are found in these systems, which we are able to search thanks to the stability of these pulsars.

The chronometric accuracy currently attained with these objects is far better than a millisecond (which is more than enough to detect a planet, even a terrestrial one). The method has the potential to detect planets with the mass of the Moon. In addition, the simplicity and accuracy of the method explain why it was the first to provide results. The first extrasolar planets detected, in 1992, were those around PSR 1257+12 (Wolszczan and Frail, 1992). But the discovery did not attract a lot of attention: the electromagnetic environment of the pulsar is, on the face of it, very hostile, and so leaves little hope that these planets could be habitable.

## 2.2.2 The Effect a Planet has on Photometry of Its Star

We have seen in the preceding section that the presence of one or more planets, orbiting a star, may be revealed indirectly by observing the movement of the star that is caused by the planet. In certain cases, we may equally hope to detect the presence of a planet by the effects produced on the luminosity of the star itself. Two principal effects may be involved: the passage of a planet in front of the star (a transit), or gravitationally induced amplification of the brightness of a background star by a multiple target.

---

[2] Unlike young pulsars, where deformation and changes in structure lead to irregularities in the period.

### 2.2.2.1 Planetary Transits

A planetary system observed edge-on (where the plane of the orbit is perpendicular to the plane of the sky), is a specific, and very interesting, case. The planets may, in fact, transit the star and cause micro-eclipses, which are detectable by continuously measuring the star's flux (Fig. 2.5). The amplitude of the extinction enables us to obtain the diameter of the planet, to a close approximation, and the duration of the phenomenon, and its periodicity, enables us to derive the planet's orbital period and hence its distance from the star.

Observation of a planetary transit is possible ONLY if the system is seen very close to edge-on. To calculate the likely probability of seeing a transit, consider Fig. 2.6

The transit will be visible only if the line of sight intercepts the cylinder, constructed on the orbit, of radius $a_p$ and height $2r_*$. Assuming that the orientation of planetary systems with respect to the plane of the sky is essentially uniform, the probability $p_T$ of actually observing a transit is thus expressed as the ratio between the surface of the cylinder that we have just described (the total number of favourable orientations of the line of sight) to the sphere of radius $a_p$ (the total number of possible orientations of the line of sight.) For a circular orbit, the following relationship may be derived:

$$p_T = \frac{2\pi a_p . 2r_*}{4\pi a_p^2} = \frac{r_*}{a_p}$$

(2.20)

where $r_*$ is the radius of the star, and $a_p$ is the semi-major axis of the planet's orbit. This relationship may equally be expressed as a function of the planet's orbital period $P$, by the application of Kepler's Third Law, and becomes:

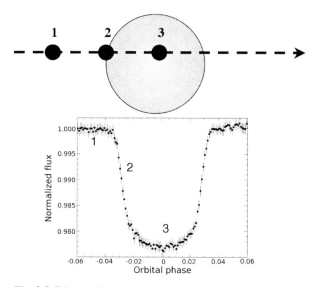

**Fig. 2.5** Diagram illustrating the principles of a planetary transit in front of a star, and the associated photometric light-curve

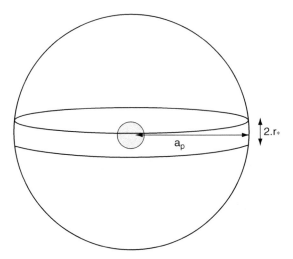

**Fig. 2.6** The geometry for observing a transit

$$p_T = \frac{r_*}{P^{2/3}} \cdot \left( \frac{4\pi^2}{Gm_*} \right)^{1/3} \tag{2.21}$$

This probability is shown in Fig. 2.7, for a star that has a radius equal to the solar radius.

It will be seen that the probability decreases rapidly with the planet's orbital distance (and thus period). Detection of planets by observing their transits of a star will, therefore, independently of any considerations of detectability, be more favourable for planets at short distances from their parent star. The probability of observing a transit of a hot Jupiter is thus about 10 per cent (with a period of 3–4 days), whereas the probability of observing the transit of a planet such as the Earth (with a period of 365 days) is only about 0.5 per cent.

To calculate the duration of a transit, we will make the simplifying assumption that the orbit is circular, and consider the geometry shown in Fig. 2.8.

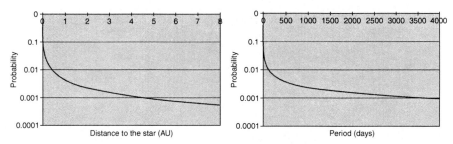

**Fig. 2.7** Probability of a transit by a planet orbiting a solar-type star as a function of its distance from the star (*left*) and its orbital period (*right*)

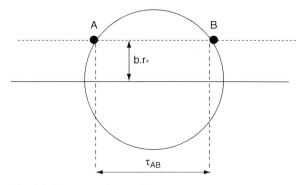

**Fig. 2.8** Geometry of a transit

The duration of a transit corresponds to the time required for the planet to traverse the chord AB (Fig. 2.9). The impact parameter (denoted $b$ and not $p$, to avoid confusion with notation used previously for the orbital period and the transit probability), is determined by the inclination of the orbit, $i$, relative to the line of sight and the relationship:

$$b.r_* = a_p.\cos(i) \tag{2.22}$$

The length of the chord AB is:

$$l = 2.r_*\sqrt{1 - b^2} \tag{2.23}$$

The time taken to traverse it, for a circular orbit of period $P$, at velocity $v = 2.\pi.a_p/P$. From this we may derive the duration of the transit: $\tau_{AB} = l/v$, which may be written, replacing $P$ by its expression as a function of $a_p$, using Kepler's Third Law, as:

$$\tau_{AB} = \frac{2.r_*\sqrt{1 - b^2}}{G.m_*^{1/2}}.a_p^{1/2} = \frac{(2\pi)^{2/3}.2\,r_*\sqrt{1 - b^2}}{(G.m_*)^{1/3}}.P^{1/3} \tag{2.24}$$

These equations may be written more simply, by taking $a_p$ in astronomical units (the mean Earth–Sun distance), $P$ in days, and $m_*$ and $R_*$ in solar units, to give:

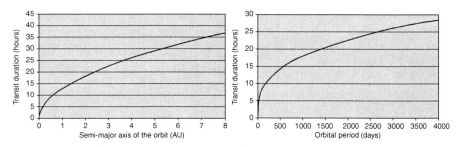

**Fig. 2.9** Duration of a planetary transit (in hours) as a function of the orbit's semi-major axis (*left*) and of the orbital period (*right*)

$$\tau_{AB} = 13.0\sqrt{1-b^2} \cdot \frac{r_*}{m_*^{1/2}} \cdot a_p^{1/2} = 1.8\sqrt{1-b^2}\frac{r_*}{m_*^{1/3}} \cdot P^{1/3} \qquad (2.25)$$

The relative amplitude of the extinction during a transit is, to a first approximation, equal to the ratio of the apparent surfaces of the planet and the star, and may be written:

$$\frac{\Delta F}{F} = \frac{r_p^2}{r_*^2} \qquad (2.26)$$

So the transit of a giant planet like Jupiter causes a photometric extinction of 1 per cent, and that of a terrestrial planet like Earth an extinction of 0.01 per cent $(10^{-4})$. To detect a planet by the transit method, it is therefore necessary to be able to carry out photometry of the star to better than the relative extinction. So, to detect a giant planet, a photometric accuracy of roughly a fraction of one per cent is required, while for a terrestrial planet, the photometric accuracy needs to be around some $10^{-5}$.

Although the detection of giant planets is possible from the ground, where one may achieve accuracies of around $10^{-3}$, limited by atmospheric turbulence, the detection of terrestrial-type planets may only be contemplated from space. The difference in the accuracy of an observation made from the ground and one made from space is illustrated in Fig. 2.10, which shows the same transit, that of a giant planet passing in front of HD 209458, observed from the ground and from space.

The method's potential is not limited to just the identification of planetary candidates. In fact, when combined with the radial-velocity method (where the indeterminacy of the angle of inclination relative to the plane of the sky is removed by the observation of a transit), detection of a transit enables us to derive:

- the size of the object (the amplitude of the extinction)
- its mass (by measurement of the radial velocities)
- and, as a result, the density of the object, thus allowing us to distinguish between a gaseous planet and a terrestrial-type planet.

In some cases, when the observation may be made with a good signal-to-noise ratio, by using differential spectroscopy before and during the transit, it is possible to determine the composition (or at least the presence of certain elements) in the planet's atmosphere. This point will be discussed in Chap. 7. We should also mention that the transit method has been used in the infrared, exploiting the fact that the planet is eclipsed when it passes behind the star (known as a secondary transit). Accurate knowledge of the ephemerides of the system allows us, by subtraction of the flux during the secondary transit (star alone) from that before the secondary transit (planet + star), to derive some spectral information about the object. To date, this method has been applied to two objects: TrEs1 and HD 209459, using the Spitzer space telescope. It set limits on the flux emitted by the objects in four spectral bands (see Chap. 7). It should become more generally available when the James Webb Space Telescope (JWST) enters service.

At present there are more than twenty, ground-based projects for the systematic observation of transits, generally with small, automated, wide-field telescopes (see

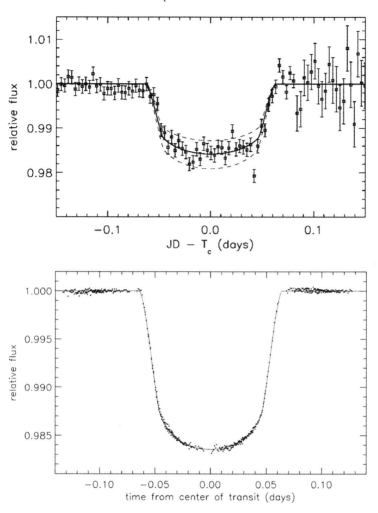

**Fig. 2.10** A transit of HD 209458b observed (*left*) from the ground (Charbonneau et al., 2000) and (*right*) from space with the HST (Brown et al., 2001)

Chap. 8). In this case, the primary difficulties are the handling of a large quantity of data and the extraction of relative photometry of the objects, when the observations have been obtained under what are sometimes extremely varied conditions (in particular, the influence of local meteorology and the succession of day and night, which do not guarantee continuity of observation). The A-STEP project, with an observing programme from Dome-C in Antarctica is an extreme example of these ground-based observational programmes. The excellent atmospheric conditions associated with the long Antarctic night appear to be serious advantages for this programme. Nevertheless, the ultimate photometric accuracy of all these ground-based programmes appears to restrict the method to the detection of giant planets.

Two space missions designed to search for planets by the transit method are envisaged for the near future, and are discussed in detail in Chap. 8:

- the CoRoT mission, led by CNES, launched in late 2006, began operation in January 2007 and should be able to identify objects of a few terrestrial radii, orbiting close (with periods of less than 75 days) to their star;
- NASA's KEPLER mission should, from 2008, be able to detect terrestrial planets with periods up to about one Earth year.

### 2.2.2.2 Gravitational Microlensing

One of the most surprising aspects of Einstein's theory of relativity is the phenomenon of the gravitational lens (Einstein, 1936). In his theory, Einstein introduced energy/matter equivalence,[3] with the consequence that the photon, the quantum of electromagnetic energy, is subject to gravitation, just like 'classical' baryonic matter, whose weight we all experience.

So a photon that passes at distance $r$ from an object of mass $M$, which is assumed to be a point, undergoes a deflection $\alpha$ relative to its direction of propagation, and this is given by the following equation:

$$\alpha = \frac{4GM}{c^2 r} = \frac{2R_s}{r} \qquad (2.27)$$

where $R_s$ is known as the 'Schwarzschild radius' of the gravitational lens caused by the point mass.

When a massive object lies between the observer and the object being observed, the image of the latter is thus deformed by the mass of the deflecting object. The amplitude of the phenomenon depends on the mass and position of the deflector. When the deflecting object is sufficiently massive – in which case $\alpha$ is greater than the resolution of the instrument being used for the observation – the effect is known as a 'macrolens': the image of the object is multiplied into an odd number of secondary images. Historically, this was the first type of lens to be observed. The image of the double quasar Q0957+561 obtained by Huchra in 1985 revealed a symmetrical multiple structure related to the presence of a galaxy on the line of sight (Huchra, 1985). When the mass of the deflecting object is low and $\alpha$ is less than the angular size of the observing instrument's diffraction disk, the situation is described as a 'microlens' and multiplication of the image is not observed (although it still exists), but the mean lensing effect effect instead, which takes the form of an overall amplification of the intensity of the source being observed. The first observation of a gravitational microlensing event was that of the quasar Q2237+0305 (Racine, 1992).

---

[3] This is the famous equation $E = mc^2$, where $E$ is the rest energy of the particle, $m$ its mass, and $c$ the velocity of light. In the case of the photon, the rest energy is $h\nu$, where $h$ is Planck's constant and $\nu$ the frequency of the wave associated with the radiation.

In a completely similar manner, if a massive object passes across the line of sight to a star, the latter's brightness is increased by the effect just described, simultaneously revealing the presence of the object that serves as a lens (Fig. 2.11). The gravitational amplification effect may, therefore, allow the transient observation of an object that cannot be detected prior to the microlensing event.

Gravitational amplification is given by the following relationship (Sackett, 1999):

$$A = \frac{u^2 + 2}{u\sqrt{u^2 + 4}} \tag{2.28}$$

where $u = \theta_S/\theta_E$, with $\theta_S$ the angular distance between the source and the deflector, and $\theta_E$ the Einstein radius, which is itself defined as:

$$\theta_E = \left( \frac{4GM}{c^2} \cdot \frac{d_{LS}}{d_L.d_S} \right)^{1/2} \tag{2.29}$$

where $G$ is the gravitational constant, $M$ the mass of the lens, $c$ the velocity of light, $d_L$ the distance between the observer and the lens, $d_S$ the distance between the observer and the source, and $d_{SL}$ the distance between the source and the lens.

The variation in the brightness of the object during the passage of the lens is shown in Fig. 2.12.

In the case of a simple lens, the variation in luminosity is symmetrical, and centred on the position of the lensing star. The overall set of the apparent positions of

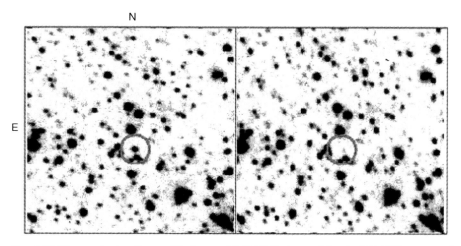

**Fig. 2.11** The gravitationally amplified event 98-SMC-01 detected by the MACHO team, and which had a gravitational amplification factor of about 100. These images show the object during the amplification phase (*left*) and then following the amplification and the passage of the lensing body (*right*). The field size is about $70 \times 70$ arcsec. This particular event has been attributed to a low-mass binary deflector in the Small Magellanic Cloud

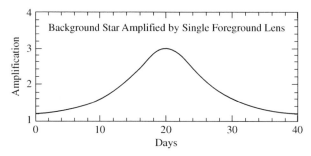

**Fig. 2.12** The theoretical gravitational amplification curve as a function of time, for a simple lens

the lens where the flux of the star is amplified by a factor of 1.34, is a circle, whose radius is precisely the Einstein radius,[4] as defined by Eq. (2.29).

This method has been used since 1995 by various groups (EROS, MACHOS, OGLE) to try to detect dark matter in the Galaxy. The theory to be verified is that dark matter consists of massive, compact, and non-luminous objects, whose overall mass could explain the difference between the mass deduced from the dynamics of the Galaxy, and that which is directly observable (stars and gas).[5]

For planet hunters, the interesting case is when the deflector is accompanied by one or more planets (a multiple lensing event). The structure of the amplification zone is then no longer symmetrical. Because of the presence of planets, a line appears between the star and the planet, known as a caustic, where the gravitational amplification is theoretically infinite for point sources such as stars. In the case of a simple lens (without a planet), the only point of infinite gravitational amplification is the position of the lens itself, i.e., when the source, lens, and observer are perfectly aligned. In the case of a multiple lens, the apparent intensity of the source undergoes significant variations when the apparent path of the source relative to the lens, approaches or crosses the caustic.[6] It is the structure of the gravitational-amplification curve (Fig. 2.13) and, more especially, the study of its artefacts (number, duration, and intensity) that enables us to determine the characteristics of the planet, or planets (mainly the mass and projected angular distance), by use of a model that incorporates a significant number of parameters that may be adjusted to the data. The error bars associated with this method are rather large.

---

[4] The Einstein radius is also the radius of the circle that is the image of the source produced by the deflecting object, when the source, the deflector, and the observer are perfectly aligned.

[5] Calculations actually reveal an enormous mass-deficit (a problem which is known in cosmology as the 'missing-mass problem'). A possible solution to this problem was to consider the existence of a vast number of brown dwarfs, which were not detectable by direct observation. This explanation, however, does not appear to be a convincing solution.

[6] The most spectacular effects occur when the apparent path of the source star crosses the caustic, giving a gravitational-amplification curve that is extremely variable, depending on the exact case. The most frequent case is that when the path approaches the caustic, but does not cross it. In such a case, we observe anomalies in the characteristic Gaussian curve of a simple lens.

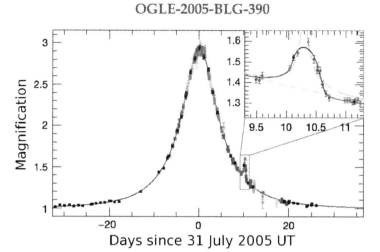

Fig. 2.13 The trace of the gravitational-amplification event OB-05-390, where the artefact indicates the presence around the deflecting star of an object of 5.8 $M_\oplus$ (After Beaulieu et al., 2006)

This method is, in principle, extremely sensitive, and enables giant planets to be detected – for example, the candidate associated with the MACHO-97_BLG-41 event, with a mass estimated as about three Jupiter masses (Bennett et al., 1999). It also allows lower-mass planets (terrestrial planets) to be detected. It was using this method that, in 2005, Beaulieu et al. detected the object OB-05-390 b, which is currently the lightest planet (5.8 $M_\oplus$ to within a factor of 2) ever detected orbiting a star other than a pulsar (Beaulieu et al., 2005). The difficulty in this case occurs in monitoring the gravitational-amplification event. The teams that are carrying out the search for planets by this method are dependent on other teams who provide details of the gravitational-amplification targets as soon as an event begins. To search for planets, as continuous a curve as possible is required so that it may be examined for artefacts. To obtain adequate sampling (about one point every 30 min) and continuous coverage of the amplification events, a collaborative effort (PLANET) has been devised, using several telescopes around the world to obtain 24-h coverage of any event.

The gravitational microlensing method is most sensitive to objects at a moderate distance (a few AU) from their parent star. It is not suitable for close objects (but there are the transit and radial-velocity methods for that). It allows the preferential detection of objects orbiting cool stars, because the stellar population of lenses is dominated by low-mass (and low-mass) objects. The object OB-05-390 b is orbiting a cold star (of spectral type M) at a distance of 2.6 AU. Given the distance of the objects, the surface temperature of this planet cannot be higher than 50 K, which means that it compares with that of a Pluto-type object in our own Solar System.

At present only the OGLE and MOA teams are continuing to issue the gravitational-amplification alerts that are likely to lead to the discovery of other planets by suitable follow-up observations. The permanence of the method thus depends, in a crucial manner, on these programmes of systematic observations.

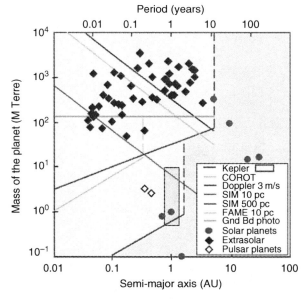

**Fig. 2.14** A diagram summarizing the domains over which the various methods of indirect detection are most sensitive

### *2.2.3 Comparison of the Different Indirect Methods*

In the previous sections we have considered the different methods for the indirect detection of extrasolar planets, their sensitivities, as well as their bias, either intrinsic, or linked to the choice of observational sample. Figure 2.14 summarizes the best observational domains for the different methods, with a diagram that plots the mass of the object against its distance from the central star (assuming the star to be of solar type).

## 2.3 Direct Detection of Exoplanets

In this section we intend to discuss direct detection of an exoplanet, in the sense of being able to detect directly photons arising from the planet, and not the effects of the latter on the central star. The objective of direct detection is to be able to separate the photons from the star from those from the planet, to enable us to make a spectroscopic analysis and to deduce details of the planet's composition, or at least that of its atmosphere. Strictly speaking, however, the transit method, when carried out in the infrared, is a direct-detection method, in particular when the secondary transit is observed (the passage of the planet behind the star). By measuring the thermal flux before and during this secondary transit, we obtain, by subtraction, a measure of the flux from the planet. This method was used on HD 209458b and TrEs-1 with the help of the Spitzer space observatory, and enabled spectral information to be gathered in four channels between 5 and 25 μm(Charbonneau et al., 2005).

Here, however, we shall restrict discussion to the collection and analysis of photons from planets.

### 2.3.1 Choice of Spectral Region

The choice of a spectral region for direct observation involves a compromise between the contrast between the star and the planet (typically between $10^4$ and $10^{10}$, depending on the nature of the companion and the spectral region chosen), and the angular resolution of the instrument, which primarily depends on the size of the collector. The limiting angular resolution is set by diffraction. The image of a point at infinity through a telescope is not a point but a disc, the form and size of which depend on the shape and size of the telescope's aperture. For a circular aperture, the diffraction disc appears as shown in Fig. 2.15.

The radius $r$ of the central diffraction disc is given by:

$$r = 1.22\frac{\lambda}{D} \tag{2.30}$$

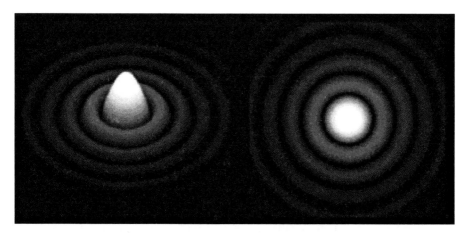

**Fig. 2.15** Form of the diffraction disc given by a circular aperture with no central obstruction

where $\lambda$ is the observational wavelength and $D$ is the diameter of the aperture (the telescope's primary mirror).

To 'image' an exoplanetary system, taking account of its spatial size, the compromise finally comes down to a choice between two domains:

- The visible and near infrared. Here, the star-and-planet pair is spatially resolved by a telescope a few metres in diameter. The difficulty lies in overcoming the contrast, which is $10^9$–$10^{10}$ for a terrestrial-type planet. In this instance, coronagraphic methods are used, possibly together with adaptive optics.
- The thermal infrared (around $10\,\mu$m). Here, the contrast is optimal ($10^7$ for a terrestrial planet), but we have to resort to interferometry to obtain the angular resolution of the star-and-planet pair, because it would require a monolithic telescope several tens of metres in diameter to resolve the system.

The luminous flux of the objects also needs to be taken into account. This point is particularly important when we want to consider the question of imaging planetary surfaces. Table 2.5 gives an estimate of this flux.

**Table 2.5** Flux received from an Earth-like planet at 10 parsecs from the Solar System

| Spectral region | Integrated flux received on Earth (omitting atmospheric absorption) |
|---|---|
| 0.3–2 $\mu$m | 0.3 photon/s/m$^2$ |
| 6–20 $\mu$m | 10 photons/s/m$^2$ |

Let us now discuss the different direct-detection techniques, together with the associated instrumentation.

## 2.3.2 Coronagraphic Methods and Adaptive Optics

### 2.3.2.1 A Historical Retrospect: The Lyot Coronagraph

The coronagraph was invented by Bernard Lyot at the beginning of the 1930s to observe the solar corona at times other than eclipses (Lyot, 1931; 1939). The principal approach – which was, nevertheless, empirical – that guided its inventor was to achieve the maximum reduction in the light scattered and diffracted by the optics. The principle of this instrument is shown in Fig. 2.16.

Lyot's trick was to re-image the primary image plane, and to place a diaphragm at the associated pupil plane to eliminate the effects of diffraction caused by the mask located as the primary image focal plane. This method, in conjunction with the use of extremely high-quality optics (with homogenous glass and exceptionally high quality polishing) enabled light scattered within the instrument to be reduced

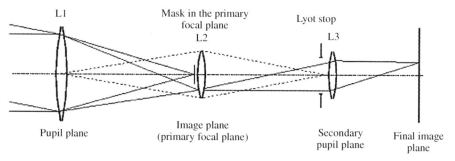

L1                    Mask in the primary                  Lyot stop
                           focal plane
                              L2                              L3

Pupil plane                Image plane                   Secondary        Final image
                    (primary focal plane)              pupil plane           plane

**Fig. 2.16** The basic principles of the Lyot coronagraph

by a factor of between 10 and 100. Lyot's coronagraph thus allowed observation of the solar corona a few arcminutes from the solar limb.

An elegant description of how Lyot's coronagraph works may be made using Fourier optics and image formation theories. A good introduction to these topics may be found in Goodman, 1996.

From the point of view of the theory of the formation of images, the principle behind Lyot's coronagraph may be described in the way shown in Fig. 2.17:

(a) at the entry pupil plane: the infinite (i.e., nominally plane) wavefront is sampled by the pupil

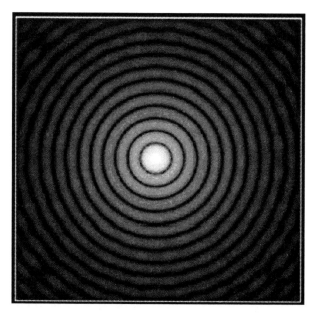

**Fig. 2.17** The Airy function (Airy disk): the diffraction pattern produced by a circular diffracting aperture

(b) at the primary image plane (the focus of the main mirror): The image of a single point at an infinite distance (which is known as the Point-Spread Function, PSF) is the square of the modulus of the pupil's Fourier transform. (The Fourier transform is a mathematical function that allows the transformation of a function into another, known as the frequency domain representation of the original function. For instance, a function of time may be transformed into one of frequency. Among the main characteristics of the Fourier transform is the fact that the operation known as a convolution, whereby two functions are manipulated to provide a third, may be simply represented in Fourier space by a single multiplication of the Fourier transform.) This point-spread function may thus be multiplied by the mask's transmission ($1-\Pi$, where $\Pi$ is the aperture function, corresponding to a mask with an occulting centre. In this image plane we therefore have:

$$\text{image} = \text{PSF} - \text{PSF} \times \Pi \qquad (2.31)$$

(c) at the secondary pupil plane: the new pupil is once again the Fourier transform of the primary image, in other words: The Fourier transform of the Airy function (i.e., the initial pupil) less the convolution product of the Fourier transform of the Airy function (the initial pupil) with the Fourier transform of the aperture function (another Airy function with a size that is inversely proportional to the size of the mask). It will be recalled that the Airy function is the result of the diffraction of light by a circular aperture (Fig. 2.17). The final transmission of a coronagraph is thus equal to 0 on the optical axis, and there remains only an annulus (of scattered light), which is eliminated by the Lyot stop at the pupil (Fig. 2.18), which may be written as:

$$\text{pupil} = \Pi - \Pi \otimes \text{Airy} \qquad (2.32)$$

where the symbol $\otimes$ represents the second magnitude convolution product.

This method, with some variants, has been employed in all solar coronagraphs, including those installed on satellites such as SOHO (Koutchmy, 1988), as well as in certain stellar coronagraphs which, when combined with adaptive-optics systems,

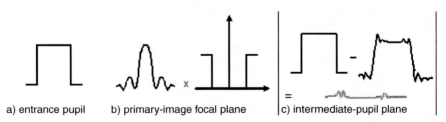

a) entrance pupil          b) primary-image focal plane          c) intermediate-pupil plane

**Fig. 2.18** Lyot coronagraph: (**a**) and (**b**) wavefront and image at the pupil planes and successive images. (**c**) in red, residual light in the plane of the secondary pupil, which is a ring of light eliminated by the Lyot stop. All photons from a source on the optical axis are therefore, in theory, completely eliminated

have attained remarkable performance (Malbet, 1996), such as the detection of a companion lying 10 Airy radii from a highly luminous star, which is $10^5$ times as bright (Beuzit et al., 1997).

The stellar version of the Lyot coronagraph, even though it may be effective down to 2 arcsec of the star that is being occulted, when used with a telescope in the 4-m class, cannot observe beyond that limit. In fact, the closer to the star that one attempts to observe, the smaller the occulting disc needs to be, and the greater the diffraction that it creates, requiring a stop whose aperture rapidly becomes very small. In practice, the Lyot stop cannot cover less than the star's Airy disk. Yet it is precisely within this region that we want to observe. Short of using a large-diameter telescope which 'super-resolves' the system and eliminates the residual contrast, Lyot's original method cannot be used, as such, to search for terrestrial planets.

### 2.3.2.2  The Phase-Mask Coronagraph

An alternative to the Lyot coronagraph was suggested by François and Claude Roddier in 1997. It replaces the circular amplitude mask in the focal plane of the Lyot coronagraph with a circular phase mask. (The mask modifies the phase of the wavefront passing through it.) This mask is smaller, covering half (of the area) of the Airy disk, and applies an achromatic phase change ($\pi$) or, at least, one that is very weakly chromatic, given the low dispersion in the chosen spectral band. Light from the star is extinguished by destructive interference by virtue of the source's symmetry relative to the mask (which must be precisely located on the image spot), and rejected outside the geometric pupil, where it is blocked by the Lyot stop.

Several variants of a phase mask have been suggested. Among the most promising are the four-quadrant mask suggested by Daniel Rouan (Fig. 2.19). The theoretical performance of such a mask suggests that extinctions of 1010 may be attained, even inside the diffraction spot. The great advantage of this mask geometry is that it is 'spatially' achromatic (unlike Roddier mask, the size of which depends on the wavelength). Here, the efficiency does not depend on the size of the image spot but, uniquely, on the ability for the phase-change $\pi$, introduced by two of the four segments of the mask, to be achromatic.

Coronagraphs, whether of the amplitude mask or phase mask type, have requisites in common: the necessity for the stellar source to be perfectly centred on the centre of the mask, and for the stellar image to be as symmetrical as possible. Any deviation from symmetry results in a stellar residue, which may mask the image of the companion that is being sought. It is therefore essential:

- to have the best possible image of the star
- to permanently monitor the telescope's alignment relative to the centre of the star.

This is why all current coronagraphic systems operating from the ground are used in conjunction with adaptive optics.

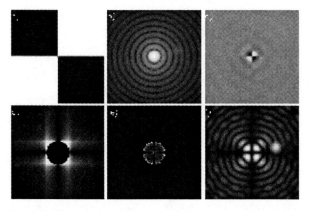

**Fig. 2.19** The principle of the 4-quadrant coronagraph: (**a**) appearance of the phase mask (white: no phase-change, black: phase-change $\pi$); (**b**) image in the primary focal plane; (**c**) the complex amplitude after passing through the mask; (**d**) secondary pupil plane before the Lyot spot; (**e**) secondary image plane; (**f**) central portion of the secondary image focal plane (After Rouan et al., 2000)

### 2.3.2.3 The Phase-Induced Amplitude Apodization Coronagraph (PIAAC)

The PIAAC instrument has been proposed by Guyon et al., and is in fact the combination of two optical devices (Guyon et al., 2005):

- a high performance coronagraph
- a pupil 'remapper' to adapt the pupil shape of the telescope to the coronagraph.

The PIAAC concept comes from the observation that the diffraction pattern produced by a classical telescope leads to the presence of light (diffraction rings) far from the on-axis direction. Such diffraction rings are classically removed or reduced by 'apodized' apertures such as Gaussian apertures that can smoothen the aperture edge effects. The main drawback of such a technique is a drastic reduction of the source flux through modification of the pupil shape and thus its transmission, because of partial occultation of the pupil. Guyon proposes apodizing the pupil by a modification of the intensity distribution within the pupil This is done by a global wavefront phase distortion, using aspheric mirrors (Guyon, 2003). This method allows apodizing the pupil without changing its overall transmission.

The PIAAC combines a first stage of apodization, which optimizes the shape of the pupil, and a coronagraph (which may be either a phase- or an amplitude-coronagraph). The principle of PIAAC is given in Fig. 2.20.

This concept has been demonstrated in the laboratory and is currently under test on the sky. It could also be adapted to a space-borne mission concept.

### 2.3.2.4 Adaptive Optics and Accurate Pointing: The Keys to Performance

Because of atmospheric turbulence, the image of a star in a large-diameter telescope is never the theoretical diffraction image as set by the limiting size of the telescope's

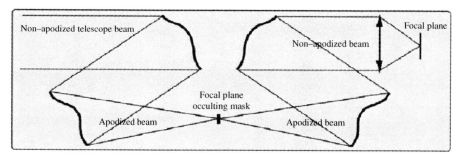

**Fig. 2.20** Principle of a Phase-Induced Amplitude Apodization Coronagraph. The beam is apodized before reaching the first focal plane where the coronagraph mask is located. The beam is then apodized backwards to revert to its initial form and to retain the initial phase information contained in the entrance pupil. It is then imaged in the detection focal plane (After Guyon, 2003)

aperture. The star's image is, in fact, the sum of multiple 'speckles', the number and size of which depend on the level of turbulence (which is a characteristic of the site and the construction of the telescope), and of the telescope's size. During a long exposure the speckles give rise to an integrated image that is distinctly spread out, and the size of which is determined by the turbulence. In terms of resolution, the degradation of the image reduces the performance of these telescopes to those of smaller instruments, typically by a few centimetres to some tens of centimetres, depending on the quality of the site and the wavelength under observation. (A thorough introduction to atmospheric optics is given by Léna, 1996). The degradation of the image into speckles results from aberration of the wavefront emitted by the star during its passage through the atmosphere. This aberration affects the phase of the wave emitted by the source. Adaptive optics (Fig. 2.21) is a system that allows:

- the analysis of the aberration of the wavefront caused by its passage through the atmosphere: this is the role of the wavefront sensor. To do this the sensor monitors a bright star within the field of view. This star serves as a point-source reference.
- the partial compensation of the degradation by means of one or more deformable mirrors, which, acting locally, will add or subtract a phase element to the wavefront in such a fashion that it will restore the wavefront's plane form that it had before passing through the atmosphere. These mirrors are actuated by a control system which uses, in real time, the information from the wavefront-sensor to calculate the correction that needs to be applied by means of the deformable mirrors.

Current 'mono-conjugate' systems correct atmospheric turbulence by assuming that it is limited to a specific atmospheric layer (and have a sole adaptive mirror). This solution proves to be inadequate as soon as one wishes to image the field around the reference object. Studies are under way to realize 'multi-conjugate' systems, which take into account turbulence at several different altitudes in the atmosphere.

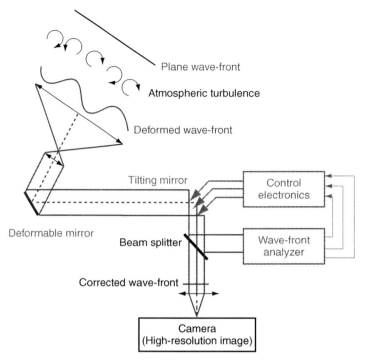

**Fig. 2.21** Schematic diagram of an adaptive-optics system. The wavefront that has been deformed by the atmosphere is analyzed, and the effects of turbulence are corrected by use of a deformable mirror (by courtesy of ONERA)

Several parameters may be used to quantify the efficiency of an adaptive-optics system. Among these, we may mention the Strehl ratio, which is defined as the ratio between the peak intensity of the image of a star (i.e., the point-spread function) and the theoretical peak intensity for an image solely limited by diffraction. A Strehl ratio of 1 corresponds to an image that is fully coherent and perfect. An image that is not corrected by adaptive optics has a low Strehl ratio, which depends on the severity of atmospheric turbulence. Current, fine adaptive-optics systems are able to attain a Strehl ratio of 0.6. Future projects for the detection and investigation of extrasolar planets from the ground will require Strehl ratios of at least 0.9.

Use of a coronagraph requires a high-quality wavefront, and which it is therefore essential to correct by adaptive optics when observations are made from the ground. In the specific case of a phase-mask coronagraph, the guiding constraints (on positioning and on centring the mask) are extremely tight (the functioning of these systems is based on the symmetry of the stellar image relative to the phase mask). These constraints are met by a fast-response, tip-tilt mirror ahead of a high-quality adaptive system which stabilizes the position of the image at the image plane.

Most 8-metre-class telescopes are now fitted with adaptive-optics systems. Some also incorporate new-generation phase-mask coronagraphs. We may mention in particular the NAOS-CONICA instrument on the VLTI. This instrument allowed Chauvin and his collaborators to obtain the first image of an exosystem (2M1207). In a few years' time, this instrument should be followed by the SPHERE instrument, which consists of a highly efficient adaptive-optics system coupled with a 4-quadrant coronagraph (see Chap. 8). These innovations, mainly developed in Europe, and more particularly in France, should, in due course, also be incorporated into American instruments.

## 2.3.3 Interferometry

As we have already seen in the section on astrometry, it is possible to measure the position and the proper motion of a star relative to a fixed reference by using interferometry. In that method, the position of the group of interference fringes and the length of the baseline are measured to derive the astrometric information. In this section we will show how analysis of the structure of the interference fringes allows us to extract information about the object itself and its environment. To understand this technique, one needs to understand some additional principles of interferometry.

### 2.3.3.1 Some Principles of Optical Interferometry

The Temporal-Spatial Coherence of Two Waves

When two sources of illumination are mutually coherent, superimposing the two waves produces a phenomenon known as 'interference'. Here, the intensity is not everywhere the sum of the intensities of the two beams taken separately, but is the result of an interference function, which we shall explain later.

Mathematically, the concept of temporal-spatial coherence is expressed by a quantity known as the 'complex degree of coherence', denoted $\gamma_{12}(\tau)$, and which is expressed as the equation:

$$\gamma_{12}(\tau) = \frac{< E(P_1, t + \tau) E^*(P_2, t) >}{\left[ < |E(P_1,t)|^2 > < |E(P_2,t)|^2 > \right]^{1/2}} \qquad (2.33)$$

In this equation, $E(P,t)$ is the electric field (complex notation) at point $P$ at time $t$; $< E(P,t) >$ is the temporal mean of this field, and $\tau$ represents the delay of wave 1 relative to wave 2. $\gamma_{12}(\tau)$ is a complex number where the modulus lies between 0 (no coherence) and 1 (complete coherence). In this expression, if $P_1 = P_2$ and $\tau \neq 0$, we are dealing with the wave's spatial coherence, and if $\tau = 0$, and $P_1 \neq P_2$, we are investigating the wave's spatial coherence.

Interferometry and Visibility

In observational astrophysics, an interferometer (Fig. 2.22) is, above all, an instrument that measures the degree of temporal-spatial coherence of the electromagnetic field emitted by a source and sampled by two telescopes.

Positions $P_1$ and $P_2$ are those of the two telescopes, the delay $\tau$ is obtained by modifying the observational configuration (the position of the object on the sky and the setting of the delay line).

If $I_1$ and $I_2$ are the intensities in the two arms of the interferometer and if $d = c.\tau$ is the path difference, then the intensity at the exit from the beam combiner may be written as:

$$I(d) = I_1 + I_2 + 2\sqrt{I_1 I_2}\,\mathrm{Re}(\gamma_{12}(d/c)) \tag{2.34}$$

Measurement of $I(d)$ allows us to measure the complex degree of coherence (or at least its real part) for each of the positions $P_1$ and $P_2$. In the case of two monochromatic plane waves of wavelength $\lambda$, spatially and temporally coherent, the preceding equation becomes:

$$I(d) = I_1 + I_2 + 2\sqrt{I_1 I_2}.\cos\left(\frac{2\pi d}{\lambda}\right) \tag{2.35}$$

This equation is interferometry's characteristic equation.

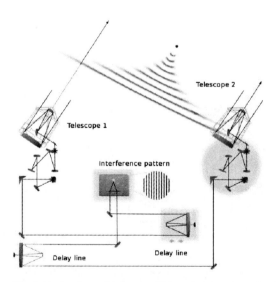

**Fig. 2.22** A schematic diagram illustrating the principle of a twin-telescope interferometer of the VLTI type. The rays of light from the two telescopes are led to the beam combiner by a series of mirrors forming the optical train. The path difference is adjusted by using a delay line. (After ESO Internet site: www.eso.org.)

In practice, one records all or part of an 'interferogram' (Fig. 2.23), obtained by fixing positions $P_1$ and $P_2$ and varying $\tau$ (by means of a delay line).

From the graph of $I(d)$ we may determine a quantity known as the 'visibility', denoted $V$, and which is defined by:

$$V = \frac{I_{max} - I_{min}}{I_{max} + I_{min}} \tag{2.36}$$

In the absence of instrumental perturbation, the visibility is a measure of the degree of mutual coherence. In practice, the experimental visibility (as measured) is the product of three terms:

- the intrinsic visibility of the object
- a term linked to the instrument function of the interferometer
- a term linked to atmospheric perturbations (for observation from the ground) or to the environment.

The last two terms are estimated by observing a 'calibration' object, which is a star whose intrinsic visibility is known, and generally a star that is not resolved with the interferometric baseline, and which is thus completely coherent (having an intrinsic visibility of 1).

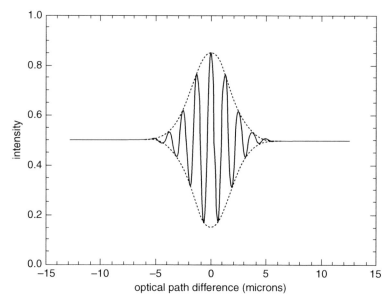

**Fig. 2.23** An example of an interferogram obtained by varying the path difference by use of a delay line. For a path difference greater than $\pm 5\,\mu$m, there is no temporal coherence. The spatial coherence is not perfect (the visibility is about 0.7)

Various studies may be found in the literature of the factors that degrade visibility, in particular, atmospheric turbulence. We shall not, therefore, discuss this aspect in detail here.

Visibility of the Source's Structure

The relationship between what may be observed (the experimental visibility) and the astrophysical source is obtained by employing the Zernicke-Van Cittert theorem, which may be expressed as follows:

> If both the linear dimensions of the source of quasi-monochromatic radiation and the distance between the two points on the screen (here, the distance between the two telescopes, i.e., the distance between $P_1$ and $P_2$) are small relative to the distance between the source and the screen (here, between the source and the Earth), then the modulus of the complex degree of coherence (the experimental visibility) is equal to the modulus of the spatial Fourier transform of the intensity of the source, normalized to the total intensity of the source.

In other words, there is a simple Fourier relationship between the spatial structure of the source and the visibility function, measured at several different baselines (several spatial frequencies) as shown in Fig. 2.24.

In practice, the opposite problem has to be tackled. The visibility curve is obtained, and then, generally by adapting the model, one can deduce the structure and parameters of the source. For example, in the case of a stellar source, the simplest model is one consisting of an evenly illuminated disk. In this case, interpolation from the measured visibility points enables one to determine the diameter of the source (by determining the point at which the visibility function first becomes zero). Obviously, this requires the source to be resolved by the interferometer; that is that the object's angular resolution should be greater than the angular resolution determined by the interferometer's baseline (the value $\lambda/B$, where $B$ is the distance between the telescopes and $\lambda$ is the observational wavelength).

It should, however, be noted that the object space (the plane of the sky, where the coordinates are denoted by $x$ and $y$), as well as the associated Fourier space (i.e., the one dealing with spatial frequencies and where the conjugate coordinates in $x$ and $y$ are denoted by $u$ and $v$, described as the $(u,v)$ plane), are two-dimensional

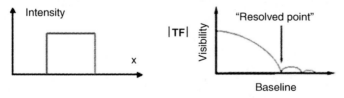

**Fig. 2.24** The relationship between an object's spatial structure and the measured visibility function. An interferometer allows the visibility to be measured as a function of the length of the baseline. By adjusting one's model to fit the experimental points, one can then obtain the structure of the source by an inverse Fourier transform

spaces that are being sampled point by point. To reconstruct a correct representation of the astronomical object, it is essential, if the second dimension is to be properly covered, to have several observational baselines (several separations between the telescopes) and, equally, several different orientations of these baselines relative to the plane of the sky. In practice, it is generally the rotation of the Earth that alters the orientation of the fixed baselines, as, in particular, it is with the VLTI. In the case of a space interferometer, several different distances between the telescopes need to be employed, as well as several different orientations of those baselines relative to the sky. (It needs to be possible to turn the interferometer or, at the very least, to change its orientation relative to the line of sight to the target.)

One of the difficulties of interferometry by measuring visibility (apart, as we have seen, from measuring the actual visibilities accurately), is the necessity for having precise models that allow for the study of the visibility and its variations, in particular as a function of the observational wavelength (chromatic models). While it is not necessary for there to be a systematic knowledge of the visibility in absolute terms, it is often useful, or even indispensable, to study the relative variations of this value as a function of wavelength (for the observation of gaps in protoplanetary disks, for example – see Chap. 7).

### 2.3.3.2 Detection of Planets via Modulation of the Visibility

A star and a planet in orbit may be considered as a combination of two mutually incoherent sources, with a high level of contrast. In this case, the visibility of this combination via interferometry is equal to the sum of the visibilities, weighted to account for the relative intensity of the sources. Mathematically, this visibility may be expressed by the equation (Bordé, 2003):

$$V^2 = \frac{V_1^2 + V_2^2 + 2\,r\,V_1\,V_2\cos\psi}{(1+r)^2} \; where \; r = \frac{I_2}{I_1} \tag{2.37}$$

$V_1$ and $V_2$ are the visibilities of the individual objects, $I_1$ and $I_2$ their relative intensities and $\psi$ denotes the angular position of the star/planet system relative to the interferometer's baseline. This visibility function is shown graphically in Fig. 2.25.

Detection of objects with very low luminosity (high contrast) therefore requires an extremely accurate measurement of visibility. The maximum accuracy over several observations is in the region of $10^{-3}$.

### 2.3.3.3 Cancellation Interferometry

Cancellation interferometry, also known as dark fringe interferometry or interference coronagraphy, differs uniquely from the method just described in its recombination technique. A diagrammatic representation of such an interferometer is shown in Fig. 2.26.

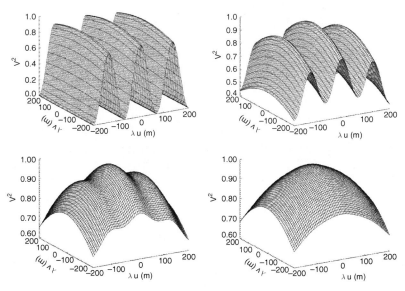

**Fig. 2.25** Visibility functions for binary systems where the contrast is 1 (*top left*), 10 (*top right*), 100 (*bottom left*) and 1000 (*bottom right*) (After Bordé, 2003.)

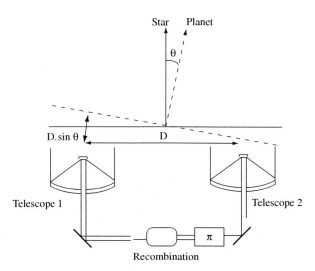

**Fig. 2.26** Schematic diagram of a dark-fringe interferometer. Light arriving from the direction in which the instrument is aimed arrives simultaneously at the two telescopes. The achromatic dephaser, $\pi$, ensures destructive interference at the recombiner. Light from a point away from the target direction (from the planet) arrives at Telescope 1 with a delay relative to Telescope 2, because it has to cover an additional distance ($D.\sin\theta$). By altering the distance between the telescopes (by varying $D$), it is possible to compensate for the $\pi$ phase shift by the additional optical path, such that the interference is constructive in the direction of the planet. We thus obtain an instrument where the transmission is zero along the direction in which the instrument is pointed, and adjustable between 0 and 1 around that axis

Let us consider two telescopes $T_1$ and $T_2$, which, because of diffraction are not able individually to resolve the star/planet pair. We point the two telescopes exactly in the direction of the star, and merge the beams of light from the two telescopes with an optical recombiner. In the direction of the star (i.e., where the two telescopes are pointing), the wavefront arrives simultaneously at $T_1$ and $T_2$. If we recombine the two beams, they will be in phase and will produce constructive interference. If we than add an achromatic dephaser, $\pi$, in one arm of the interferometer (for example, the path from $T_2$), the light from the two telescopes will be recombined with opposite phases. In other words, the interference will be destructive and everything coming from the star's direction (in particular the star's flux) will be extinguished. In the direction of the planet (which typically lies at an angle $\theta \sim 0.1$ arcsec for a terrestrial analogue orbiting a star at a distance of 10 pc), we introduce a delay for $T_1$ relative to $T_2$ that is equal to $D.\sin(\theta)$, where $D$ is the distance between the two telescopes. If $D$ is altered (by moving the telescopes), this may be done so that at an average wavelength (for example at the centre of a spectral band, or at two wavelengths so chosen as to maximize the spectral coverage) the difference in the path-lengths $D.\sin(\theta)$ compensates for the dephasing that $\pi$ introduces into the arm incorporating $T_2$. This results in constructive interference in the direction of the planet. To summarize, such an instrument allows the observation of a faint object that lies outside the optical axis, where the theoretical transmission is zero. The response obtained with such an instrument is shown in Fig. 2.27. In its original version, the Bracewell interferometer was rotated such that the signal from the planet was modulated rel-

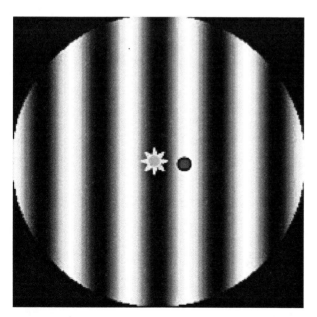

**Fig. 2.27** Plot of the transmission of a dark-fringe interferometer with two telescopes (a Bracewell interferometer). The star is located on the dark fringe (it is extinguished), and the planet on a bright fringe

ative to the leakage of light from the star (because of the non-zero size of the star and of the finite size of the zone with zero transmission, the star is not perfectly extinguished).

In practice, a dark-fringe interferometer and its associated detection system (which may, in principle, be a single-pixel detector), behaves as a photometer with the spatial transmission shown in Fig. 2.27. In particular, measurements obtained with this instrument are measurements of the flux, which consists of several individual contributions:

- the off-axis object (for example, the planet),
- leaks from the star (because the central fringe is not perfectly dark and the star is not a perfect point), and the associated noise and fluctuations,
- other contributions related to the source, for example, a portion of the zodiacal disk in a planetary system,
- thermal emission from the instrument itself (in cases where such wavelengths are a problem).

The specific remedy that needs to be applied is to try to separate these different contributions to be able to isolate the component linked to the off-axis object that one is trying to detect and analyze. One classic technique consists of taking the geometry of the signals into account (the emission from a disk of zodiacal light is overall centrally symmetrical with respect to the plot of transmission, whereas that from the planet is not). To do this, the system is observed at several orientations of the interferometer, enabling discrimination of the different contributions. Another possibility is to use an internal modulation between several subsidiary interferometers.

This concept of a dark-fringe interferometer is the basis of the proposal for the DARWIN space mission and its American counterpart TPF (Terrestrial Planet Finder), described in Chap. 8.

It is interesting to note that from the instrumental point of view, the two interferometric concepts: the 'measurement of visibility' and dark-fringe interferometry, may, potentially, be combined in a single instrument. In particular, the achromatic dephaser does not prohibit the measurement of the experimental visibility. In such a case, it is necessary to limit exploration of the interference fringes to a small number of fringes, the characteristics of which (the amplitude and the phase) have been determined by a method that is precisely comparable with the classic method of measuring the visibility.

## 2.3.4 Interferometry and Imagery: Hypertelescopes

We have seen that the spatial resolution of a telescope is intrinsically limited by diffraction and, in practice, by the diameter of the telescope. The size of instruments is limited by current technology to:

- telescopes between 10 and 42 metres in diameter (ESO's current Extremely Large Telescope projects). The largest current telescope is a multi-mirror system with an elliptical pupil ($10 \times 11$ m): the Hobby Eberly Telescope, which is slightly larger than the two Keck telescopes (10 m in diameter). The largest monolithic telescopes are the two mirrors of the Large Binocular Telescope (each 8.4 m in diameter);

- space telescopes of a few metres in diameter. (The JWST is to have a diameter of about 6.5 m. This is the largest space instrument observing in the visible or infrared that has ever been constructed or is in the course of construction.)

However, if we are to contemplate the observation of planetary systems in the thermal infrared, or even direct imagery of exoplanetary surfaces in the visible, we will require diameters that are distinctly larger, and also with significantly greater collecting surfaces, because the flux from planets is weak. We soon encounter the limit for monolithic instruments – or at least with unified pupils (all points of the pupil are available without leaving the pupil).

Table 2.6 gives the minimal baseline (diffraction limited) as well as the collecting surface required to image the surface of a planet such as the Earth at a distance of 10 parsecs, with the corresponding spatial resolution (given as the number of pixels). The assumptions used for these calculations are:

- image of a planet like the Earth orbiting a solar analogue in the visible/near IR spectral region (0.1 photon/s/m$^2$ at the Earth, in the 0.3–1 $\mu$m spectral region, in three colours, to obtain chromatic information i.e., to differentiate vegetation, continents, oceans, etc.);
- the integration time should be sufficiently short to 'freeze' the planet's possible rotation and avoid a blurred image. In this calculation, the integration time is limited to one hour, but could be adapted according to the target if the rotation period were known;
- observation of the planet is limited by photon noise with a signal-to-noise ratio of 10 per pixel. As a result, an average flux of 100 planetary photons is required per bright pixel. (This relatively stringent assumption enables an estimate of the observation's minimal characteristics.);
- at 10 parsecs, an Earth-like body would have an angular diameter of $8.5 \times 10^{-6}$ arcsec;
- the overall transmission of the photometric chain (including the telescope and the detector's efficiency) is assumed to be equal to 0.2.

**Table 2.6** Parameters for imaging a terrestrial planet at a distance of 10 parsecs

| Image size (pixels) | Collecting surface (m$^2$) | Minimum baseline (km) | Integration time |
|---|---|---|---|
| $16 \times 16$ | 20 000 | 450 | 3 min |
| $32 \times 32$ | 20 000 | 900 | 12 min |
| $128 \times 128$ | 20 000 | 3600 | 50 min |
| $256 \times 256$ | 320 000 | 7200 | 50 min |
| $512 \times 512$ | 1 300 000 | 15 000 | 50 min |
| $1024 \times 1024$ | 5 000 000 | 30 000 | 1 h |

The minimum size of instruments required for planetary imagery clearly shows that we will have to abandon the idea of monolithic instruments and instead examine the possibility of imaging via interferometry.

The concept of imaging via interferometry is quite old, and several different configurations have been suggested. However, to obtain a field around the target, only the so-called Fizeau form is suitable. It is, nevertheless, very difficult to implement because it requires an optical system that guarantees homothetic mapping of the entry pupil to the exit pupil, and the accuracy of which increases as the spatial resolution increases (i.e., as the number of pixels in the image increases). If this homothetic mapping is not attained, the field is limited to the target direction (the centre of the image).

In addition, if we consider the baselines required to image planets (several 100 km), we are faced with the necessity of using small-sized telescopes (a few metres in diameter), separated by great distances (unless we use a large number of telescopes). As a result, if we want to preserve the homothetic mapping between the instrument's entry and exit pupils, the image will be very poor, and not capable of being exploited in terms of spatial frequencies. (Coverage of spatial frequencies is given by the pupil's autocorrelation function, which introduces many gaps in the frequency space if we use small telescopes separated by large distances) In other words, the image will be of very poor quality because it will contain few spatial frequencies.

An elegant solution for obtaining very-high-spatial-resolution imagery has been suggested by Labeyrie (1996), and this consists of combining the imaging capabilities of classic telescopes with the angular resolution given by interferometers. This concept, which has been called a 'hypertelescope' (Fig. 2.28), consists of several small telescopes that lie on the fictional mirror of a larger telescope (like Fizeau interferometers). An arrangement called a 'densified pupil' enables the frequency

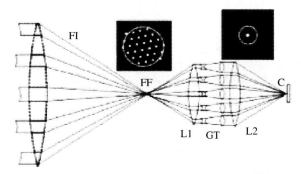

**Fig. 2.28** A diagrammatic representation of the principle of a hypertelescope. Several small telescopes synthesize a larger aperture (shown schematically by a single lens). The intermediate image (FF) lacks spatial frequencies because the pupil is only weakly sampled. The densified-pupil combiner ($L_1 + GT + L_2$) enables a PSF to be attained that is comparable to that of a monolithic telescope with a size equal to the sum of those of the small telescopes. This gain is at the expense of the field that is covered. A compromise must be made between field size and frequency coverage (After Labeyrie, 1996)

coverage to be increased by assuming an equivalence between the sub-pupils of the hypertelescope's entry pupil. Admittedly, one then loses the homothetic mapping of the pupil (and thus reduces the field observed), but the whole interest in this arrangement is to find the best compromise between field and frequency coverage. The other special feature of hypertelescopes is that the (virtual) primary mirror that is covered with the small, individual telescopes is spherical. (The reason being that it simplifies the problem.) An optical corrector (a Mertz corrector) is then used to re-establish a stigmatic configuration at infinity, and to restore a large field.

The hypertelescope concept is currently being investigated. The Carlina project (see Chap. 8) should, in a few years' time, allow it to be validated and allow the first actual instruments to be defined, with initial baselines of a few hundred metres. The extension of this concept to baselines measured in kilometres or even more, and particularly within the framework of a space-borne implementation, will need to be accompanied by complementary efforts into positioning in precise formation. A hypertelescope in space, with long baselines, cannot be designed as a rigid structure. The sampling telescopes will therefore need to be autonomous in operation and positioning. The overall instrument will need to be maintained in a precise position, most especially when being aimed at a target, and during the observation. Several studies are under way, most notably within the framework of the DARWIN project (a dark-fringe interferometer), to validate ideas about formation positioning. Be that as it may, imaging the surface of a terrestrial-like exoplanet remains an extremely difficult task, which will require several generations of instruments before it becomes a reality. Intermediate stages will be necessary, such as imaging a whole planetary system. Luckily, it is not necessary to image the surface of a planet to obtain spectroscopic information about the planet itself, and thus be able to study it. This less sensitive method will therefore be favoured initially.

### 2.3.5 Detection by Radio

The problem of direct detection of exoplanets is generally described in terms of contrast and angular separation. This is, after all, the way in which we have initially attacked the subject. There is one spectral region where the contrast between the stellar flux and the planetary flux may be very low: this is a radio wavelengths, in particular at decametric wavelengths. In fact, planets with magnetic fields produce non-thermal auroral emissions (i.e., interactions between charged particles and the magnetic field in the polar regions as in the aurorae that occur on Earth). The intensity of these emissions is comparable, at these frequencies, with the emission from the star itself. The auroral emissions are very specific (Zarka et al., 1997):

- they have a typical duration of 30–300 ms on the planets in our Solar System,

- their frequency spectrum is relatively uniform and broad (from less than 20 kHz to more than 40 MHz), with a typical spectral power that lies between 0.1 and 100 W Hz$^{-1}$, corresponding to a flux density that varies between 0.4 and 400 Jy.[7]

Detecting auroral emission from the Solar System's planets or from exoplanets is, however, rather tricky because of the presence of two main sources of parasitic signals:

- fluctuations in the sky background, of galactic origin, which have a very high brightness temperature (30 000–50 000 K at 25 MHz)
- parasitic signals of terrestrial (and human) origin. Such signals have the characteristic of being limited to relatively narrow frequency bands; and this obviously depends on the properties of the transmitter. Take, for example, CB[8] signals, which completely saturate a frequency band around 27 MHz. On the other hand, the bands allotted to these emissions are numerous, spread out, and constantly being developed (Denis and Zarka, 1996).

In addition, at these wavelengths, the angular resolution is very low: the antenna's lobe is relatively broad (about 1° for an antenna 1 km in diameter, so the signals from both the star and the planet are detected simultaneously within the lobe). This breadth tends to increase the contribution of the galactic background in the signal that is detected. It is interesting to note that in this case it is not the noise in the detection chain that limits the sensitivity of the method, but the galactic contribution and any parasitic signals. Any detector, for example, can work at ambient temperatures.

Radio astronomers have suggested an observational strategy appropriate to these signals, and its specific features are as follows (Zarka et al., 1997):

- a receiving surface with a large area, so that the angular size of the antenna's lobe is reduced. In practice, the UTR-2 array at the Radio Astronomy Institute at Kharkov (Ukraine), which is currently the largest, and which is being used pending the commissioning of the future LOFAR (LOw Frequency ARray) and SKA (Square Kilometric Array) systems. The north-south arm of UTR-2 has an effective area of about 50 000 m$^2$;
- the use of an acousto-optic spectrograph, which would enable time-frequency diagrams to be obtained with a temporal resolution (integration time) of about 250 ms, and spectral resolution of a few dozen kHz. The short duration of auroral emissions forbids the use of long integration times, because these would distinctly increase the contributions from parasitic signals and from the galactic background in every sample. It appears more sensible to integrate for short periods with durations that are comparable with those of the emission, and to eliminate empty samples so that only samples containing a signal are retained for integration;

---

[7] 1 Jy = 1 Jansky = $10^{-26}$ W m$^{-2}$ Hz$^{-1}$.

[8] Radio transmitting equipment, generally employed on vehicles. The development of mobile telephony (in a higher frequency region), however, is tending to marginalize their use, which appears to be restricted to professional drivers.

- handling the data with thorough observational and computational procedures, designed to eliminate parasitic signals (Zarka et al., 1997). In crude terms, this means detecting signals of terrestrial origin by observing two neighbouring areas of the sky, one on axis (and thus containing the astrophysical target), and the other off-axis. This technique may be used at Kharkov because the array's set of delay lines enables the array to be pointed in two different directions with an angular separation of $1°$, and observations carried out in both lobes simultaneously.

Given the level of the background and parasitic signals, and using the method just described, it may be shown that, using UTR-2, it would be possible to detect Jupiter at about 0.2 parsec. This is, unfortunately, insufficient to even offer the hope of detecting extrasolar planets (remember that the star closest to the Sun, Proxima Centauri, lies at a distance of about 1.3 parsecs). On the other hand, Zarka has also shown that, in the case of giant planets close to their parent star (i.e., hot Jupiters), one could expect a mechanism that generated radio emission $10^3$–$10^5$ times as strong as that produced by Jupiter. Under such circumstances, we could hope to detect emissions out to about 20–25 parsecs, which makes the method distinctly more attractive.

A series of observations has been made with UTR-2, but the complex processing to which the data are subject means that, to date, no candidates have yet been identified. When the SKA array is commissioned, it should increase the sensitivity of the method by reducing the size of the antenna's lobe and, as a result, the level of the galactic background.

# Bibliography

Beaulieu, J.-P., et al., 'Discovery of a cool planet of 5.5 Earth masses through gravitational microlensing', *Nature*, **439**, 437–440 (2006)

Bennett, D. et al., 'Discovery of a planet orbiting a binary star system from gravitational microlensing', *Nature*, **402**, 57–59 (1999)

Beuzit, J.-L., Mouillet, D., Lagrange, A.-M. and Paufique, J., 'A stellar coronograph for the COME-ON-PLUS adaptive optics system', *A&ASS*, **125**, 175–182 (1997)

Bordé, P., *Détection et caractérisation de planètes extrasolaires par photométrie visible et interférométrie infrarouge à très haute précision*, PhD thesis, Université de Paris-VI (2003)

Bracewell, R.N., 'Detecting nonsolar planets by spinning infrared interferometer', *Nature*, **274**, 780 (1978)

Brown, T.M., Charbonneau, D., Gilliland, R.L., Noyes, R.W. and Burrows, A., 'Hubble space telescope time-series photometry of the transiting planet of HD 209458', *Astrophys. J.*, **552**, 699–709 (2001)

Charbonneau, D., Brown, T.M., Latham, D.W. and Mayor, M., 'Detection of planetary transit across a sun-like star.', *Astrophys. J.*, **529**, L45–L48 (2000)

Charbonneau, D. et al., 'Detection of thermal emission from an extrasolar planet', *ApJ*, **626**, 523–529 (2005)

Davis, M.M., Taylor, J.H., Weisberg, J.M., Backer, D.C., 'High-precision timing observations of the millisecond pulsar PSR 1937 +21', *Nature*, **315**, 547–550 (1985)

Denis, L., Zarka, P., Interference problems at the Nançay Decameter Array and studies towards a better immunity, JCE Symposium "Interference problems in Radio Astronomy and Communications – or Cosmic Ecology", U.R.S.I. XXVth General Assembly, Lille, France, p. 752 (1996)

Einstein, A., *Science*, **84**, 506 (1936)

Goodman, J.W., *Introduction to Fourier Optics*, 2nd edn, McGraw Hill, Columbus, Ohio (1996)

Guyon, O., 'Phase-induced amplitude apodization of telescopes pupils for extrasolar terrestrial planet imaging', *Astron. Astrophys.*, **404**, 379–387 (2003)

Guyon, O., Pluzhnik, E.A., Galicher, R. and Martinache, F., 'Exoplanet imaging with a phase-induced amplitude apodization coronagraph. I. Principle', *ApJ*, **622**, 744–758 (2005)

Huchra, J., Gorenstein, M., Kent, S., Shapiro, I., Smith, G., Horine, E., Perley, R., '2237 + 0305 – A new and unusual gravitational lens', *Astronom. J.*, **90**, 691–696 (1985)

Koutchmy, S., 'Space-borne coronagraphy', *Space Sci. Rev.*, **47**, 95 (1988)

Labeyrie, A., 'Resolved imaging of extra-solar planets with future 10–100 km optical interferometric arrays', *A & A*, **118**, 517–524 (1996)

Léna, P., 'Observational Astrophysics', sec. Ed, Springer, Berlin (1998)

Lyot, B., 'Photographie de la couronne solaire en dehors des éclipses', *C.R. Acad. Sci. Paris*, **193**, 1169 (1931)

Lyot, B., 'A study of the solar corona an prominences without eclipses', *Mon. Not. R. Astron. Soc.*, **99**, 580 (1939)

Malbet, F., 'High angular resolution coronography for adaptive optics', *A&ASS*, **115**, 161–174 (1996)

Racine, R., 'Continuum and semiforbidden C III microlensing in Q2237 + 0305 and the quasar geometry', *ApJ*, **395**, L65–L67 (1992)

Rouan, D., Riaud, P., Boccaletti, A., Clénet, Y. and Labeyrie, A., 'The four-quadrant phase-mask coronagraph. I. Principle', *PASP*, **777**, 1479–1486 (2000)

Roddier, F. and Roddier, C., 'Stellar coronograph with phase mask', *PASP*, **109**, 815–820 (1997)

Sackett, P. 'Searching for unseen planets via occultation and microlensing' in *Planets Outside the Solar System: Theory and Observation*, Mariotti, J.-M and Alloin, D. Eds, NATO-ASI Series, Kluwer, 189–227 (1999)

Wolszczan, A. and Frail, D., 'A planetary system around the millisecond pulsar PSR1257+12', *Nature*, **255**, 145–147 (1992)

Zarka, P., Queinnec, J., Ryabov, V., Shevchenko, V., Arkhipov, A., Rucker, H., Denis, L., Gerbault, A., Dierich, P., Rosolen, C., 'Ground-based high sensitivity radio astronomy at decameter wavelength', in *Planetary Radio Emissions IV*, (eds) Ruckern, H.O. Bauer, S.J. and Lecacheux, A., Austrian Acad. Sci. Press, Vienna (1997)

# Chapter 3
# Extrasolar Planets, 12 Years After the First Discovery

The first detection of an extrasolar planet orbiting a solar-type star, that was announced as such, took place in 1995. Some 13 years after this major discovery, more than three hundred planets have been discovered, primarily by the method used to detect 51 Peg b – radial-velocity determinations – but also by the transit method, which is beginning to bear fruit. These detection methods are 'indirect' (we do not 'see' the planet, but instead deduce its presence from the effect that it has on its central star). They do, however, allow us (from knowledge of the characteristics of the central star) to estimate certain planetary parameters, in particular:

- the mass of the body (or, more precisely, the product $m.sin(i)$, where $i$ is the angle at which the system is observed ($i = 90°$ if the system is seen edge-on)
- the semi-major axis of the orbit (deduced from Kepler's Third Law – cf. Appendix)
- the eccentricity of the orbit
- the possibility that the system may be multiple
- the radius of the planet when the system is observed simultaneously by radial-velocity measurements and in transit.

In this chapter we shall discuss these planets, and determine, from a statistical point of view, the preliminary information that these objects provide. All the graphs and statistical representations shown in this chapter have been prepared from the list of objects available on the site: exoplanet.eu, which is maintained by J. Schneider.

Figure 3.1 shows a histogram plot of the annual totals of planetary discoveries. It will be seen that the annual totals have been practically constant since about 2000, which indicates that the detection methods currently employed have got into their stride. It will undoubtedly be necessary to implement new methods to refine our knowledge of the properties of the objects that have already been detected, and to discover others with different characteristics, in particular, less massive bodies.

M. Ollivier et al., *Planetary Systems*. Astronomy and Astrophysics Library,
DOI 978-3-540-75748-1_3, © Springer-Verlag Berlin Heidelberg 2009

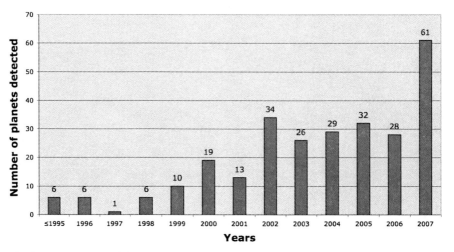

**Fig. 3.1** Annual totals of detected exoplanets. It is obvious that the yearly total has increased a lot since 2007 after five years where it was more or less constant

## 3.1  Exoplanets and Exoplanetary Systems

By the beginning of September 2008[1], 306 planets have been detected, including those around the pulsar PSR 1257 + 12. They fall into 262 planetary systems, of which 31 are multiple systems that may include as many as five planets (the 55-Cancri system, for example). Most of these planets are giant planets, because the methods used to detect them (mainly radial-velocity measurements and/or transit measurements from the ground) are method which are particularly sensitive to this kind of object. However, microlensing and pulsar timing methods as well as radial-velocity measurements around low mass stars have enabled several objects of a few Earth-masses to be detected close to their parent star. The current statistical picture will therefore be rendered complete only when it has become possible to identify planets with lower masses. The arrival of satellites for the observation of transits from space (primarily COROT and KEPLER), should enable us to achieve this goal. Given the length of time that observations have been made (some ten years at most), it has only been possible to confirm the existence of planets with periods that do not exceed a few years (again, up to about ten years).

## 3.2  The Mass-Distribution of Exoplanets

Searching for companion bodies by the measurement of radial velocities (i.e., the measurement of the Doppler shift in the spectrum of the parent star), is not specific to the detection of exoplanets. It enables us to identify all types of multiple system,

---

[1] The date on which these numbers were updated was 1 September 2008

and thus companions, whether they are stellar (double or multiple stars, known as 'spectroscopic' doubles), sub-stellar (brown dwarfs), or planetary. By measuring $M.sin(i)$ (where $i$ is the angle at which the system is viewed; with $i = 90°$ the system is viewed side-on), we obtain a lower limit for the mass of the companion. Before concentrating specifically on exoplanets, it is interesting to examine the distribution of the masses of companions, both stars and planets. Figure 3.2 shows this distribution for some one hundred objects.

It is immediately obvious that the mass distribution is bimodal, with planets $(M < 0.01\ M_\odot)$ on the left[2], and stars $(M < 0.08\ M_\odot)$ on the right. Between the two there are very few objects. This zone, known since the 1980s, is called the 'brown dwarf desert'. It was revealed by the first programmes devoted to the systematic search for sub-stellar companions, carried out on a few tens of objects, and which indicated the relative non-existence of this type of object (Campbell et al., 1988, Marcy and Benitz, 1989). Current programmes for monitoring radial velocities have only allowed the identification of about 20 brown-dwarf candidates (where the value of $M.sin(i)$ that has been measured lies within the range of brown-dwarf masses).

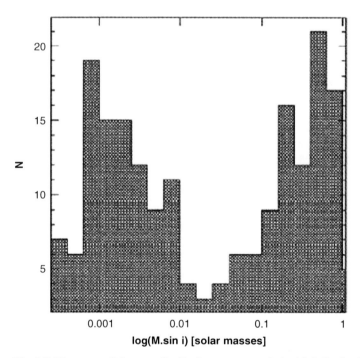

**Fig. 3.2** Histogram of the mass-distribution – more precisely, $M.sin(i)$ - for the stellar and sub-stellar companions discovered by the radial-velocity method. It is obvious that the distribution is bimodal with planets on the left, and stars on the right. The poorly populated central zone between the two peaks (where $0.01\ M_\odot < M.sin(i) < 0.08\ M_\odot$) is known as the 'brown dwarf desert' (After Santos et al. (2002))

---

[2] $1\ M_\odot$ = mass of the Sun

Use of data from the European astrometry satellite Hipparcos has, however, enabled seven of the candidates detected by the ELODIE instrument (11 candidates in the histogram shown in Fig. 1.2), to be definitely excluded[3] (Halbwachs et al. 2000). This particular result tends to show that the 'brown dwarf desert' is not only real, but, in addition, that the brown-dwarf candidates detected by radial-velocity measurements are, probably in the majority of cases, objects with significantly greater masses, where $sin(i)$ is low. This significant separation between planets and stars may be explained – or at least understood – when the standard models for the formation of stellar and planetary bodies are considered. Stars form by the gravitational collapse of an even more massive nebula (of some $100\,M_\odot$), whereas planets, even giant ones, start in the form of embryonic nuclei and then grow by the subsequent accretion of matter. In the first case, the process is one of fragmentation from high-mass objects to less massive ones, whereas in the second case, the embryonic nucleus becomes a planet through the processes of accumulation and accretion. These two processes involve various physical phenomena which have different governing parameters, and which explain the different scale limits.

If we now consider just exoplanets, we can derive the same type of histogram for the 306 planets currently known (Fig. 3.3).

Examination of Fig. 3.3 clearly shows several points:

- although the sensitivity of the radial-velocity method (which has currently provided more than 90 per cent of candidate planets) increases as the mass of the object increases, almost half of the planets have a mass less than, or comparable with that of Jupiter. High-mass planets (several Jupiter masses) therefore seem to be rare, which gives us an indication of planetary systems' formation processes. It should, however, be mentioned that the distribution in Fig. 3.3 shows the product of the mass of the object and the sine of the angle of inclination to the line of sight (which may be derived from the radial-velocity measurement). The mass that is deduced is therefore a lower limit for the real mass of the objects. However, and making allowance for the size of the sample, it is possible to deconvolve the histogram (Santos et al., 2002), and this confirms the real tendency shown in the histogram. High-mass planets are, therefore, truly less numerous;
- The mass-distribution between 0.1 and 2 Jupiter masses is more-or-less uniform, which tends to show that, for planets, the whole range is covered. So it is obvious that there was segregation by mass over this range when the formation processes took place for these objects;
- There are more than a dozen objects whose mass is less than one tenth of a Jupiter mass. This observation is the result of the detection of objects with lower and lower masses (masses comparable with those of Uranus and Neptune), and tends to suggest that the range of masses extends as far as the domain of terrestrial masses. This result is confirmed by the first detections of several Earth-mass objects by microlensing and radial velocity measurements.

---

[3] The astrometric data from Hipparcos enable the uncertainty over the value of $sin(i)$ to be resolved.

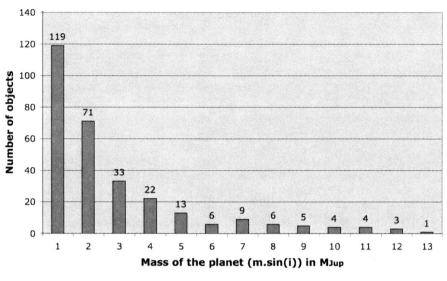

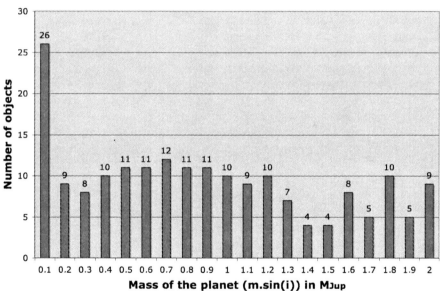

**Fig. 3.3** Mass distribution of exoplanets detected to date. top, the total population; bottom, detail between 0 and 2 Jupiter masses (the mass of Jupiter is about $10^{-3}$ solar masses)

## 3.3 The Distance-Distribution of Exoplanets

The semi-major axis of a planetary orbit may be deduced from the measurement of
the orbital period by applying Kepler's Third Law, provided one can estimate the
mass of the central star, which is derived from measurement of the spectral type
(cf. Appendix). The distribution of the semi-major axes of exoplanets is given in
Fig. 3.4.

This statistical analysis provides various items of information:

- Nearly half of the objects orbit at distances less than 0.4 AU (which is the distance between the Sun and Mercury in the Solar System). Among these, about
  one third orbit at distances that are extremely close to their stars, typically 0.05
  AU, and thus with orbital periods of just a few days. The first planet detected
  by radial-velocity measurements (51 Pegasi) is the prototype for these planets,
  which were completely unknown until the first observation, and which have been
  called 'hot Jupiters' because of their size (being giant planets) and distance from
  their parent stars. Objects with a lower mass (a mass comparable with that of
  Neptune or Uranus) have since been detected in orbits that are as close to their
  parent star, and have thus been described as belonging to the 'hot Uranus' or
  'temperate Uranus' class, as a result of their distance from the star and thus of
  the temperature that the object's thermal equilibrium implies. Apart from the fact
  that this type of object is completely unknown in the Solar System, the discovery of the first hot giant planets poses many questions about their formation and
  evolution. In fact, current models of protoplanetary nebulae make it impossible
  for these planets to have formed where they are currently found, primarily because the temperature would have been too high to allow the grains of the core
  to condense, and also because of the density of the disks. We must, therefore,
  envisage that these objects formed at greater distances from their stars and that
  the parameters of their orbits have evolved over the course of time, so that they
  attained their current position through 'orbital migration'. The question of orbital migration is considered in Chap. 6. The consequences for a planet of having
  an orbit close to its parent star are numerous, because the planet is permanently
  strongly irradiated, and the spectrum of these objects is completely different from
  that of a classical Jupiter-type planet, such as we know from the Solar System. In
  addition, taking account of the proximity of the star, the orbit of the planet is generally practically circular (or of low eccentricity), and the period of revolution is
  probably equal to the period of rotation (i.e., there is spin-orbit synchronization).
  This point is considered in Chap. 7.
- With the exception of the hot objects, the distribution of objects with distances
  between 0 and 5 AU is relatively uniform. These observations do not contradict
  all the theoretical approaches, which show that certain orbits are more stable
  than others in a given planetary system. The diverse range of parent stars, most
  notably with regard to mass, tends to smooth out the distance distribution.
- The detail shown in Fig. 3.4 for distances between 0 and 1 AU indicates that the
  distribution of distances is not uniform when seen at that scale. There does, in

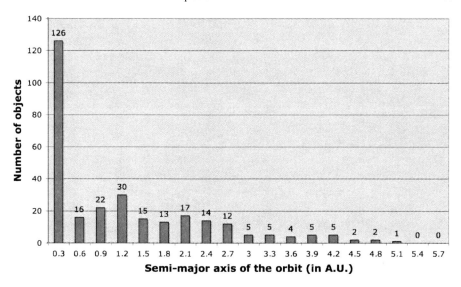

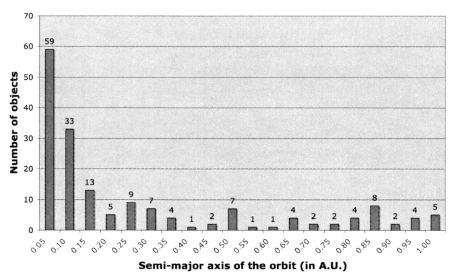

**Fig. 3.4** Distribution of the semi-major axis of the orbits of exoplanets. Bottom: detail of the region between 0 and 1 AU

fact, appear to be a relative deficit of objects around 0.4 AU. This limit seems to correspond to a distinction between objects that have migrated, and whose orbits are therefore not primitive ones, and other objects which may, potentially, have formed at the location where they are currently found. It should be pointed out that this observation is not the result of bias, because the orbital parameters are directly derived from the radial-velocity measurements, which are greater (and measurement easier), the larger the mass of the object, and the closer it is to the parent star.

Study of certain multiple systems has also revealed that various planets may occupy resonant orbits (cf. Appendix for the definition of resonance). This is particularly the case for the system Gliese 876, where three planets have been identified. One of these (Gl 876 b) has a period of 60.94 days, while another (Gl 876 c) has a period that is practically half that, 30.1 days, and the third component (which is more controversial) is a hot object with a period that is less than 2 days. Resonance effects are discussed in detail in Chap. 6.

## 3.4 The Relationship Between the Mass of Exoplanets and Their Distance from Their Star

In the Solar System, the most massive planets, the four giant planets, lie at great distances from the Sun, and the less massive ones are closer[4]. Is this separation found in exoplanetary systems? The mass/distance distribution for roughly the first 100 planets discovered and for the 209 currently known is shown in Fig. 3.5.

Analysis of the distribution for the first 100 objects clearly revealed several specific points [After Zucker and Mazeh (2002); Udry et al. (2003)]:

• the absence of high-mass objects at small distances (on the face of it, 'hot super-Jupiters' do not appear very numerous)
• the absence of any objects with masses below that of Jupiter between 0.5 and 5 AU.

At the time, any suggestion of observational bias was ruled out by statistical arguments based on the sensitivity of the method of detection by means of measurements of the radial velocity of the parent star (which, it may be recalled, is the more sensitive, the greater the mass of the planet and the closer it is to the star).

The analysis may be repeated for the current state of detection (Fig. 3.5, *right*), and this shows that the situation is not really any different. Apart from a few exceptions, which may be explained by the fact that measurement of the mass is marred by an unknown in the inclination of the planetary system (which only gives a lower

---

[4] According to its new definition, established at the General Assembly of the International Astronomical Union, in August 2006, the Solar System consists of just 8 planets, 4 terrestrial ones close to the Sun (Mercury, Venus, Earth and Mars), and 4 giants at greater distances (Jupiter, Saturn, Uranus and Neptune), with minor planets (asteroid) more-or-less scattered throughout the system.

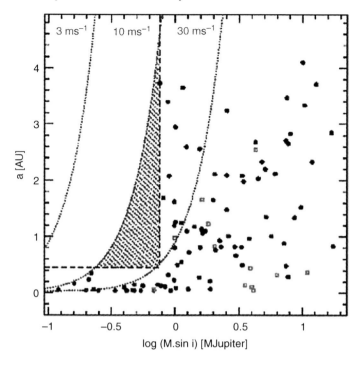

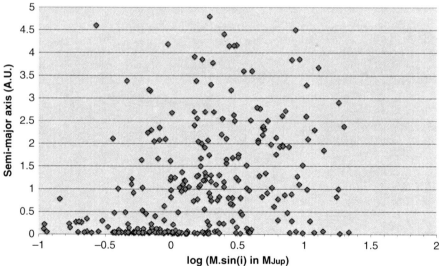

**Fig. 3.5**  The relationship between planetary mass and distance from the star: (top) for the first 100 objects [After Udry et al. (2003)]; the limit for detection at 10 m.s$^{-1}$ corresponds to the sensitivity limit for instruments in 2003; (bottom) for the objects currently listed

limit), and by the fact that the sensitivity of our instruments has increased, we are forced to admit that the observations in 2002–2003 remain completely valid.

An explanation of these observations may derive from the analysis of the effectiveness of the processes of migration as a function of the mass of the objects (even if in the type II migration process – i.e., that involving massive planets (cf. Sect. 6.3) – it is solely the viscosity of the protoplanetary disk that is involved, and not the mass of the planet). For the most massive objects, with a greater inertia, the migration process is less effective, and as a result they tend to remain in their initial position, at a great distance from the star. By contrast, objects of low mass where the migration process is, on the face of it, more efficient, are statistically more abundant at small distances. There is a detailed discussion of this point in Udry et al. (2003). A complete description of the planetary migration process is given in Chap. 6.

It should also be noted that a new generation of instruments has begun operation, and which has a sensitivity better than $1 \text{ m.s}^{-1}$ (for example, the HARPS instrument on the ESO 3.6-m telescope at La Silla), and that certain new objects, notably of very low mass, have been discovered by this means.

## 3.5 Orbital Eccentricity Among Exoplanets

Determination of the eccentricity of the orbit of a planet, detected using the radial-velocity method, is carried out by analyzing the deviations of the radial velocity from the sine wave displayed by the radial velocities (cf. the previous chapter). Figure 3.6 shows the eccentricity of exoplanets as a function of their orbital period.

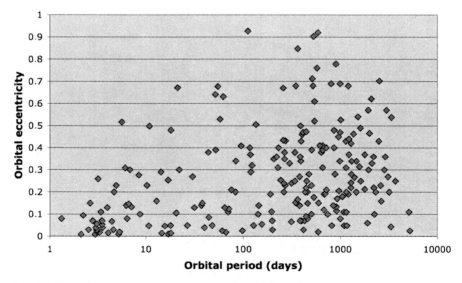

**Fig. 3.6** Eccentricity of exoplanets as a function of their orbital period

Figure 3.6 basically shows three points:

- The short-period objects (with periods shorter than a few days) have practically circular orbits
- Above a few days, there is no obvious correlation between the period and the orbital eccentricity
- Unlike our Solar System, where eccentricities are low[5], some planetary systems exhibit eccentricities that are extremely large (and may even attain 0.92)
- The distribution of eccentricities as a function of orbital period is perfectly comparable with those observed in multiple stellar systems (in binary stars, for example).

These observations may be explained by recourse to the following arguments: at short distances, the tidal effects are greater. The gravitational-potential gradient is stronger than at greater distances, which creates a differential attraction between the inner side and the outer side of the planet, which results in the formation of a bulge on the exoplanet. This bulge leads to the appearance of a torque exerted by the star on the planet, causing the dissipation of energy through friction in the planetary mantle. The dissipation of energy tends to circularize the orbit of the planet. At the same time, there should also be progressive synchronization of the periods of the planet's revolution and rotation[6] as with the Moon in its orbit around the Earth. It may be shown that the time required to circularize an object whose orbital period is 7 days, amounts to about 1000 million years (Halbwachs et al., 2005). When the average age of the objects observed is taken into account, this explains the limit at a few days seen in Fig. 3.6. Above this figure of a few days, the objects' orbits are not yet completely circularized, and the distribution of eccentricities becomes more dispersed. (It should be noted that no object with a period less than 30 days has an eccentricity greater than 0.5.)

To explain the fact that the distribution of eccentricities as a function of orbital period does not display any obvious correlation for periods about a few days – unlike the case with multiple stellar systems – the following argument may be used. At a greater distance, the tidal effects (differential attraction) become negligible relative to the interaction between the planet and the protoplanetary disk in which it is forming (a theory proposed by Goldreich and Sari in 2003), or relative to the interactions between the planet and its planetary and stellar environment (a theory by Marzari and Weidenschilling in 2002). The latter interactions, however, tend to increase the orbital eccentricity, which was originally low at the time of the planet's formation, through the transfer of angular momentum (see Chap. 6).

---

[5] The planet with the greatest eccentricity (after the eccentricity: 0.246, of Pluto, which has just lost its status as a planet), is Mercury with an eccentricity of 0.206. The other planets all have eccentricities less than 0.1, which is very low when compared with what is observed with certain exoplanets.

[6] This is probably not without consequences for the planet's climate, because it always turns the same face towards the star. This point is discussed in Chap. 8.

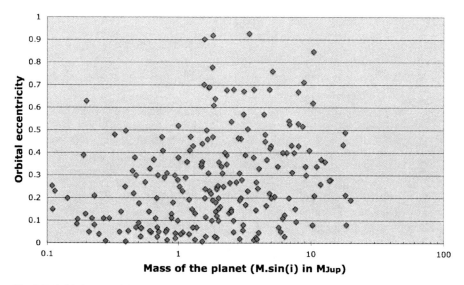

**Fig. 3.7** Orbital eccentricity of exoplanets as a function of their mass

The absence of massive objects in the immediate vicinity of the Sun may be invoked to explain the fact that eccentricities in the Solar System have remained small.

It should also be noted that certain theoretical studies predict that the eccentricity of planetary orbits increases with the mass of the planet. Figure 3.7, which represents the eccentricity of exoplanets as a function of their mass, does not, however, reveal any obvious correlation. So there is no observational support for these theoretical conclusions.

## 3.6 Exoplanets and Their Parent Stars

The diversity of parent-star types is one of the highlights of the currently known distribution of planetary systems. Planetary systems have been identified around a wide range of objects:

- main sequence stars for spectral types F to M
- giant (e.g. HD 122430); sub-giant (e.g. HD 47536) dwarf (e.g. HD 209458), and sub-dwarf stars (e.g. V391 Pegasi)
- stellar remnants (e.g. pulsar PSR 1257+12)
- young objects (e.g. 2MASS 1207)
- multiple stars (e.g. Gl86))
- protoplanetary disks
- pulsating stars
- ...

Let's consider now the metallicity of parent star. To astronomers – to put it rather dramatically – the periodic classification of the elements has, in the end, come down to just three categories:

- hydrogen, whose relative abundance (the partial fraction) by mass is denoted by $X$,
- helium, whose relative abundance by mass is denoted by $Y$,
- all the other heavier elements, known generically as 'metals', whose overall relative abundance is denoted by $Z$ or $Fe$.

By definition, we have $X + Y + Z = 1$, and the 'metallicity' of a star is the value of $Z$. For the Sun, for example, $X = 0.73$, $Y = 0.25$, and $Z = 0.02$. (These abundances are known as 'cosmic abundances'.) For convenience, the logarithm of the abundance of iron in a star, relative to the abundance of iron in the Sun is also known as the metallicity (with no risk of confusing it with $Z$). It is denoted [Fe/H] and given by the ratio:

$$[Fe/H] = \log\left(\frac{[Fe]/[H].}{[H]/[Fe].}\right) \tag{3.1}$$

This value is zero for stars with the same metallicity as the Sun, negative for objects of lower metallicity, and positive for stars with a greater metallicity. Stars that have a higher metallicity than the solar metallicity (or cosmic abundances) are said to exhibit a 'metallicity excess'.

The metallicity of a star is measured spectroscopically in the visible or even the near UV region. The metal abundances are obtained by comparing the depths of the spectral lines associated with certain electronic transitions of the metallic atoms (Fe, Ti, etc.).

Figure 3.8 shows the distribution of metallicity of the stars that have been identified as having planets.

In examining Fig. 3.8, we can see that stars that have exoplanets are generally stars with a metallicity excess (about two-thirds of the stars in this histogram have a metallicity that is greater than the metallicity of the Sun).

This result, first published in 2002, was initially subject to some discussion, because some people saw the excess metallicity as a sign of bias in the choice of the stellar sample that had been used for the radial-velocity method. After comparison and study of various stellar samples (defined, as far as M. Mayor's team were concerned, as being all the stars of one or several, given, spectral types out to a certain distance from the Sun), this result is now accepted by the general astronomical community.

This observation is not really surprising, because in the standard model for the formation of planetary systems, planets form within a disk of dust and gas around a young star (cf. Chap. 5). The composition of this disk is similar to the composition of the star. To form planets, it is initially necessary to create rocky kernels that will form the cores of future planets. Now, these kernels consist of the metallic elements (as we have defined them). So it is not surprising that we should find planets around metallic stars.

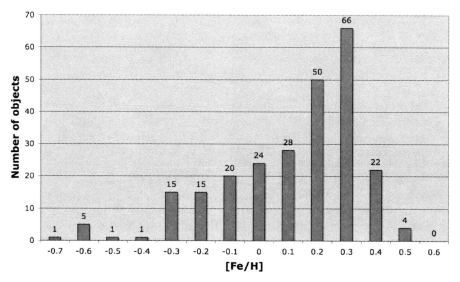

**Fig. 3.8** Metallicity of stars with exoplanets

Another interpretation that is possible from these data rests on the fact that systems of high metallicity preferentially form more massive planets, which are thus easier to detect with current methods. Finally, we should note that the Solar System is actually located in the metallicity zone where the probability of finding planets is low ...

Another theory proposed to explain the excess metallicity is the presence of planets that have been engulfed and increased the metallic concentration of the star. Even if the hypothesis of the primordial composition of the protoplanetary disk is generally preferred, the hypothesis of planets being engulfed is not completely eliminated, and has been reconsidered using recent models of the convective layers in stars.

## 3.7 Mass/Diameter Ratio

Some objects (a good dozen) have been observed by two different methods: the radial-velocity method, which gives the mass of the object; and the transit method, which gives the diameter[7]. This enables us to obtain information about the average density of the planet. In certain cases, the transit has been observed by space telescopes (HST, Fuse, Spitzer, etc.), allowing accurate and chromatic measurements of the planet's diameter, leading to information about its atmospheric structure and composition (as described in the next paragraph). Figure 3.9 shows the relationship between the masses of planets observed in transit and their radii. Theoretical studies

---

[7] When a planet is observed in transit, the planetary system is seen from the side, so that the value of the radial velocity measured gives the mass directly, because $(sin(i) = 1)$.

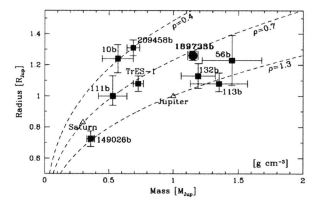

**Fig. 3.9** Mass/radius diagram for giant planets observed in transit

of the internal structure of these objects have enabled us to estimate the mass of the rocky cores. A complete discussion of this point is given in Chap. 7.

## 3.8 Characteristics of Extrasolar Planetary Atmospheres

Even if, 2MSS 1207 excepted, no direct detections *sensu stricto* have been achieved, the observation of transiting extrasolar systems during transits both in front of, and behind their stars (*see* Chap. 3) has enabled the identification of atmospheres and several compounds within them. The first identification of an atomic element was that of sodium in the atmosphere of HD 209458b by Charbonneau et al. in 2002. Using the FUSE satellite data, Vidal Madjar et al. (2004) then announced the observation of atomic hydrogen, carbon, and oxygen, and concluded that the atmosphere was evaporating. Several observations using the HST provided constraints on the radius of several planetary candidates as a function of the wavelength, thus probing the structure of their atmospheres. Recently, the observation of several objects in the thermal infrared using the Spitzer telescope had enabled several molecules to be identified in certain atmospheres and, in particular, $H_2O$ in the atmosphere of HD 209458b (Berman, 2007). Richardson et al. (2007) published the identification (by the same techniques) of silicate features in the same object, interpreted as the presence of clouds in the atmosphere. The detection of $CH_4$ in the atmosphere of HD 189733b has also been announced recently by Swain et al. (2008)

To conclude, the study of the objects detected in the last 10 years allows us to gain an initial statistical view of giant exoplanets orbiting stars close to the Sun. This picture will be rendered complete by:

- the exploration of more distant regions of the Galaxy to confirm the (implicit) theories that the distribution is spatially homogeneous
- study and analysis of terrestrial-type objects.

These two points are the focus of the next five years of research, in particular with the advent of dedicated observations from space, such as those provided by COROT and KEPLER.

# Bibliography

Barman T., 'Identification of absorption features in an extrasolar planet atmosphere', *Astrophys. J.*, **661**, L191–L194 (2007)

Campbell, B., Walker, G.A.H. and Yang, S., A search for substellar companions to solar-type stars, *Astrophys. J.*, **331**, 902–921 (1988)

Charbonneau, D., Brown, T.M., Noyes, R.W. Gilliland, R.L., 'Detection of an extrasolar planet atmosphere', *Astrophys. J.*, **568**, 377–384 (2002)

Goldreich, P. and Sari, R, Eccentricity evolution for planets in gaseous disks, *Astrophys. J.*, **585**, 1024–1037 (2003)

Halbwachs, J.-L., Arenou, F., Mayor, M. Udry, S. and Queloz, D., Exploring the brown dwarf desert with Hipparcos, *Astron. Astrophys.*, **355**, 581–594 (2000)

Halbwachs, J-L., Mayor, M., Udry, S., 'Statistical properties of exoplanets. IV. The period-eccentricity relations of exoplanets and of binary stars', *A & A*, **431**, 1129–1137 (2005)

Marcy, G.W. and Benitz, K.J., A search for substellar companions to low-mass stars, *Astrophys. J.*, **344**, 441–453 (1989)

Marzari, F. and Weidenschilling, S.J., Eccentric extrasolar planets: the jumping Jupiter model, *Icarus*, **156**, 570–579 (2002)

Mayor, M. and Queloz, D., A Jupiter-mass companion to a solar-type star, *Nature*, **378**, 355 (1995)

Richardson, L.J., Deming, D., Horning, K., Seager, S., Harrington, J., 'A spectrum of an extrasolar planet', *Nature*, **445**, 892–895 (2007)

Santos, N.C., Mayor, M., Queloz, D. and Udry, S., Extra-solar planets, *The Messenger*, **110**, 32–38 (2002)

Swain, M.R., Vasisht, G., Tinetti, G., 'The presence of methane in the atmosphere of an extrasolar planet', *Nature*, **452**, 329–331 (2008)

Udry, S., Mayor, M. and Santos, N.C., Statistical properties of exoplanets: I. The period distribution: constraints for the migration scenario, *Astron. Astrophys.*, **407**, 369–376 (2003)

Vidal-Madjar, A., Désert, J-M., Lecavelier des Etangs, A., Hébrard, G., Ballester, G.E., Ehrenreich, D., Ferlet, R., McConnell, J.C., Mayor, M., Parkinson C.D., 'Detection of oxygen and carbon in the hydrodynamically escaping atmosphere of the extrasolar planet HD 209458b', *Astrophys. J.*, **604**, L69–L72 (2004)

Zucker, S. and Mazeh, T., On the mass-period correlation on the extrasolar planets, *Astrophys. J.*, **568**, L113–L116 (2002)

# Chapter 4
# What we Learn from the Solar System

As we have mentioned earlier (Sect. 1.4.3), it has been the study of the Solar System that has led us to favour the model of star formation that envisages the collapse of an interstellar nebula. In this chapter we intend to examine in more detail how a coherent scenario for stellar and planetary formation has been reached. This takes overall account of the observations that have been accumulated about the objects around us. These observations concern not only the dynamics of the objects (their orbits, and the way in which their paths have evolved), but also their physical properties (mass, density, atmospheric composition, nature of their surfaces, etc.). More comprehensive discussions of the Solar System may be found in Lewis (1997), Pater and Lissauer (2001), and Encrenaz et al. (2004).

## 4.1 Observational Methods

For several centuries, observation of the planets was limited to visual observation, which allowed the orbits of the brightest planets to be determined. At the beginning of the 17th century, Galileo's use of the refractor that has since carried his name brought a decisive advance in the observation of planets and their systems (the Galilean satellites being discovered in 1610). Planetary exploration was carried out and advanced by the observations of Cassini and Huygens, in particular, and by the construction of major observatories. A century later (1781) with Herschel, the arrival of large telescopes brought the discovery of a new planet, Uranus, invisible to the naked eye, at a distance of about 20 AU from the Sun. In 1846, Le Verrier, based on perturbations in the orbit of Uranus, deduced the existence of an eighth planet, and calculated its orbit. Neptune was discovered near the position indicated, at a distance of nearly 30 AU from the Sun. In 1930, Pluto was discovered by C. Tombaugh; its semi-major axis is approximately 40 AU. It was classed as a planet, and remained one until 2006.

Do other planets exist beyond Pluto? This question, which remained open throughout the 20th century, took on a new significance in the 1990s with the discovery by D. Jewitt and J. Luu of a new population of objects, the TNOs (Trans-Neptunian Objects). These occupy the Kuiper Belt, most lying between 30 and

M. Ollivier et al., *Planetary Systems*. Astronomy and Astrophysics Library, DOI 978-3-540-75748-1_4, © Springer-Verlag Berlin Heidelberg 2009

**Fig. 4.1** Artist' view of the Cassini probe (© NASA/ESA)

50 AU. Nowadays, it seems that Pluto is one of the largest representatives of this new class of objects. In 2003, another TNO (2003 UB 313) was discovered, with a larger diameter. Other more distant objects will probably be discovered, thanks for the use to ever more powerful telescopes. In parallel with imaging and photometric observations, spectroscopy, which has been carried out since the beginning of the 20th century, has provided information about the chemical composition of the planets and their atmospheres.

Alongside telescopic observations, the exploration of space, which started at the beginning of the 1960s, has revolutionized our knowledge of the planets in the Solar System. After the Moon, which was explored in-situ by the Apollo missions, and whose success is well-known, Venus (with the Venera, Pioneer Venus and Magellan probes), and Mars (with Mariner 9, Vikings 1 and 2, Mars Global Surveyor, Mars Odyssey, Mars Express, and the Spirit and Opportunity rovers) were the various space agencies' favoured targets. Despite several failures, the various space missions have returned a rich harvest of data on the nature of the surface and atmosphere of the terrestrial planets (Fig. 4.1). As for the giant planets, they have also greatly benefited from space exploration. After the fly-bys of Jupiter and Saturn by Pioneers 10 and 11, the Voyager 1 and 2 probes flew past the four giant planets between 1979 and 1989, completely altering our knowledge of these objects. More recently, the Galileo and Cassini missions, orbiting Jupiter (1995–2003) and Saturn (from 2004, Fig. 4.2), have carried out an in-depth investigation into the properties of the two planets, with that for Saturn still in progress.

**Fig. 4.2** The planet Saturn as imaged by the Cassini probe (image credit: courtesy NASA, JPL Space Science Institute)

## 4.2 The Observational Data

In this section we intend to summarize the observational features that form the basis for the model of the Solar System's formation. These relate to the orbital properties of the planets; their physical properties; the properties of the small bodies; and the Solar System's age.

### 4.2.1 Orbits that are Essentially Co-Planar and Concentric

The first, and most obvious, feature, is the regularity of the orbits of the planets: most of them are co-planar, quasi-circular, and close to the plane of the ecliptic (defined as the plane of the Earth's orbit). The only exception is the eccentricity of Mercury (which exceeds 0.2, whereas all the others are less than 0.1), and its inclination (7°, whereas all the others are less than 4° – see Table 1.1 in Chap. 1). With the exception of Venus and Uranus, all rotate in a direct sense, which is also the direction of rotation of the Sun itself, as seen from the north pole of the ecliptic. This regularity of the orbits is also found in many of the asteroids and TNOs. It also occurs among the orbits of the regular satellites of the giant planets. We shall see that this feature is also relevant to the formation of these planets.

### 4.2.2 Terrestrial Planets and Giant Planets

The planets in the Solar System fall into two categories, the terrestrial planets and the giant planets. The terrestrial planets (Mercury, Venus, the Earth, and Mars) are dense and small in diameter (Table 1.2, Chap. 1); they either have no satellites or very few. The giant planets (Jupiter, Saturn, Uranus, and Neptune) by contrast are very large in size, have low densities, have systems of rings, and a large number of satellites, many of which occupy circular orbits in the planets' equatorial planes. The giant planets therefore appear to be Solar Systems in miniature.

**Fig. 4.3** The planet Mercury, photographed by the Mariner 10 probe (© NASA)

Mercury (Fig. 4.3), close to the Sun and of low mass, does not have a sufficiently strong gravitational field to preserve a stable, neutral atmosphere. The atmospheres of the other terrestrial planets basically consists of $CO_2$ and $N_2$, with $H_2O$ being also present on the Earth and Mars. In the case of Earth, the $CO_2$ is held in the oceans, and $O_2$ has appeared following the development of life. In contrast, the atmospheres of the giant planets (Fig. 4.4) is dominated by hydrogen and helium. Other elements are primarily present in reduced form ($CH_4$, $NH_3$, $PH_3$, $H_2S$, etc.). We shall see how measurements of the abundances of the elements provides constraints on models for the formation of the giant planets.

### 4.2.3  The Small Bodies

The main concentration of asteroids (or minor planets) extends between the orbits of Mars and Jupiter (Fig. 4.5). With diameters of less than 1000 km, they main occur

**Fig. 4.4** The planet Jupiter, photographed by the Cassini probe during its fly-past in December 2000 (image credit: courtesy NASA)

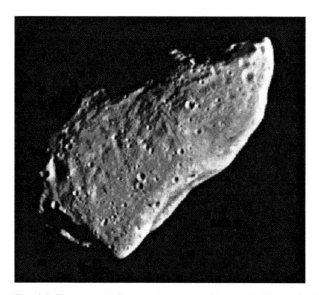

**Fig. 4.5** The asteroid Gaspra, photographed by the Galileo probe (© NASA)

within the Main Belt, which lies between 2.0 and 3.5 AU in heliocentric distance. We will see that, in the model for the formation of the Solar System, their presence is interpreted as being the residue of a disk of planetesimals, that has been dispersed by direct or indirect gravitational perturbations by Jupiter. The chemical composition of the asteroids varies with their heliocentric distance, with the most dense closer to the centre. Simulations of the dynamical history of the asteroids provide constraints on the age of the Main Belt and the date of Jupiter's formation.

A wide range of small bodies populate the outer reaches of the Solar System. Some have been known since antiquity: these are the comets, icy nuclei, whose diameters do not exceed a few tens of kilometres, with extremely eccentric orbits. When their paths approach perihelion, the surface sublimes, resulting in the emission of gas and dust in a coma that may occasionally be extremely spectacular. Comets may have two different origins. Some, with a wide range of inclinations, originate in the Oort Cloud, a vast shell lying at the limits of the Solar System, at a heliocentric distance of some 40 000 AU. The gravitational perturbations of the giant planets, particularly Jupiter, have ejected them into this vast reservoir (the existence of which was established by the work of Jan Oort and, subsequently, of Brian Marsden), and which may contain some $10^{11}$ comets. A tiny fraction of these may, by chance, be returned towards the inner Solar System, and eventually established in stable orbits, once again through planetary gravitational perturbations. In particular, this is the case with Comet Halley, which has been known since antiquity, which has a period of 76 years. The second class of comets is distinguished by a low inclination to the ecliptic and a shorter period. These are believed to originate in the Kuiper Belt, the reservoir of trans-Neptunian objects that has recently been discovered.

The existence of comets, of TNOs, of the Kuiper Belt, and of the Oort Cloud are all simply explained in the planetary formation model in which the planets formed within a protoplanetary disk, by accretion of solid particles subject to multiple collisions and mutual gravitational interactions.

### 4.2.4 Dating the Solar System Through Radioactive Decay

Some of the isotopes trapped in interplanetary grains at the time of their condensation are unstable. These isotopes, or 'parent nuclei', have decayed into 'daughter nuclei' since the date at which they were incorporated into the grains. The excess abundance of the latter isotopes may be measured, which allows the formation of the grains to be dated, once the radioactive decay-rate constant has been determined. This is the inverse of the time $T_e$, the time after which the number of parent isotopes has been divided by $e$. To measure the age of the Solar System use is made of long-period 'clocks', in particular the ($^{40}$K, $^{40}$Ar), ($^{87}$Rb, $^{87}$Sr), and ($^{238}$U, $^{238}$Pb) pairs, which have constants less than $10^{-10}$/year. Measurements made of these elements in meteorite samples have shown that the age of the Solar System is 4.55 $\times$ $10^{10}$ years. In addition, measurements made on plutonium-244 and iodine-129 (with shorter radioactive half-lives), show that $10^8$ years at most passed between the separation of the protosolar material from the interstellar medium and the formation of the planets. These results show that the Sun and the material in the protosolar disk have the same origin.

## 4.3 The Emergence of a 'Standard Model'

The observers of the 17th century were well aware that the Solar System's planets all orbited the Sun in the same direction, and on quasi-circular orbits, close to the plane of the ecliptic. These fundamental properties prompted the concept of a primitive nebula, first proposed by Immanuel Kant and the by Pierre-Simon Laplace. According to this idea, which is the basis of the model accepted today, a primitive, rotating nebula collapsed into a disk, within which the planets and the small bodies subsequently formed. Other models for the formation of the Solar System have been proposed, however, both before and after the nebular model.

### 4.3.1 The Nebular Theory

The model of a primitive nebula in rapid rotation that collapses into a disk, first proposed by Kant and Laplace in the 17th century, has the merit of explaining simply the essential orbital properties of the objects in the Solar System. These formed through condensation within the disk itself. The principal objection to this theory was put forward at the end of the 19th century, and concerned the conservation of angular momentum in the system. In fact, almost all of the angular momentum is located in the giant planets, whereas the Sun represents 99.8 per cent of the overall mass. If it had retained most of its original angular momentum, it would rotate extremely rapidly (in about half a day), which is obviously not the case: its rotation period is 26 days at the equator. How could the Sun have transferred its angular momentum to the planets? For a long time the question remained without an answer, which encouraged the development of alternative theories, such as tidal theories. Nowadays we know the answer to this question: The Sun formed from material that transferred its angular momentum to material that remained in the protoplanetary disk. In addition, the solar magnetic field acted to slow the Sun's rotation (*see also* Sect. 6.1.4) If the particles are transported out to a distance of 10 solar radii, a mass-loss of just 0.003 $M_.$ is sufficient to explain the Sun's current rate of rotation, and such mass-losses are actually measured for young, rapidly rotating stars with strong magnetic fields.

With the principal objection to the nebular model having been discounted, the model has slowly been accepted and refined over recent decades. We see the emergence of two main classes of model:

- the massive nebula model, developed by A.G.W. Cameron in particular, in which the mass of the disk is about one solar mass. The planets form directly within it through gravitational instabilities. The remnants of the disk are either accreted by the Sun, or ejected from the system by the solar wind. As we shall see, this model cannot account for the formation of the planets (both terrestrial and giant) in the Solar System, but it may apply to other planetary systems;

- the low-mass nebular model, notably as suggested by V. Safronov, in which the mass of the disk is about one-hundredth of that a solar mass. The disk forms through gravitational collapse, and cools. The solid material gathers together into planetesimals and then into planetoids through the action of multiple collisions. The latter have the effect of regularizing the orbits, which tend to lose their eccentricity and their inclination to the plane of the disk. During a phase of intense solar activity, the gas is ejected from the system, carrying dust with it, and leaving only objects with diameters larger than about one kilometre. This model is currently the basis for our understanding of the formation of the Solar System. It is still subject to many further developments, which aim to take better account of all the phenomena that are observed.

## 4.3.2 The Standard Model: The Chronology of Events

The model is based on a disk of dust rapidly settling out within a low-mass nebula of gas. The solid particles accumulate by condensation and coalescence, forming objects, planetesimals, the size of which was about one kilometre. The later stages were dominated by gravitation: the accretion of more massive bodies, planetoids, occurred through multiple collisions and gravitational interactions. The young Sun then passed though a phase of intense activity during which the extremely intense solar wind stripped the protosolar disk of its gas and remaining dust.

In this section we summarize the principal characteristics of the standard model.

### 4.3.2.1 Collapse of the Nebula

Modern observations show that stars of the solar type form collectively, through the propagation of an internal instability, deep within molecular clouds that are 2–5 pc in diameter (*see* Chap. 6). The collapse phase lasts between $10^5$ and $10^6$ years. A protostar forms at the centre of an accretion disk. A fraction of the material falling onto the star is ejected in jets that are observed following the protostar's axis of rotation (Figs. 4.6 and 5.9). The disk subsequently cools over a few hundred thousand years (Fig. 5.19), which causes material within the disk to condense, beginning with refractory elements.

### 4.3.2.2 The Condensation Sequence

This section describes the simple case of a dynamically static, cooling nebula. A more realistic case would be a dynamically evolving nebula (an accretion disk with radial mass transport). The sequence in which the elements in the protoplanetary

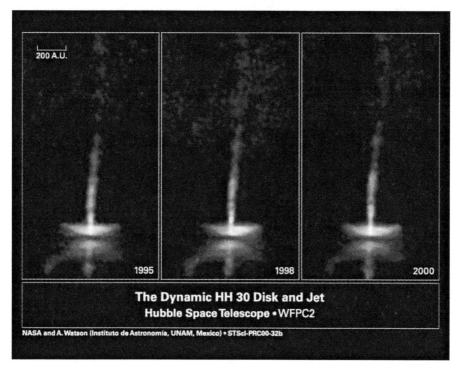

**Fig. 4.6** The bipolar flux and accretion disk around the young object HH 30, photographed by the Hubble Space Telescope (© NASA)

disk condensed is known through laboratory studies of meteorites. Close to the Sun, metals condensed first (between 4.56 and 4.55 thousand million years ago), then, at a greater heliocentric distance, silicate materials (between 4.55 and 4 thousand million years ago). Beyond a certain distance from the Sun, the temperature was sufficiently low to allow the condensation of ices, starting with that of water. The most abundant elements (O, C, N, etc.) were also present in solid form, which considerably increased the solid:gas ratio in the disk. We shall see that this limit, known as the 'ice line', does in fact represent the boundary between the region in which the terrestrial planets formed and that in which the giant planets arose. According to the standard model, this boundary lies at a distance of about 4–5 AU.

The condensation sequence (Fig. 2.8) may be calculated, assuming chemical equilibrium, from the measured abundances of elements in the Sun, based on a series of equations that incorporate factors governing all the condensation products that are capable of being formed from a given molecular gas. The calculations show that the condensation sequence revealed depends exceptionally closely on the state of oxidation of the initial gas. The results of the work of Grossman and Larimer (1974) are shown in Fig. 4.7, for a C:O value of 0.55, in accordance with the solar abundances of C and O, measured at that time (Prinn and Owen, 1976; however, it should be noted that these measurements are still the subject of debate).

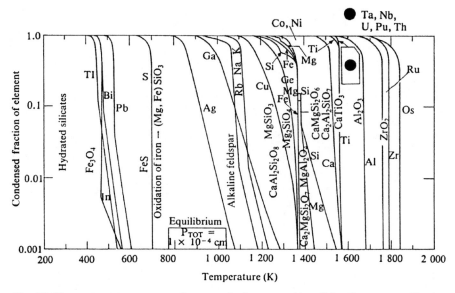

**Fig. 4.7** The condensation sequence for a gas of solar composition (After Grossman and Larimer, *Rev. Geophys. Space Sci.* 12, 71, 1974)

According to these investigations, the order in which the elements condensed is as follows: Al, Ti, Ca, Mg, Si, Fe, Na, and S. The abundances of the main elements, as with the stable phases around 140 K, are in extremely good agreement with the measured abundances in refractory inclusions in C3 chondrites (for example, in the Allende meteorite), which are the most ancient samples that we have at our disposal.

It should, however, be noted that the complex composition of meteorites, where components that condensed at high and low temperatures are in close association, suggests changes in heating that were very localized in both time and space, and possibly cold zones, outside the plane of the disk, that were subject to sudden surges in solar activity. The measurement of certain isotopic anomalies (such as those of oxygen in the Allende meteorite, and those of magnesium in other meteorites) appears to indicate the presence of presolar grains, formed in another nucleosynthetic environment, and which have survived without being vaporized when they were incorporated into the protosolar nebula.

It is possible to estimate the growth rate of grains through condensation, assuming that whenever there is a collision between a grain and a molecule, the molecule remains trapped on the grain. For a solid:gas ratio of $10^{-2}$ and a mean molecular mass of 20, calculations (carried out without allowing for turbulence) show that objects several centimetres across may form rapidly, and then collapse into the disk with characteristic times of $10^3$ to $10^6$ years, depending on whether the particles are some millimetres across or less than a few microns in size. Within the disk, increasing numbers of collisions allow bodies tens of kilometres across, the planetesimals, to form. The chemical composition of these varies with heliocentric distance

**Fig. 4.8** The distribution of minor-planet classes within the Main Belt. The asteroids with the highest densities (types E and S) occur close to the Sun, whereas primitive asteroids (type D), occur at the greatest heliocentric distances (After Bell, 1989)

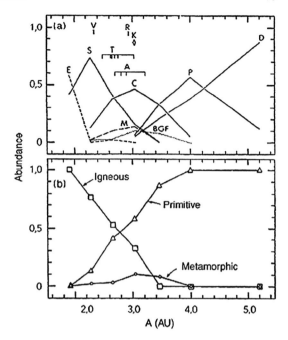

(Fig. 4.8): close to the Sun, it is dominated by silicates and heavy metals, whereas beyond 5 AU, ices predominate. The planetesimals are more massive beyond the ice line, because there solid material is far more abundant.

### 4.3.2.3 The Burst of Accretion and the Formation of the Giant Planets

The growth of the protoplanets and small bodies, beyond a diameter of 10 km, is governed by gravitational attraction. To take account of the phenomena involved, numerical simulations use a statistical approach. According to the models, two bodies may agglomerate if they impact on one another with a collision velocity that is less than, or comparable to, their escape velocity. Two classes of solutions emerge from these calculations: (1) an orderly growth resulting in the formation of a few large bodies, with a power-law distribution for the mass of the smallest; (2) a burst of accretion, linked to the way in which a given body sweeps up surrounding material, and which ceases then the zone available for accretion becomes empty. This second mechanism allows rapid accretion of the giant planets, in particular Jupiter, in a time compatible with the model's other constraints (the existence of a minor-planet belt, and dispersal of the disk by the Sun's T-Tauri phase, about one to a few million years after the collapse of the nebula).

Jupiter's great mass, compared with that of the other giant planets, is very probably explained by its position relative to the ice line. Lying just outside the latter, it was able to sweep up the increase in solid material generated by the condensa-

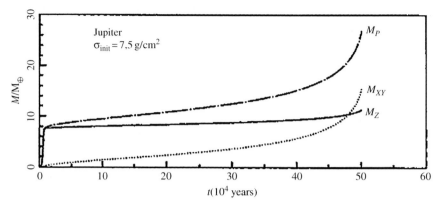

**Fig. 4.9** A simulation of the formation of Jupiter through the accretion of solid material (*continuous line*), and gases (*dotted line*) (After J.B. Pollack et al., 62, 1996)

tion of ices. (Dynamical models indicate that Jupiter's orbital migration must have remained very limited.) Saturn also benefited, to a lesser extent, from the increased availability of material, as well as the other two giant planets, Uranus and Neptune. All thus acquired a core, primarily consisting of ices, with a sufficient mass (some ten Earth masses) to induce gravitational accretion of the surrounding protosolar gas. According to the model by Pollack et al. (1996), the accretion processes for the solid core and for the gas were not entirely decoupled. The formation of the giant planets took place in three phases: (1) accretion of the solid core; (2) simultaneous accretion of gases and solid material; (3) a surge in the accretion of gas (Fig. 4.9).

### 4.3.2.4 Gas Giants and Ice Giants

It remains for us to explain why Uranus and Neptune (Fig. 4.10) have masses significantly lower than that of Jupiter and Saturn (see Table 4.2). Whereas the masses of Jupiter and Saturn are, respectively, 318 and 95 Earth masses, those of Uranus and Neptune are only 15 and 17 Earth masses, respectively, which means that more than 50 per cent of their mass consists of their initial core material. They certainly experienced the phase that saw the collapse of the surrounding gas, as is shown by their numerous regular satellites (i.e., those that lie in the planet's equatorial plane and have quasi-circular orbits), but the amount of gas accreted was not at all abundant. What is the explanation for this effect? Uranus and Neptune, being more distant from the Sun, possibly took longer (between a few million to ten million years) to form an initial core that had the critical mass to capture the surrounding gas. Phase 3 described above (a burst of gas accretion) must have occurred after the Sun's T-Tauri phase, during which almost all of the gas and planetoids were swept out of the protosolar disk. So there was very little material available for the final phase in the formation of Uranus and Neptune. Other possible factors that should be taken into account are gaseous loss to the Sun and photo-evaporation.

(a)

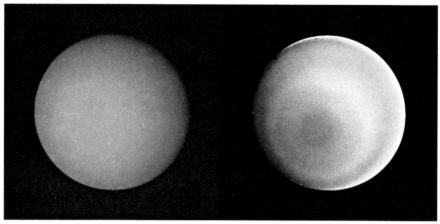

(b)

**Fig. 4.10** Uranus (**a**) and Neptune (**b**), photographed during the Voyager 2 fly-bys in 1986 and 1989, respectively (image credit: courtesy NASA, JPL)

#### 4.3.2.5 Rings and Satellites in the Outer Solar System

The nucleation model for the formation of the giant planets naturally explains the existence of systems of rings and regular satellites – i.e., that lie in the equatorial plane – around these planets. These bodies must have formed within the sub-nebula that collapsed around the nucleus in the final phase of accretion. The rings lie in the immediate proximity to the planets, within the Roche limit, where the tidal forces are too strong to allow the particles to accrete into satellite. (The Roche limit generally lies at a distance of about 2.5 planetary radii.) The rings may also be fed by bodies captured by the planet that are broken up within the Roche limit. Beyond the Roche limit, any captured bodies become the giant planets' irregular satellites, which are characterized by high inclinations and strong eccentricities.

#### 4.3.2.6 Interaction of the Giant Planets with the Small Bodies in the Outer Solar System

Recent numerical-simulation models show that the giant planets may interact with the residual disk of planetesimals and with the other giant planets, and that this may result in migration relative to the location at which they formed. The presence of Pluto and Kuiper-Belt objects in a 3:2 resonance (see Appendix A.5) with Neptune, for example, suggests a process of mutual perturbations, where Jupiter has migrated slightly inwards, whereas the other giant planets have moved outwards. Neptune would have trapped Pluto and numerous other TNOs (the Plutinos), forcing them to migrate as well, which would explain their high eccentricities. Numerical simulations seem to indicate that at a certain time, Saturn and Jupiter were in a 2:1 resonance, which must have produced significant perturbations of the orbits of the asteroids, the other giant planets, and the TNOs. This event appears to be the source of the massive bombardment that occurred 3,800 million years ago, that is, 800 million years after the formation of the giant planets.

#### 4.3.2.7 Formation of the Terrestrial Planets

Whereas the duration of the formation of the giant planets is reckoned in millions of years, that for the formation of the terrestrial planets may have been, according to numerical simulations, ten to one hundred times as long. The major reason is the smaller quantity of material available within the ice line. The planets condensed within a narrow zone, the initial agglomeration phase having ended with some one hundred embryonic bodies the size of Mercury. From this situation, the numerical models end up, in an unpredictable manner, with a small number (less than ten) of planets with sizes comparable with those of the terrestrial planets. Once formed, the planets continued to be subject to a bombardment by planetesimals, notably originating from the outer regions of the Solar System, which (with the exception of Mercury) seems to be the principal source for the formation of their atmospheres.

Mercury, very close to the Sun, does not have a stable atmosphere. On the day side the temperature may reach $700\,K$, and the average molecular velocity is too high, when compared with the escape velocity at the surface (less than $5\,ms^{-1}$), to allow a neutral atmosphere to be trapped – even one consisting of heavy gases. Mercury's density implies an abnormally high metal:silicate ratio, which is difficult to explain within the framework for the condensation of the protosolar disk. It is possible to explain this paradox by invoking a collision between a proto-Mercury that was already differentiated, with a body one fifth of its mass. The material lost (primarily silicates) would have been ejected towards the Sun.

The method by which the terrestrial planets formed, in which there is no accretion of surrounding gas, explains the absence of rings and the low number of satellites orbiting these planets. In the case of Mars, the small satellites Phobos and Deimos are probably asteroids captured by the planet a long time after is formation. As for the Earth-Moon system (Fig. 4.11), its origin remained unexplained for a very long time. A model has come to be accepted recently, thanks to developments in the numerical simulation of chaotic and hydrodynamic situations. In this scenario, the Earth-Moon pair is the result of a glancing impact on the proto-Earth by a body of about one tenth the mass, which led to the ejection into Earth orbit of a portion of the mantles of both bodies, accompanied by the merger of the heavy elements in the two cores. The fragments ejected into Earth orbit re-accreted to form the Moon. This scenario has the advantage of accounting for the Moon's low density, as well as for the Moon's initial orbit, which was strongly elliptical and highly inclined, as is deduced from the current rate of lunar recession (4 cm per year). The history of Mercury, like that of the Earth-Moon pair, illustrates the importance of collisions in the early phases of planetary formation. This type of collision may also be responsible for the deviation of Uranus' axis of rotation to lie almost in the plane of the ecliptic.

**Fig. 4.11** The Earth-Moon system, photographed by the Galileo probe. Dynamical models show that the presence of the Moon may be explained by a collision between the proto-Earth and a body with one tenth of the mass (© NASA)

## 4.4  The Physical and Chemical Properties of Solar-System Objects

In this section, we examine the physical and chemical properties of the different classes of object in the Solar System, stressing those that provide information on formation and evolutionary processes. This listing will also enable us to highlight the extreme diversity of planets and satellites, even within any given family.

### 4.4.1  The Electromagnetic Spectrum of the Objects in the Solar System

Study of the physical and chemical properties of celestial bodies depends greatly on observation of their electromagnetic spectrum. Objects in the Solar System receive energy from the Sun, part of which is reflected or scattered back to space, and the remainder of which is absorbed and converted into thermal energy. The effective temperature (that is, that of a black body emitting the same thermal energy) depends on the surface albedo (i.e., the fraction of the solar energy that is reflected) and their heliocentric distance, as expressed in the following equations:

$$\frac{F}{D^2}\pi R^2(1-A) = \sigma T^4 4\pi R^2 \tag{4.1}$$

for an object in rapid rotation (as with the giant planets), or

$$\frac{F}{D^2}\pi R^2(1-A) = \sigma T^4 2\pi R^2 \tag{4.2}$$

for an object rotating slowly (such as Venus).

In these equations, $F$ is the solar flux at a heliocentric distance of 1 AU, $D$ is the heliocentric distance, $R$ is the radius of the object, $A$ is its albedo, and $\sigma$ is Stefan's constant.

In CGS units, the effective temperature may be expressed as:

$$T = (1-A)^{1/4}\frac{273}{D^{1/2}} \tag{4.3}$$

for a rapidly rotating object, and

$$T = (1-A)^{1/4}\frac{324}{D^{1/2}} \tag{4.4}$$

for a slowly rotating object.

For the planets in the Solar System, the albedo is of the order of 0.3. The effective temperatures vary from 700 K (maximum) for Mercury to about 50 K for Neptune. The maxima for the corresponding thermal flux range from 4 µm

(Mercury) to $70\,\mu m$ (Pluto). The radiation from Solar-System objects therefore primarily lies in the visible for the reflected component and in the infrared for the thermal component. The infrared region is also a favourable region for the observation of neutral molecules in planetary atmospheres, because it is the spectral region where the strongest rotational and vibrational transitions are found.

## 4.4.2 Planetary Atmospheres

### 4.4.2.1 The Chemical Composition of Planetary Atmospheres

An important factor in the chemical composition of the various planetary atmospheres is the escape velocity, which is a function of the mass of the body concerned, and its radius:

$$V_{esc} = \left[ \frac{2GM}{R} \right]^{1/2} \tag{4.5}$$

An atmospheric molecule has an average thermal velocity, $V_{th}$, that may be estimated from the Maxwell velocity distribution:

$$V_{th} = \left[ \frac{2kT}{m} \right]^{1/2} \tag{4.6}$$

The probability that a molecule will escape from any given atmosphere depends on the relationship between $V_{esc}$ and $V_{th}$: a molecule escapes more easily the lighter it is (which really goes without saying), the smaller the planet, and the higher its temperature. This explains the absence of a permanent atmosphere on Mercury, as well as the atmospheric composition of the giant planets, which is dominated by hydrogen and helium: their gravitational field is such that even hydrogen has not been able to escape over the whole lifetime of the Solar System. In contrast, the gravitational fields of the terrestrial planets are insufficient to retain hydrogen and helium: their chemical composition is dominated by the elements C, N, and O (as $CO_2$, $N_2$, $H_2O$, and $O_2$).

In what form did the carbon, nitrogen and oxygen occur in the protosolar nebula? Assuming thermodynamic equilibrium, the abundance of molecules containing C and N is governed by the following reactions:

$$CH_4 + H_2O \leftrightarrow CO + 3H_2 \tag{4.7}$$

$$2NH_3 \leftrightarrow N_2 + 3H_2 \tag{4.8}$$

which tend to form $CH_4$ and $NH_3$ at low temperatures and high pressures, and CO and $N_2$ at high temperatures (a few hundred K) and low pressures. When CO and $N_2$ predominate, CO reacts with $H_2O$ to form $CO_2$:

$$CO + H_2O \leftrightarrow CO_2 + H_2 \tag{4.9}$$

Based on thermodynamic equilibrium, therefore, to a first approximation we may expect to find an atmosphere rich in $CH_4$ and $NH_3$ in the giant planets, and one that is dominated by CO, $CO_2$, and $N_2$ (following the escape of hydrogen) in the terrestrial planets. Overall, this model agrees with observations. It does not, however, explain the atmospheres of Titan, Triton, and Pluto, which are rich in $N_2$ and $CH_4$. Lewis and Prinn (1984) have suggested that CO and $N_2$ that formed in the protosolar nebula at its beginning, when the temperature was sufficiently high, could have been trapped when the disk cooled, if this took place sufficiently quickly. In the case of Titan, it is possible that nitrogen was present originally in the form of ammonia, $NH_3$, which was subsequently altered into $N_2$ through photolysis in the atmosphere.

### 4.4.2.2 The Thermal Structure of the Planetary Atmospheres

An atmosphere is the gaseous envelope that separates the body from its external environment (interplanetary medium, solar wind, or magnetosphere). The processes that take place there arise partly through this interaction with the outer environment (photolysis of neutral molecules, dissociation and ionization, interaction with the solar wind and, where appropriate, with the magnetic field), and partly its interaction with the surface (volcanism, tectonics, and greenhouse effect), or with the deeper layers (internal dynamics). The same mechanisms are likely to be present in the atmospheres of exoplanets (*see* Chap. 7).

Let us remind ourselves of the general nomenclature (Fig. 4.12): the lower region of an atmosphere, where all the non-condensable components are uniformly mixed, is the homosphere. Above this, and separated from it by the homopause, is the heterosphere where gases diffuse independently of one another as a function of the atomic or molecular mass. The homosphere is divided into several regions, depending on the dominant energy-transfer process within each. The lowest layer of an atmosphere is the troposphere. In the terrestrial planets, this layer is in direct contact with the surface. Within it the temperature decreases as altitude increases. Above the troposphere (at an altitude of approximately 12 km for the Earth), lies the stratosphere. The temperature increases as altitude increases, because of the absorption of solar radiation by the molecules and aerosols in the atmosphere as a function of the photochemical reactions that take place. (In the case of the Earth, this is the region where the ozone layer forms though photochemical changes to oxygen; in the giant planets, methane photochemistry dominates within this layer.)

In general terms, the parameters of an atmosphere (temperature, pressure, and density) are linked by the following equations:

- the hydrostatic equilibrium law, which is an exact balance between the pressure-gradient force and the gravitational force:

$$dP = -\rho g dz \tag{4.10}$$

where $P$ is the pressure, $\rho$ the density, $z$ the altitude, and $g$ gravity;

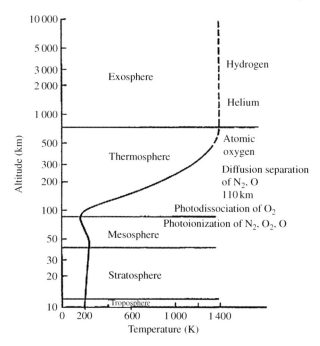

**Fig. 4.12** The thermal profile of the Earth atmosphere (Encrenaz et al., 2004)

- the perfect gas law:

$$P = \frac{R\rho T}{\mu} \tag{4.11}$$

where $R$ is the perfect gas universal constant, and $\mu$ the average molecular mass.

Equation (4.10) may be written:

$$\frac{dP(z)}{P(z)} = \frac{\mu g \, dz}{RT(z)} \tag{4.12}$$

Over the altitude range of planetary atmospheres, the gravity $g$ may be assumed to be constant. Equation (4.12) is analytically integrable only if the temperature is constant or is a simple function of altitude. As a first approximation, we assume the atmosphere to be isothermal. In this case we obtain:

$$P(z) = P(z_0) \exp^{-\frac{(z-z_0)}{H}} \tag{4.13}$$

where $H$ is the scale height:

$$H = \frac{RT}{\rho g} \tag{4.14}$$

and where $z_0$ may be the surface or, in the case of the giant planets, a given reference level (that of a cloud layer, for example).

At the surface of the terrestrial planets, the scale height is 8 km for the Earth, 14 km for Venus, and 10 km for Mars. It is equal to 20 km on Jupiter, and 40 km on Saturn at a pressure level of 0.5 bar, which for Jupiter is close to the $NH_3$ cloud layer.

The temperature structure of a planetary atmosphere mainly depends upon two mechanisms of transport, convection and radiation (*see* Sect. 5.2.6). Convection, which is dominant in dense atmospheres ($P \geq 0.1$ bar), where the temperature decreases as the altitude increases, is caused by density gradients resulting from temperature differences. Using the equation of hydrostatic equilibrium, it may be shown (*see* e.g., de Pater and Lissauer, 2001) that the temperature gradient follows the dry adiabatic lapse rate:

$$\frac{dT}{dz} = -\frac{g}{C_p} \tag{4.15}$$

where $C_p$ is the thermal heat capacity at constant pressure.

The radiation mechanism dominates in the planet's upper troposphere and stratosphere, where the quantity $e^{-\tau}$ is neither negligible nor equal to 1, with $\tau$ being the optical depth of the gas:

$$\tau_\nu = \int_z^\infty \kappa_\nu \rho\, dz \tag{4.16}$$

where $\kappa_\nu$ is the absorption coefficient of the gas and $\rho$ the density.

The radiative transfer equation is written (see e.g., Encrenaz et al., 2004):

$$\mu \frac{dI_\nu}{d\tau_\nu} = I_\nu - J_\nu \tag{4.17}$$

where $\mu = \cos\theta$, $\theta$ being the angle between the line of incidence and the vertical; $I_\nu$ is the specific intensity (a function of the frequency, $\nu$) and $J_\nu$ is the source function. In planetary atmospheres, for pressures above about 1 mbar, collisions are usually sufficient to produce local thermodynamical equilibrium (LTE): the physical properties of the medium depend only on the temperature; the velocity distributions of the atoms and molecules are Maxwellian; the populations at different energy levels obey Boltzmann's Law; and the source function is the Planck function. Each atmospheric layer radiates like a blackbody, following Kirchhoff's Law:

$$J_\nu = B_\nu(T) \tag{4.18}$$

The temperature at a level z is given by:

$$\int_\nu \int_z \int_\mu B(T) e^{-\tau/\mu} d\nu d_z d\mu = \sigma T^4 \tag{4.19}$$

The opacity, for a given frequency, depends upon the spectroscopic properties of the atmospheric gases. Condensates are also to be taken into account: $NH_3$ and $NH_4SH$ in Jupiter and Saturn, and additionally $CH_4$ and hydrocarbons in Uranus and Neptune.

### 4.4.2.3 Thermal Emission from a Planetary Atmosphere

The thermal radiance emerging from a planetary atmosphere may be calculated assuming LTE, assuming the radiative transfer equation (*see* e.g., de Pater and Lissauer, 2001; Encrenaz et al., 2004):

$$I_\nu = \int_\mu \int_{z_0}^\infty B(z,\nu)\, e^{-\tau_\nu(z,\mu)} \tag{4.20}$$

where $\mu$ is the cosine of the angle of incidence, $B(z,\nu)$ is the blackbody emission at the altitude $z$ and frequency $\nu$, and $\tau_\nu(z,\mu)$ is the optical depth above the altitude $z$, as defined above (Eq. 4.15).

A convenient parameter is the brightness temperature $T_B(z,\nu)$, which is the temperature that a blackbody would emit the same radiance at the given frequency. It may be estimated using the Barbier-Eddington approximation:

- In the case of the specific intensity emitted along the vertical ($\mu = 1$), the brightness temperature is, to a first approximation, the temperature of the atmospheric layer for which the optical depth is equal to 1.
- In the case of the flux emitted over the whole planetary disk (integrated over $\mu$), the brightness temperature is, to a first approximation, the temperature of the atmospheric layer for which the optical depth is equal to 2/3.

The thermal spectrum of a planetary atmosphere thus depends on two main quantities: the temperature and the optical depth at each level. The optical depth, in turn, depends on the density of the absorber (i.e., the mixing ratio of the atmospheric constituent) and its absorption coefficient. The absorption coefficient is known from spectroscopic data. When an atmosphere has a temperature gradient that changes sign with altitude, the infrared spectrum shows molecular lines either in emission (when the gradient is positive) or in absorption (when the gradient is negative), depending on the intensity of the line, and depending on the abundance an vertical distribution of the absorbing gas. Thermal spectra of Jupiter and Saturn beyond $5\,\mu m$ typically show a combination of emission features (probing the stratosphere) and absorption features (arising from lower, tropospheric levels), as shown in Fig. 4.21.

## *4.4.3 The Terrestrial Planets*

### 4.4.3.1 Variation Among the Terrestrial Planets

The four terrestrial planets – Mercury, Venus (Fig. 4.13), Earth, and Mars (Fig. 4.14), in order of their increasing distance from the Sun – are notable for their relatively small size, their high density, and a low number of satellites (or even none). Their surface properties are in sharp contrast. We have seen that Mercury, the planet closest to the Sun, has no stable atmosphere. The other three terrestrial planets have

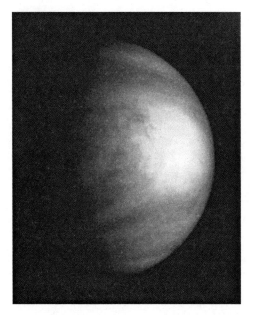

**Fig. 4.13** The planet Venus as observed by the Galileo probe during its fly-by in December 1989 (image credit: courtesy NASA, JPL Galileo Project)

**Fig. 4.14** The planet Mars as observed by the Mars Global Surveyor probe (image credit: courtesy NASA, JPL)

atmospheres. On Venus the surface pressure is close to 100 bars at a temperature of 730 K; on Mars, in contrast, it is only 6 millibars on average, with a surface temperature that swings between 150 K (on the southern polar cap) and 300 K at the equator. The Earth occupies a position between these two extremes. The extreme conditions that the terrestrial planets experience, even though the bodies arose under what were essentially neighbouring conditions, pose one of the major challenges to current planetology.

By contrast, from the point of view of their chemical composition, the atmospheres of Venus, the Earth, and Mars have remarkable similarities. For Venus and Mars, carbon dioxide predominates and nitrogen, $N_2$, is present at just a few per cent. The Earth's primitive atmosphere probably had a similar composition, but the presence of abundant liquid water allowed $CO_2$ to be deposited at the bottom of the oceans in the form of calcium carbonate, $CaCO_3$. Oxygen appeared following the development of life, to give the Earth's current atmospheric composition (78 per cent $N_2$, 21 per cent $O_2$).

An important characteristic of the terrestrial planets is their severe depletion of all volatile elements, compared with solar abundances. The properties of their primitive atmospheres were probably strongly affected by massive collisions during their late stages of formation.

### 4.4.3.2 The Thermal Structure of the Terrestrial Planets

The three terrestrial planets with atmospheres exhibit a layer, near the surface, where the temperature decreases with increasing altitude: this is the troposphere (*see* Sect. 4.4.2.2). In the case of Venus (Fig. 4.15) and Mars (Fig. 4.16), above this region above about 50 km, there is the mesosphere, a layer that is more-or-less isothermal. On Earth, the tropopause (the upper limit to the troposphere) lies at an average altitude of about 12 km. Above it lies the stratosphere, where the temperature increases with height, thanks to the absorption of ultraviolet solar radiation by molecular oxygen with the formation of the ozone layer.

In most cases, the infrared spectra of the terrestrial planets exhibit atmospheric lines in absorption, because the surface temperature is generally higher than that of the lower atmosphere immediately above it. Because carbon dioxide has the property of absorbing infrared radiation emitted by the surface, the lower atmosphere is correspondingly heated, in turn causing an increase in the temperature of the surface. This is the greenhouse effect, which is particularly strong on Venus. There are regions on Mars and on the Earth where the surface temperature is lower than that of the lower atmosphere: these are the polar caps. Over them, the atmospheric lines appear in emission, with an absorption core because the temperature decreases again in the upper atmosphere. When instrumental methods permit, it will be possible, by observing the thermal spectra of exoplanets, to detect lines in emission or in absorption, and thus determine constraints on their thermal structure.

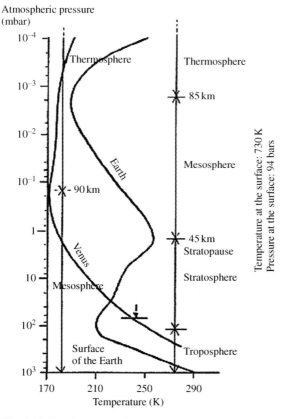

Atmospheric pressure
(mbar)

Fig. 4.15 The thermal profile of Venus' atmosphere, compared with that of the Earth (Encrenaz et al., 2004)

### 4.4.3.3 Comparisons Between the Evolution of the Terrestrial Planets

Why have the terrestrial planets evolved in such different ways? Mercury's case is simply explained: the escape velocity at its surface is too high (its gravity being weak and the temperature too high) for the planet to retain a permanent atmosphere. Initially, the other three planets had an atmosphere consisting of $CO_2$, $N_2$, and $H_2O$. On Earth, a vast quantity of water is present in liquid form, but it appears, as water vapour, in only trace amounts in the atmospheres of Venus and Mars. How do we know that abundant water was originally present? That information comes from measurement of the D:H ratio, determined by infrared spectroscopy from the $HDO:H_2O$ ratio. Observations have revealed an important result: the D:H ratio on Venus is enriched by a factor close to 120 relative to the terrestrial value, and the same ratio on Mars is about 6 times the terrestrial value. Interpretation of this enrichment is as follows: on Venus, water, which was probably extremely abundant initially, has largely escaped; heavy water, HDO, escaping less easily than $H_2O$, has

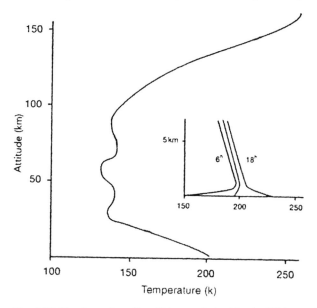

**Fig. 4.16** The thermal profile of Mars as derived by the Viking probes (Encrenaz et al., 2004)

accumulated over the course of the planet's history. The same reasoning applies to Mars to a lesser extent, and seems to indicate that the primitive atmosphere of Mars was denser than it is today. The presence of traces of fluid flow on the surface of Mars is another indicator of the presence of liquid water during the planet's history.

A broad outline of the history of the terrestrial planets may thus be given. The primitive atmosphere of Venus, dominated by $CO_2$ and $H_2O$, was sufficiently hot (because of the planet's heliocentric distance, which is slightly less than that of the Earth) for water to be in the form of vapour. As $CO_2$ and $H_2O$ are both particularly effective greenhouse gases, the temperature of the surface and of the lower atmosphere rose and, in the absence of any regulatory mechanism, there was a runaway greenhouse effect, leading to the surface conditions that we see today. The water vapour has disappeared, probably through photodissociation and subsequent loss of hydrogen. The Earth's heliocentric distance was such that water occurred primarily in liquid form. Most of the carbon dioxide was therefore absorbed, and the greenhouse effect remained at a moderate level over the Earth's lifetime.

The case of Mars is different from those of Venus and the Earth, because the planet, farther away from the Sun and thus colder, is also only about one tenth of the mass. Its internal energy sources and its gravitational field are thus much weaker: two factors that prevented it from acquiring a primitive atmosphere that was as dense at those of Venus and the Earth. The planet appears to have had a magnetic field early in its history, but that this decayed within the first thousand million years. The primitive atmosphere must have been denser than today (without ever being comparable with that of Venus), but this has not been confirmed. What happened to the water on Mars? Although it is almost completely absent from the atmosphere, it seems that a significant quantity may be trapped beneath the surface, in the form of ice in the polar caps, and perhaps in the form of permafrost at lower latitudes. Understanding the past climate of Mars, from its origin to the present day, is one of the major issues to be tackled by the exploration of Mars.

### 4.4.4 The Giant Planets

We have seen that the giant planets may be divided into two classes, the gas giants (Jupiter and Saturn) and the ice giants (Uranus and Neptune), and that this classification has a simple explanation in terms of the formation of the giant planets by nucleation. We will now summarize the principal physical and chemical properties of these four planets.

The giant planets have no surfaces in the sense that we understand the term as applied to the terrestrial planets and the asteroids. Measurements of density and gravity, coupled with theoretical models, suggest the existence of a central core consisting of heavy elements, surrounded by a mixture of hydrogen and helium under great pressure, which is probably metallic in Jupiter and Saturn, and molecular in Uranus and Neptune (Fig. 4.17). The pressure at the boundary would be about

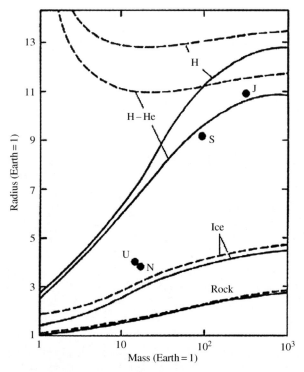

**Fig. 4.17** The mass:radius relationship for a self-gravitating body, for different compositions. After Stevenson, 1982

40 Mbars in Jupiter, 13 Mbars in Saturn, and 6 Mbars in Uranus and Neptune, with temperatures – still according to the models – of 23 000 K, 12 000 K, 3000 K, and 3000 K, respectively (see Chap. 7). Only the outermost layer of the atmospheres of the giant planets is accessible to observation. Remote sounding measurements have enabled the tropospheres to be studied down to a depth of a few bars; in addition, in December 1995, the Galileo descent probe provided measurements of Jupiter's troposphere down to a level at which the pressure was 20 bars. The chemical composition of the giant planets is dominated by hydrogen and helium, with any other elements bound to hydrogen in a reduced form ($CH_4$, $NH_3$, $PH_3$, $H_2S$, etc.).

With the exception of Uranus, the giant planets have an internal energy source. The ratio between the energy radiated to space and the amount of solar energy received is 1.7, 1.8, and 2.6, for Jupiter, Saturn and Neptune, respectively. It is less than 1.1 for Uranus. The origin of this energy is probably the contraction and cooling of the planets, following the initial collapse of the gas onto the core, and the heating this produced. Another possible contribution in the case of Jupiter and Saturn comes from the predicted condensation of helium within the metallic hydrogen phase (see Sect. 7.1.4.2). With Uranus, the absence of an internal source of heating remains an enigma.

#### 4.4.4.1  The Thermal and Cloud Structure of the Giant Planets

A feature of the thermal profiles of the four giant planets (Fig. 4.18) is a troposphere where the gradient is adiabatic (about 2 K km$^{-1}$) below the tropopause, which in each case lies at a pressure level that is about 0.1 bar. This similarity is explained by the generally similar composition of the four planets, and which is dominated by hydrogen and helium. In contrast, however, the temperature at the tropopause decreases as the heliocentric distance increases (110 K for Jupiter, 90 K for Saturn, and about 50 K for Uranus and Neptune). The tropopause therefore acts as a very

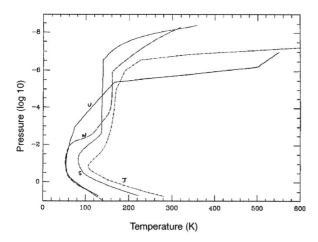

**Fig. 4.18**  The thermal structure of the giant planets (Encrenaz et al., 2004)

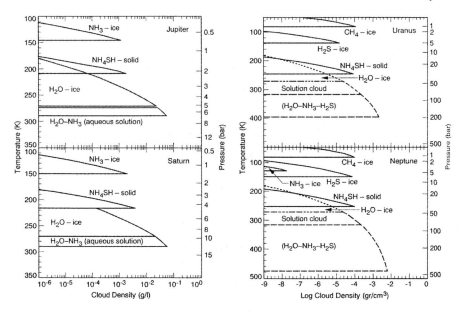

**Fig. 4.19** The cloud structure of the giant planets (After R.A. West, 1999)

efficient cold trap, where the products of the dissociation of methane, produced at a higher altitude, condense. In the stratosphere, the temperature again increases with altitude in a manner that varies greatly from one planet to another, which is a sign of different heating mechanisms in each case. (The heating being caused by the absorption of solar radiation by the aerosols and the dissociation products.) These differences increase in the upper stratosphere, where other heating mechanisms (gravity waves, and energetic particles) intervene.

There is a striking contrast between our excellent knowledge about the thermal structure of the giant planets and current uncertainties as to their cloud structure. The basic reason for this is undoubtedly because the spectroscopic determination of the nature of the condensates is far more difficult that for gases. The spectral signatures of solids are often ambiguous. So, for example, we still do not know with any degree of certainty the nature of the components of the Great Red Spot, a large-scale dynamical structure, discovered three centuries ago, and which has proved to be stable since then. It is possible to derive the structure of an atmosphere theoretically, assuming a given chemical composition, by determining the photochemical equilibrium of the predicted molecular compounds, and by determining the condensation sequence associated with each molecule (Fig. 4.19).

### 4.4.4.2  Chemical Composition

The chemical composition of the atmospheres of the giant planets is shown in Table 4.1. The components may be grouped into two major categories: tropospheric components and stratospheric components.

**Table 4.1** The atmospheric composition of the giant planets (After Encrenaz et al., 2004)

| Species | Jupiter | Saturn | Uranus | Neptune |
|---|---|---|---|---|
| $H_2$ | 1 | 1 | 1 | 1 |
| HD | $3.6 \times 10^{-5}$ | $4.6 \times 10^{-5}$ | $11 \times 10^{-5}$ | $13 \times 10^{-5}$ |
| He | 0.136 | $0.11 - 0.16$ | 0.18 | 0.23 |
| $CH_4$ (trop) | $2.1 \times 10^{-3}$ | $4.4 \times 10^{-3}$ | $2 \times 10^{-2}$ | $4 \times 10^{-2}$ |
| $CH_4$ (strat) | $2.1 \times 10^{-3}$ | $4.4 \times 10^{-3}$ | $3 \times 10^{-5} - 10^{-4}$ | $7 \times 10^{-4} (0.05 - 1\,mb)$ |
| $^{13}CH_4$ (trop) | $2 \times 10^{-5}$ | $4 \times 10^{-5}$ | | |
| $CH_3D$ (trop) | $2.5 \times 10^{-7}$ | $3.2 \times 10^{-7}$ | $10^{-5}$ | $2 \times 10^{-5}$ |
| $CH_3D$ (strat) | | | | $2.2 \times 10^{-7}$ |
| $C_2H_2$ | $9 \times 10^{-7} (0.3\,mb)$ | $3.5 \times 10^{-6} (0.1\,mb)$ | $2 - 4 \times 10^{-7} (0.1\text{-}0.3\,mb)$ | $1.1 \times 10^{-7} (0.1\,mb)$ |
| $^{12}C^{13}CH_2$ | * | $2.5 \times 10^{-7} (mb)$ | | |
| $C_2H_6$ | $4.0 \times 10^{-6} (0.3 - 50\,mb)$ | $4.0 \times 10^{-6} (< 10\,mb)$ | | $1.3 \times 10^{-6} (0.03 - 1.5\,mb)$ |
| $CH_3C_2H$ | * | $6.0 \times 10^{-10} (< 10\,mb)$ | | |
| $C_4H_2$ | | $9.0 \times 10^{-11} (< 10\,mb)$ | | |
| $C_2H_4$ | $7 \times 10^{-9}$ | | | |
| $C_3H_8$ | $6 \times 10^{-7}$ | | | $3 \times 10^{-7} (1\,\mu b)$ |

**Table 4.1** (continued)

| Species | Jupiter | Saturn | Uranus | Neptune |
|---|---|---|---|---|
| $C_6H_6$ | $2 \times 10^{-9}$ | * | | |
| $CH_3$ | | $0.2 - 1 \times 10^{-7}(0.3\,\mu b)$ | | $2 - 9 \times 10^{-8}(0.2\,\mu b)$ |
| $NH_3$ (trop) | $2 \times 10^{-4}(3-4b)$ | $2 - 4 \times 10^{-4}(3-4b)$ | | |
| $^{15}NH_3$ | $8 \times 10^{-7}$ | | | |
| $PH_3$ (trop) | $6 \times 10^{-7}$ | $1.7 \times 10^{-6}$ | | |
| $GeH_4$ | $7 \times 10^{-10}$ | $2 \times 10^{-9}$ | | |
| $AsH_3$ | $3 \times 10^{-10}$ | $2 \times 10^{-9}$ | | |
| CO (trop) | $1.5 \times 10^{-9}$ | $2 \times 10^{-9}$ | | |
| CO (strat) | $1.5 \times 10^{-9}$ | $2 \times 10^{-9}$ | $10^{-6}$ | |
| $CO_2$ (strat) | $4 \times 10^{-10}(<10\,mb)$ | $3 \times 10^{-10}(<10\,mb)$ | | $5 \times 10^{-10}(<5\,mb)$ |
| $H_2O$ (trop) | $1.4 \times 10^{-5}(3-5b)$ | $2 \times 10^{-7}(>3b)$ | | |
| $H_2O$ (strat) | $1.5 \times 10^{-9}(<10\,mb)$ | $2 - 20 \times 10^{-9}(<0.3\,mb)$ | $5 - 12 \times 10^{-9}(<0.03\,mb)$ | $1.5 - 3.5 \times 10^{-9}(<0.6\,mb)$ |
| HCN | | | | $3 \times 10^{-10}$ |
| $H_3^+$ | * | * | * | |

The tropospheric components are, in descending order of abundance: $H_2$, He, $CH_4$, and $NH_3$, as well as certain hydrogenated components ($PH_3$, $GeH_4$, $AsH_3$, $H_2O$, and $H_2S$) as shown in Fig. 4.20. For Uranus and Neptune, these products (with the exception of $H_2$, He, and $CH_4$) are not detectable because they condense at levels that are too deep to be accessible for observation.

The stratospheric species may be of two different origins. Some of them arise through the photodissociation of methane, the most abundant being $C_2H_6$ and $C_2H_2$. Others, discovered at the end of the 1990s by the ISO satellite (Fig. 4.21), have an external origin. These are $H_2O$ and $CO_2$, whose presence betrays the existence of an oxygen flux, which may be of local (from the rings or satellites), or interplanetary (meteoritic, micrometeoritic, or cometary) origin. The collision of Comet Shoemaker-Levy 9 with Jupiter in July 1994 showed that such events could actually occur, despite their low probability.

A certain number of isotopic species have also been detected in the giant planets: HD, $CH_3D$, $^{13}CH_4$, $^{12}C^{13}CH_2$, and $^{15}NH_3$. These measurements have been used to determine the D:H, $^{15}N{:}^{14}N$, and $^{13}C{:}^{12}C$ ratios. Together with the element ratios, these factors provide important constraints on models for the formation and evolution of the giant planets.

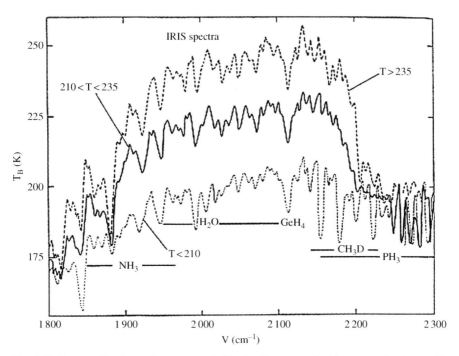

**Fig. 4.20** Spectra of Jupiter at 5 μm as recorded by the Voyager probe. This spectral region enables the tropospheric layers of the planet to be examined (After Drossart et al., 1982)

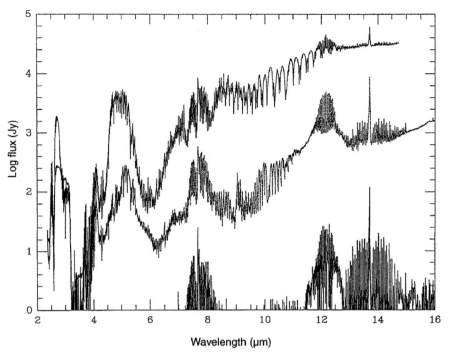

**Fig. 4.21** The infrared spectra of the giant planets, Jupiter (*top*), Saturn (*middle*) and Neptune (*bottom*) as recorded by the SWS instrument on the ISO satellite. Stratospheric lines appear in emission, and tropospheric lines in absorption (After Encrenaz et al., 2004)

### 4.4.4.3 Element and Isotopic Ratios in the Giant Planets

#### a. Element ratios

In the nucleation model that we have described earlier, the giant planets formed initially as a core of heavy elements ($A > 10$), with a mass of 10–12 Earth masses. This mass was sufficiently large to cause the collapse of the surrounding nebula. According to cosmic abundances, the mass fraction of heavy elements is about 2 per cent. Assuming that all the heavy elements are equally trapped in ices, and that the elements became homogenized within the planets' interiors through the heating caused by the accretion phase, it is possible to calculate, from the total masses of the giant planets, their enrichment in heavy elements relative to hydrogen, by comparison with solar values. In the absence of any initial core, the enrichment would be 1. For a planet consisting solely of its initial core of heavy elements, the enrichment would be 50. Table 4.2 lists the enrichments predicted for the four giant planets, assuming a mass of 12 Earth masses for the initial core of heavy elements.

Table 4.2 shows a remarkable agreement between predictions and the observational measurements that are available. The most striking example is that of Jupiter, where precise measurements have been obtain by the mass spectrometer on the

**Table 4.2** Enrichment in heavy elements predicted (using the nucleation model, assuming an initial core of $M_C$ of 12 Earth masses) and observed in the giant planets (After Encrenaz, 2005)

| Planet | $M_T$ | $M_{PS}$ | $M_{TPS}$ | $E_t$ | $E_{obs}$ |
|--------|-------|----------|-----------|-------|-----------|
| Jupiter | 318 | 6 | 18 | 3 | 3 (GPMS-Galileo) |
| Saturn | 95 | 2 | 14 | 7 | $6 \pm 2$ (CH$_4$) |
| Uranus | 15 | 0.06 | 12.06 | 40 | 20–50 (CH$_4$) |
| Neptune | 17 | 0.1 | 12.1 | 36 | 20–50 (CH$_4$) |

$M_T$: Total mass of the planet (in Earth masses)
$M_{PS}$: Mass of heavy elements in the protosolar ($M_{PS} = M_T \times 0.02$) (in Earth masses)
$M_{TPS}$: Total mass of heavy elements in the planet ($M_{TPS} = M_C + M_{PS}$) (in Earth masses)
$E_t$: Enrichment in heavy elements (relative to the solar value) predicted assuming an initial core of 12 Earth masses: $E_T = M_{TPS}/M_{PS}$
$E_{obs}$: Observed enrichment in heavy elements

Galileo probe. Apart from three elements (Ne, He, and O) for which various explanations may be found, overall the elements measured (Ar, Kr, Xe, C, N, and S) exhibit an enrichment by a factor of $3 \pm 1$, relative to hydrogen (Fig. 4.22). For Saturn, Uranus and Neptune, we are only able to use remote sounding measurements of CH$_4$. The enrichments measured are 6 for Saturn, and between 20 and

**Table 4.3** A list of parent molecules detected in comets. (After Encrenaz et al., 2004)

| Molecule | | Abundance | Method of Observation | Note |
|----------|--|-----------|-----------------------|------|
| $H_2O$ | Water | = 100 | IR | |
| CO | Carbon monoxide | 2–20 | UV, radio, IR | Extended source |
| $CO_2$ | Carbon dioxide | 2–6 | IR | |
| $CH_4$ | Methane | 0.6 | IR | |
| $C_2H_6$ | Ethane | 0.3 | IR | |
| $C_2H_2$ | Acetylene | 0.1 | IR | |
| $H_2CO$ | Formaldehyde | 0.05–4 | Radio | Extended source |
| $CH_3OH$ | Methanol | 1–7 | Radio, IR | |
| HCOOH | Formic acid | 0.1 | Radio | |
| HNCO | Isocyanic acid | 0.07 | Radio | |
| $NH_2CHO$ | Formamide | 0.01 | Radio | |
| $CH_3CHO$ | Acetaldehyde | | Radio | |
| $HCOOCH_3$ | Methyl formiate | 0.1 | Radio | |
| $NH_3$ | Ammonia | 0.5 | Radio, IR | |
| HCN | Hydrogen cyanide | 0,1–0.2 | Radio, IR | |
| HNC | Hydrogen isocyanide | 0.01 | Radio | |
| $CH_3CN$ | Cyanomethane | 0.02 | Radio | |
| $HC_3N$ | Cyanocetylene | 0.02 | Radio | |
| $N_2$ | Nitrogen | 0.02–0.2 | Visible | Indirect, from $N_2^+$ |
| $H_2S$ | Hydrogen sulphide | 0.3–1.5 | Radio | |
| $H_2CS$ | Thioformaldehyde | 0.02 | Radio | |
| $CS_2$ | Carbon disulphide | 0.1 | UV, radio | Indirect, from $N_2^+$ |
| OCS | Carbonyl sulphide | 0.4 | Radio, IR | |
| $SO_2$ | Sulphur dioxide | $\approx 0.2$ | Radio | |
| $S_2$ | Sulphur | 0.05 | UV | |

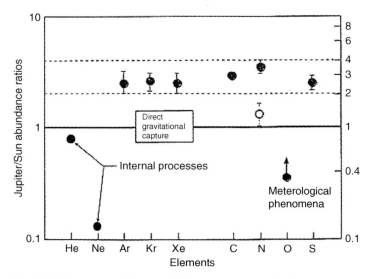

**Fig. 4.22** Abundances of elements in Jupiter, as measured by the mass spectrometer on the Galileo probe (After Owen et al., 1999)

50 for Uranus and Neptune. Overall, these measurement do, therefore, give decisive support to the nucleation model for the giant planets. In addition, it seems to indicate that the giant planets formed from planetoids of solar composition.

There is, however, one significant, unanswered question. The calculation just made assumes trapping of the elements in the form of ice, whatever the temperature considered. Laboratory experiments, however, have shown that certain elements, in particular nitrogen and argon, cannot be trapped at temperatures above approximately 30 K. So did the planetoids that later formed Jupiter agglomerate at a temperature less than 30 K? Models for the evolution of the protosolar nebula favour temperatures that are significantly higher, at least equal to 100 K. The question is currently unanswered.

### b. The D:H ratio

Valuable information is provided by isotopic ratios, especially the D:H ratio. Laboratory measurements have shown that ices are enriched in deuterated molecules, because of the ion-molecule and molecule-molecule reactions that come into play at low temperatures. The enrichment is greater, the lower the temperature. So a measurement of the D:H ratio in various bodies in the Solar System provides information on the temperature at which the planetoids formed, and from which the bodies we are investigating later accreted.

In the giant planets, the deuterium came from two sources: the protosolar nebula (the most important reservoir for Jupiter and Saturn), and the ices in the initial core (the most important reservoir for Uranus and Neptune). Jupiter's D:H ratio should, therefore, reflect the protosolar value; Saturn's value should be slightly higher, while the D:H ratio for Uranus and Neptune should be significantly enriched.

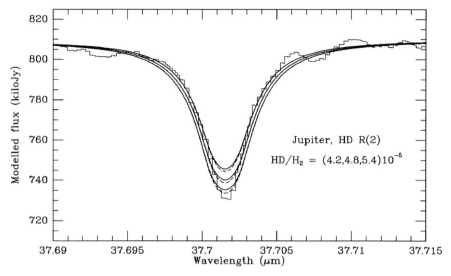

**Fig. 4.23** Measurement of D:H in Jupiter and Saturn from ISO observations (After Lellouch et al., 2001)

The predictions were confirmed by the D:H measurements carried out by the ISO satellite, notably from the $HD:H_2$ ratio (Fig. 4.23). The D:H enrichment in Uranus and Neptune is about 3, relative to the protosolar value (Fig. 4.24), which constrains the value in the proto-Neptunian ices which served as building blocks for the initial cores. A remarkable fact is that this value is less, by a factor of 2, than the D:H ratio measured in comets, and this sets constraints on the conditions under which the different types of objects formed. In fact, the D:H ratio is an indication of the temperature at which the body being observed formed (the former is lower, the higher the ratio), and thus of the heliocentric distance at the time of its formation.

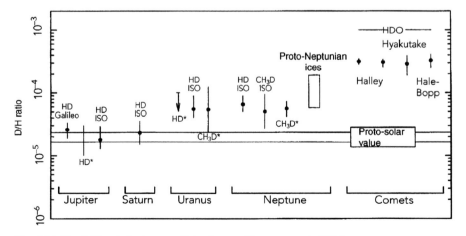

**Fig. 4.24** The D:H ratio in the outer Solar System (Encrenaz et al., 2004)

## 4.4.5  Rings and Satellites in the Outer Solar System

The formation of the giant planets, following the collapse of the protoplanetary disks around their initial cores, possibly suggests an analogy with the Solar System itself, the result of the collapse of the presolar nebula. Numerous satellites of the giant planets were thus formed in their equatorial planes (known as regular satellites), just as the planets formed in the plane of the ecliptic. Such a comparison soon reveals its limits, however: whereas Mercury, the closest planet to the Sun, lies at about 80 solar radii, Io, the closest Jovian satellite, is at only about 6 planetary radii from the planet. The existence of rings in the equatorial planes of the four giant planets, in their immediate proximity, bears witness to the presence of material that is even closer. Below a limit at about 2.5 planetary radii (the Roche limit), the tidal forces to which the regular satellites are subject, created by the planet's gravitational field, lead to the destruction of the former. This fragmented material is the reason for the existence of planetary rings.

Another family of satellites exists around the giant planets, alongside the regular satellites. These are satellites that have been captured by the planetary gravitational field. These characteristically have high orbital inclinations relative to the planet's equatorial plane, and high eccentricities.

### 4.4.5.1  The Rings of the Giant Planets

The rings consist of a myriad grains and rocky blocks, whose dimensions may vary from a millimetre to several tens of metres. They occupy quasi-circular orbits, in planes that are very close to the equatorial plane of the parent planet. Each particle follows its own individual path, and permanently interacts with the particles in neighbouring orbits.

The rings have been studied by analyzing sunlight that they scatter. The observations made by the Voyager and later Cassini missions (Fig. 4.25) have shown the great complexity of the rings' dynamics. Their behaviour does not just depend on the planet's gravitational field, but also on that of the small satellites that are located within the outer rings.

The particles in planetary rings are subject to two effects:

- solar-radiation pressure which acts to push them away from the Sun, and which is proportional to the surface area of the grain; because gravitational attraction is proportional to volume, radiation pressure is particularly effective when acting on small-sized grains, about a micrometre across;
- the Poynting-Robertson effect, which arises when a particle in orbit around the Sun receives the solar radiation along the radius vector, but radiates it away isotropically. The result is a braking force which constantly decreases the eccentricity of the grain's orbit, and leads to a spiral path heading towards the Sun. This effect is particularly important for grains that are a few centimetres in size.

**Fig. 4.25** Saturn's rings as observed by the Cassini probe (© NASA)

Particles in rings are thus subject to two types of forces that tend, either to brake them, or to eject them from the system. Calculations show that their lifetimes are very short, around a few million years. Rings are therefore systems that are permanent over time, but which consist of transient bodies. They are permanently fed by the small, inner satellites, while part of their material escapes from the system.

One cannot fail to be surprised by the extraordinary diversity of the ring systems surrounding the four giant planets. While Saturn's has been known since Galileo's time, the other three ring systems, far more tenuous, were discovered only in the last few decades. Jupiter's rings, discovered by the Voyager 1 probe in 1979, are extremely tenuous and contain heavy elements (Si, S, etc.) derived from Jupiter's innermost small satellites Saturn's rings primarily consist of water ice, as was shown by spectroscopic observations carried out in the infrared from Earth some twenty years ago. Observations made by the Cassini probe, which has been orbiting Saturn since 2004, have enabled the rings to be studied in immense detail and for their chemical composition to be analyzed as a function of their distance from Saturn. Finally, the rings around Uranus and Neptune, initially detected from observations of stellar occultations from Earth in 1977 and 1984, respectively, and subsequently analyzed by the Voyager 2 probe during its fly-bys of 1986 and 1989, have an extremely low albedo. This suggests the presence of organic material, which might result from the irradiation of ices containing hydrocarbons. The presence of ring systems around the four giant planets in the Solar System seems to suggest that giant exoplanets might also possess such ring systems. The diversity in the rings observed in the Solar System suggests an equal diversity of rings around exoplanets.

### 4.4.5.2  The Satellites of the Giant Planets

Even within the regular satellites of an individual planet, there is an extremely wide variety of objects. Most of the satellites are without a permanent atmosphere. The most notable exception is Titan, the largest satellite of Saturn, which has a nitrogen atmosphere whose surface pressure (1.5 bar) is close to that of the Earth. The number of satellites known continues to increase, as a result of advances in instrumentation on ground-based telescopes. Currently, more than one hundred are known. The diameters of the smallest are no more than about ten kilometres.

The four largest satellites of Jupiter (known as the Galileans after their discovery by Galileo in 1610) exhibit major differences in composition and morphology, which may be explained primarily through their different distances from Jupiter. Io (Fig. 4.26), the satellite closest to Jupiter, is subject to intense tidal forces, which, in association with Io's elliptical orbit that is created by the interaction with the other Galilean satellites, leads to continuous remodelling of the surface and active volcanism. This volcanism feeds a permanent, but extremely tenuous, atmosphere, mainly consisting of $SO_2$. The surface pressure is around ten nanobars at the equator, and it decreases significantly towards the poles as a result of the condensation of sulphur dioxide. The other three Galilean satellites (Europa, Ganymede, and Callisto) have surfaces that primarily consist of water ice. In the case of Europa, which is also subject to Jupiter's tidal effects, the satellite's internal energy might be sufficient to maintain a liquid ocean beneath the surface. The depth of the surface layer of ice

**Fig. 4.26** The satellite Io (image credit: courtesy NASA, JPL Caltech)

may reach several tens of kilometres. The surfaces of the Galilean satellites are less evolved, the farther they are away from Jupiter.

Most of the regular satellites of Saturn are also covered with water ice, and without atmospheres. Lying at a distance of four Saturn radii, Enceladus is exceptional since the discovery by the Cassini probe of active cryovolcanism at the south pole. The other exception, already mentioned, is Titan (Fig. 4.27). Orbiting at a distance of 25 Saturn radii, Titan has a dense atmosphere of molecular nitrogen, with a

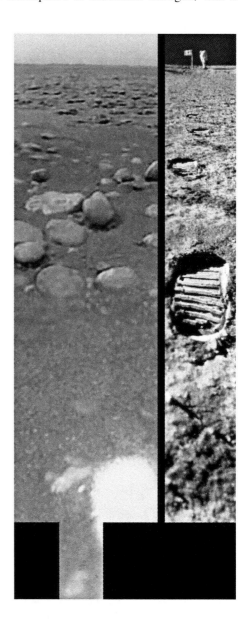

**Fig. 4.27** The surface of Titan as seen by the Huygens probe (image credit: courtesy ESA; NASA, JPL University of Arizona)

few per cent of methane. The methane is dissociated in the upper atmosphere and forms complex hydrocarbons that fall back onto the surface. The nitrogen is also dissociated by the energetic particles in the planet's magnetosphere, and becomes incorporated into nitriles. The methane, continuously photodissociated, is constantly regenerated in the atmosphere by an internal source. It is probably the same for $NH_3$, degassed from the surface by cryovolcanism (that is, by being ejected as liquid or gas which immediately solidifies on the surface), and transformed into $N_2$ by photodissociation. Unlike those of the giant planets, Titan's atmosphere was not produced from the surrounding nebula, but has been degassed from the body itself.

At a greater heliocentric distance from the Sun, the satellites of Uranus are darker than those of Jupiter and Saturn. As with the rings, this dark colour is undoubtedly the result of irradiation of ices containing hydrocarbons by energetic particles. Among Neptune's satellites, the most remarkable is Triton, an irregular satellite, whose features closely resemble those of Pluto when the latter, which has a very eccentric orbit, is at a comparable heliocentric distance. It has a tenuous atmosphere of molecular nitrogen, with a surface pressure of a few microbars, and methane is present at rate of about one per cent. In the light of the discoveries made in the last ten years about the trans-Neptunian (TNO) family, of which Pluto is one of the largest representatives, it seems that Triton itself is a TNO that has been captured by Neptune.

## 4.4.6 Small Bodies in the Solar System

Several classes of objects orbit the Sun in interplanetary space: the asteroids, comets, and the trans-Neptunian objects. Investigation of them provides information about the formation process and dynamical evolution of the outer Solar System. Comets have been known since antiquity, because their highly eccentric orbits may bring them close to the Sun and the Earth, when they become observable. The asteroids were first discovered at the beginning of the 19th century, and the trans-Neptunian objects at the end of the 20th century. As with the outer satellites, the number of small bodies discovered continuously increases, thanks to the commissioning of deep probes using increasingly powerful cameras in conjunction with large telescopes on the ground.

### 4.4.6.1 The Asteroids

The asteroids are objects with no atmospheres, with diameters less than 1000 km, most of which have quasi-circular orbits at heliocentric distances between 2 and 3.5 AU; this is the Main Belt (Fig. 4.28). Their accumulation within this region suggests that they are the remnants of a swarm of planetesimals, which were not able to accrete into a single planet, because of Jupiter's immense gravitational field. Other families of asteroids lie inside and outside the main asteroid belt: The NEA

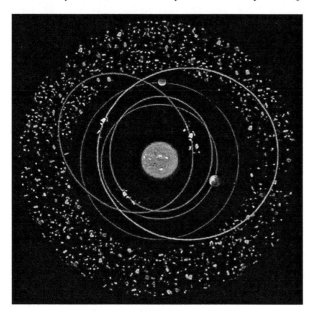

**Fig. 4.28** The orbits of several asteroids

objects (Near-Earth Asteroids) have one specific feature in that they come close to the Earth's orbit; the Trojans lie on Jupiter's orbit at the L4 and L5 Lagrangian points, 60 degrees ahead of, and behind, Jupiter. Farther from the Sun, the Centaurs are on more eccentric orbits, between the orbits of Jupiter and Neptune. Some of these objects are surrounded by a transient gaseous envelope, resembling those found around comets.

We know the chemical and mineralogical composition of the asteroids through spectroscopic observations in the near infrared, made from the ground. Several spectral types are recognized: the principal ones are the silicaceous asteroids (S), the metallic (M), and the carbonaceous (C). Their distribution as a function of heliocentric distance reveals that the densest asteroids (S and M) are, on average, closer to the Sun that the carbonaceous ones, the latter being more primitive. We find stratification resulting from the condensation sequence, in agreement with the model of planetary formation. A few asteroids have been the targets of fly-bys by space probes, and it has therefore been possible to study their surface properties in detail. Gaspra and Ida were examined by the Galileo probe, Mathilde and Eros by NEAR, Braille by Deep Space 1, Hayabusa by MUSES-C, etc.

Another source of information about asteroids comes from laboratory studies of meteorites, whose parent bodies are the asteroids. Such measurements have enabled the D:H ratio in particular to be measured for asteroids. We have seen that this ratio is greater, the lower the temperature, because low temperatures favour the formation of deuterated molecules. It has thus been possible to show that the D:H ratio in the terrestrial oceans (SMOW, Standard Mean Ocean Water) is close to that in asteroids of type D, primitive carbonaceous objects populating the outer region of the Main

Asteroidal Belt. The water in terrestrial oceans thus primarily stems from that population, with a minor component (with a lower D:H ratio than SMOW), originating from degassing by the globe, and another tiny component (with a higher D:H ratio than SMOW) originating in comets (*see below*).

### 4.4.6.2  Comets

Comets are objects with diameters generally less than some ten kilometres, orbiting the Sun on very eccentric orbits. Far from the Sun, they consist of a nucleus of ice and dust. Their periods range from a few years for the closest comets (such as Encke) to several thousand years (e.g., Hale-Bopp). When a comet approaches the Sun, its surface sublimes and we observe the ejection of gas and dust: this is the coma, which reflects and scatters sunlight, rendering the comet visible from Earth, sometimes even with the naked eye. The parent molecules that are ejected from the nucleus, observable from the ground in the infrared and millimetric regions, are, in turn, themselves dissociated into daughter molecules, radicals and ions. These species are detectable in the visible and ultraviolet regions through their fluorescent emissions.

Cometary physics has benefited greatly from the exploration of Comet Halley, with a period of 76 years, known since antiquity, and which is particularly bright. When it returned in 1986, several space probes were directed to encounter it, allowing the first in-situ analysis of a cometary nucleus. Since Halley's return, other very bright new comets (Hyakutake in 1996 and Hale-Bopp in 1997) (Fig. 4.29) have been the subject of profound study, in particular thanks to infrared and millimetric spectroscopy.

**Fig. 4.29**  Comet Hale-Bopp

Comets consist of 80 per cent water, but numerous other molecules have also been detected within them, generally at abundances that are less than a few per cent: $CO$, $CO_2$, $HCN$, $CH_3OH$, $HCO$, etc. The degassing rate for these molecules depends on the heliocentric distance. Degassing of $CO$, at several AU (11 AU in the case of Hale-Bopp) is probably the first sign of a comet's activity. Degassing of water occurs at around 2 AU from the Sun. The list of parent molecules (Table 4.3) contains more than twenty. All have similarly been observed in the interstellar medium. The nucleus of a comet has a very low albedo (less than 0.10). It appears to be covered in carbonaceous material, probably the result of irradiation of hydrocarbon-rich ices by cosmic rays. This carbonaceous material has been observed to be abundant in cometary dust released from the nucleus. Finally, isotopic ratios measured in comets, highly enriched in deuterium, bear witness to their formation at extremely low temperatures. The D:H ratio measured for water in particular, is twice as high as the terrestrial ratio.

All these signs bear witness to a common parentage between cometary material and interstellar material. Comets are some of the most primitive objects in the Solar System. Their small size has protected them from any effects of differentiation. Because they evolved for most of the time far from the Sun, in a cold and extremely tenuous environment, they also avoided thermal effects and collisions. Comets are therefore precious witnesses to the conditions surrounding the formation of the outer Solar System.

We have seen (Sect. 4.2.3) that the study of the orbits of comets has enable us to determine that high-inclination comets originate in the Oort Cloud, a vast shell lying some 40 000 AU from the Sun. Injected into this reservoir by perturbations by Jupiter, at the very beginning of their history, on rare occasions they return as the result of other local perturbations and may, in certain cases, stabilize on orbits within the Solar System following other planetary perturbations. Certain comets approach sufficiently close to the Sun to disappear into it. The SOHO satellite has shown that more than one hundred comets are swallowed up by the Sun every year. We may note the similarity here with that of the regular fall onto the star β Pictoris of comets from its disk of debris (*see* Chap. 6.) Finally, alongside comets from the Oort Cloud, other comets, with low inclinations and short periods, originate in another reservoir, the Kuiper Belt, the location of the trans-Neptunian objects.

### 4.4.6.3 The Trans-Neptunian Objects

Around the middle of the 20th century, the astronomers K. Edgeworth and G. Kuiper suggested, on the basis of dynamical studies, the existence of a toroidal belt of small bodies lying beyond Neptune – known as the trans-Neptunian objects (TNOs) – where low-inclination comets originated. In 1992, D. Jewitt and J. Luu detected the first member of this new family of objects. About 15 years later, more than a thousand are known, and the number continues to increase, with objects that are even farther away becoming detectable with the constant improvements in observational techniques (Fig. 4.30).

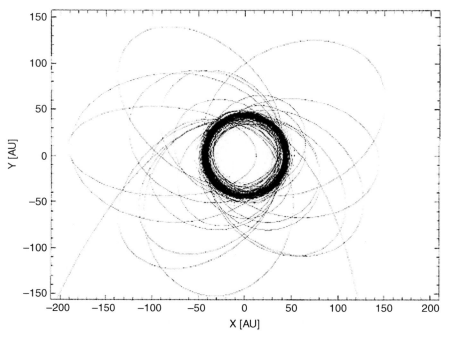

**Fig. 4.30** The paths of a number of TNOs (After Encrenaz et al., 2004)

The TNOs are large icy bodies, with an average diameter of a few hundred kilo-metres, and whose albedo is higher than that of comets. At present we know little about their physical and chemical properties, because spectroscopy of these faint objects is at the limits of current techniques. The first spectra obtained seem to indi-cate a resemblance to the spectra of Triton, Pluto, or with the Centaur, Pholus, with the presence of water ice in some cases, and of hydrocarbon (methanol) ices. The difference in composition may reflect different stages in the alteration of the surface, caused by different rates of bombardment by cosmic rays.

In contrast, we know more about the dynamical characteristics of the TNOs. The latter may be divided into three major classes:

- 'classic' objects, which includes more than half of these bodies, have quasi-circular orbits with low inclinations. Their average heliocentric distance is 42 to 47 AU, and they are not associated with any resonances;
- resonant objects (between 10 and 15 per cent of the total population) are, like Pluto, in the 3:2 resonance with Neptune, at a distance of 39.2 AU; these are known as Plutinos. This property confirms that Pluto, once classed as the ninth planet in the Solar System, is in fact one of the largest representatives of the TNO family.
- scattered objects have a perihelion outside the orbit of Neptune and a high ec-centricity. It is in this category that astronomers may well expect to make new discoveries of objects that are more and more distant and larger in size, such as the object 2003 UB313, discovered by M. Brown, which has a diameter larger than that of Pluto.

The Kuiper Belt therefore has a well-defined inner boundary that is determined by the 3:2 resonance with Neptune, but its outer boundary is far less well-defined.

To conclude, the Kuiper Belt may be considered, together with the comets in the Oort Cloud, as remnants of the initial protoplanetary disk. Studies of its dynamical families appears to show that the giant planets have migrated to a certain extent (*see* Sect. 4.3.2.6 and Chap. 6). Continuing exploration of the TNOs over the next few years should enable us to gain a better understanding of the overall dynamical history of the outer Solar System.

## 4.5 Conclusions: The Solar System Compared with Other Planetary Systems

What lessons may we draw from our study of the Solar System that are relevant to our exploration of planetary systems? We may obtain two forms of information from this comparison. The first concerns the scenario covering the formation of the Solar System: it is the exception or the rule? At present, do we have sufficient information on planetary systems to answer that question? The second lesson to be extracted from study of the Solar System concerns the study of the different classes of objects, with all their diversity. Can we gain a better idea of the atmospheres of exoplanets by studying planetary atmospheres? Finally, apart from exoplanets, what other bodies might be detectable in other planetary systems, and to what extent can we carry out remote sensing to investigate them?

### *4.5.1 The Scenario for the Formation of the Solar System*

We have seen that the Solar System formed through the collapse into a disk of a protoplanetary nebula, and that the planets formed by nucleation within this disk, starting with solid material. The observational facts on which this model is based are primarily:

- the quasi-coplanar and quasi-concentric orbits of the planets
- the enrichment on the giant planets in heavy elements, which is a factor that is strongly in favour of their formation through nucleation
- the existence of systems of rings and numerous regular satellites in the equatorial planes of the giant planets, and close to the latter, which bear witness to the collapse of surrounding subsidiary nebulae onto an initial solid core
- the existence of the main asteroid belt, of the Kuiper Belt, and the Oort Cloud, all vestiges of the accretion process. They represent the overall population of planetesimals that were not incorporated into more massive objects.

How do these pieces of data compare with the data that we currently have regarding exoplanetary systems? Their properties are, on the face of it, very different:

- Many giant exoplanets occur in close proximity to their star. The model of the formation of the Solar System does not explain their presence at such a location, unless they formed farther away and subsequently migrated to the immediate neighbourhood of their parent star;
- Many giant exoplanets, farther away from their parent star, have very eccentric orbits, which does not agree well with their formation from a disk that was produced by the collapse of a protostellar nebula;
- The number of exoplanets detected around a star seems to increase with the latter's metallicity rate, which seems to favour the theory of planetary formation by nucleation; however, the metallicity of the Sun is not particularly high, and the Solar System does not conform to this property.

Must we therefore conclude that the Solar System is an exceptional case? It is probably too soon to maintain this view. There are giant exoplanets at considerable distances from their parent star, in orbits with very low eccentricities. Nothing precludes the existence of exo-Earths or small exoplanets in the inner regions of a planetary system. We do not currently have the means to detect them. From the ubiquity of disks around young stars, we can infer that the method by which the Solar System formed is probably not unique, but that it does not seem to be widespread in the solar neighbourhood. Another possibility is that the method by which the Solar System formed might, under limiting conditions, evolve into very different systems. An important step in understanding this problem will be the detection of exoplanets of the terrestrial type. We will have to await the results of the CoRoT and KEPLER space missions to know if small exoplanets exist, and where they are located.

## 4.5.2 Objects in the Planetary Systems Observable from Earth

Our knowledge of the physical and chemical properties of exoplanets is still very poor. A comparative study of the planets in the Solar System should enable us to understand them better. In addition, apart from exoplanets, other bodies belonging to planetary systems are observable, or could become so with improvements in methods of observation. These include comets (by their possible fall into their parent stars), the Kuiper Belt (through the large quantity of water that it contains), the rings and satellites of exoplanets (by observation of transits or gravitational microlensing).

### 4.5.2.1 The Atmospheres of Exoplanets

Study of the planets in the Solar System teaches us one lesson. This is the astonishing diversity of objects that may be found within a single family. The properties of each planet are the result of the way in which it formed, but also of the evolutionary processes it has undergone. The most striking example is the diversity of the surface conditions of the terrestrial planets. When we are able to detect exo-Earths, we can thus expect to find a similar diversity. The atmospheric composition of an exoplanet will be identifiable from its infrared spectrum. An exoplanet having an abnormally high temperature (like Venus) could be detected from its night side through the thermal radiation. A surface temperature less than that of the atmosphere would be detectable through a thermal emission spectrum. Finally, numerous studies, based on models of stellar and planetary atmospheres, have been carried out to model the composition and thermal structure of an exoplanet as a function of its distance from its parent star (*see* Chap. 7).

### 4.5.2.2 Comets and the Kuiper Belt in Planetary Systems

We have seen that numerous comets that approach the Sun end their lives by dis-integrating and becoming swallowed up within it. This phenomenon would be ob-servable from outside the Solar System, through the observation, in the visible and UV regions, of transient, fluorescent emission lines, created when these object are volatilized. Observations of β Pictoris have shown transient phenomena that sug-gest the fall of comets onto the central star (*see* Sects. 5.3.2 and 6.5). Beta Pictoris is surrounded by a disk of dust, and the presence of a gap in this disk suggests the existence of at least one exoplanet. Taking account of the relatively late age of the star, the disk may not correspond with the planetary formation stage, but may con-sist of a residue of planetesimals that did not lead to the formation of planets. It would thus be the equivalent – albeit with a much greater volume – of our Kuiper Belt. Other observations of disks of dust (such as HR 4796A and Epsilon Eridani) suggest the presence of a Kuiper Belt around other stars. Similarly, observation of an abnormally large quantity of water vapour around a late star might be explained by a massive vaporization of its Kuiper Belt, in which the star has swollen as it passes into the red-giant stage. We will then be witnessing the scenario that is scheduled to take place when the Sun finally dies, in a few thousand million years.

## Bibliography

Bell, J.F., Davis, D.R., Hartmann, W.K. and Gaffey, M.J. 'Asteroids: the big picture', in *Asteroids II*, (eds.) Binzel, R.P., et al., 921, University. of Arizona Press, Tucson (1989)

De Pater, I. and Lissauer, J., *Planetary Sciences*, Cambridge University Press, Cambridge (2001)

Drossart, P., Encrenaz, T., Kunde, V., et al., 'An estimate of the $PH_3$, $CH_3D$ and $GeH_4$ abundances on Jupiter from the voyager IRIS data at 4.5 μm', *Icarus*, **49**, 416–426 (1982)

Encrenaz, T., 'Neutral atmospheres of the giant planets: an overview of composition measure-ments', in *The Outer Planets and Their Moons*, (eds.) Encrenaz, T., Kallenbach, R., Owen, T. and Sotin, C., Springer-Verlag, Heidelberg (2005)

Encrenaz, T., Bibring, J.-P. and Blanc, M., et al., *The Solar System*, 3rd edn, Springer-Verlag, Heidelberg (2004)

Grossman, L. and Larimer, J.W. Early chemical history of the solar system, *Rev. Geophys. Space Res.*, **12**, 71 (1974)

Lellouch, E., Bézard, B. and Fouchet, T., et al., 'The deuterium abundance in Jupiter and Saturn from ISO-SWS observations', *Astron. Astrophys.*, **370**, 610–622 (2001)

Lewis, J.S., *Physics and Chemistry of the Solar System*, rev. edn, Academic Press, New York (1997)

Lewis, J.S. and Prinn, R.G., *Planets and Their Atmospheres: Origin and Evolution*, Academic Press, New York (1984)

O'Leary, B., Chaikin, A.L. and Kelly Beatty, J., *The New Solar System*, rev. edn, Cambridge University Press, Cambridge (1999)

Owen, T., Mahaffy, P. and Niemann, H., et al., 'A low-temperature origin for the planetesimals that formed Jupiter', *Nature*, **402**, 269–270 (1999)

Pollack, J.B., Ubickyj, O. and Bodenheimer, P., 'Formation of giant planets by concurrent accretion of solids and gas', *Icarus*, **124**, 62–85 (1996)

Prinn, R.G. and Owen, T., 'Chemistry and spectroscopy of the jovian atmosphere', in *Jupiter*, (ed.) Gehrels, T., 319–371, University of Arizona Press, Tucson (1976)

Stevenson, D.J., 'Interiors of the giant planets', in *Ann. Rev. Earth and Plan. Sci.*, **10**, 257–295 (1982)

West, R.A., 'Atmospheres of the giant planets', in *Encyclopedia of the Solar System*, (eds.) Weissman, P.R., McFadden, L.-A. and Johnson, T.V., 315–338, Academic Press, New York (1999)

# Chapter 5
# Star Formation and Protoplanetary Disks

Observation of stars at different stages of their life has enabled us to retrace the processes of their formation and of the different stages in their evolution. Stars are born within interstellar condensations that are inhomogeneous and turbulent, as a result of the gravitational collapse of a rotating cloud. After a contraction phase, they reach a long period of equilibrium during which they move along the Main Sequence in the Hertzsprung-Russell diagram, which defines the evolution of their luminosity as a function of the temperature (see Appendix A.4 and Fig. 11.2). The collapse process of the initial cloud leads to the formation of a disk, perpendicular to the cloud's axis of rotation, within which planets may form by the accretion of solid particles.

In this chapter, we summarize the first stages of star formation, from the collapse of the interstellar cloud until the star arrives at the Main Sequence, and we describe the evolution of protoplanetary disks as well as their spectral signature. Finally, we describe the physical mechanisms governing the formation of planetesimals and planetary embryos inside the disk. The dynamical evolution of planetary systems and the dynamical interaction between planets and disks are studied in Chap. 6.

## 5.1 The First Stages in Star Formation

### 5.1.1 Properties of the Interstellar Medium

The existence of an interstellar medium that was 'not empty' was suspected at the end of the 18th century, when the British astronomer William Herschel observed dark areas among the stars, and which might correspond with material that absorbed visible radiation (Fig. 5.1). At the beginning of the 20th century, this theory was confirmed by spectroscopic detection of narrow absorption lines, which could be attributed to the presence of gas along the line of sight to the star being observed. Following predictions made by C. Van de Hulst and J. Oort, the neutral hydrogen transition at 21 cm was observed in the 1950s. This transition, although intrinsically very faint, is observable only because of the long optical path-lengths available

M. Ollivier et al., *Planetary Systems*. Astronomy and Astrophysics Library,
DOI 978-3-540-75748-1_5, © Springer-Verlag Berlin Heidelberg 2009

**Fig. 5.1** The Horsehead Nebula (B33) in the Orion Nebula. The dark silhouette is caused by the absorption by the dust in a dark molecular cloud of the light from the nebulosity lying behind the cloud (IC 434) (image credit: courtesy NOAO, AURA, NFS)

for observation in the Galaxy. Detection of the transition proved the existence of significant quantities of gas in the form of atomic hydrogen. In the 1940s, the CH and CN radicals and the $CH^+$ ion were detected in the centimetric region, followed, in the 1960s, by OH $H_2O$, and $NH_3$, and finally, beginning in the 1970s, CO and numerous molecules, detected in the millimetric region. The interstellar medium does not contain just atomic hydrogen, but also radicals, ions, and molecules. Beginning in the 1970s, observations from space in the UV region (from the Copernicus and IUE – International Ultraviolet Explorer) satellites allowed intense UV sources to be detected, together with absorption lines in their spectra. Infrared observations (with IRAS – InfraRed Astronomical Satellite – then ISO – Infrared Space Observatory, and most recently, Spitzer) have opened access to the study of interstellar dust and circumstellar envelopes. Infrared spectroscopy is a valuable tool for the study of cool regions, such as dense molecular clouds, because the thermal flux from dust peaks in that region, where the spectral signatures from molecules are also the most intense.

What have we learned from observations? The interstellar medium consists, by mass, of 99 per cent gas (of which about 76 per cent is hydrogen, and 22 per cent helium), and about 1 per cent dust. It is characterized by various phases under very different physical conditions (see, for example, Acker, 2005 and Lequeux, 2005): (1) a cold phase (about 80 K) that is neutral and relatively dense (approximately $40\,H\,cm^{-3}$); (2) a hot phase (8000 K) that is neutral and very tenuous ($0.4\,H\,cm^{-3}$);

(3) an ionized phase, with similar physical parameters; and (4) an extremely hot, ionized phase (where T is around $10^6$ K, and the density $0.003\,H^+\,cm^{-3}$). The extreme diversity of these physical conditions is explained by the variety of sources of radiation in which they are bathed (stellar radiation, cosmic rays, etc.). In addition to these four phases, there are dense molecular clouds, with average temperatures of 10 K and with densities $H > 300\,cm^{-3}$. Even though they occupy only a small volume, they contain a significant fraction (about 50 per cent) of the overall mass of interstellar material.

Probes of the molecular gas within galaxies have shown that the gas is often concentrated in the spiral arms, within vast molecular complexes that are about one kiloparsec across, and whose mass may reach $10^7$ solar masses. These complexes contain giant molecular clouds, some hundred parsecs in diameter, with masses of one million solar masses. They may, in turn, contain massive concentrations, several parsecs across and containing several thousand solar masses, and also small dense clouds, about 0.1 parsec across and of a few solar masses (Larson, 2003). The interstellar medium appears to exhibit a hierarchical structure, that is probably fractal in nature (see, in particular, Pfenniger and Combes, 1994; Pfenniger et al., 1994; Larson, 1995; and Elmegreen et al., 2000). A summary of the different stages of stellar evolution and the evolution of planetary disks may be found in Najita (2001) and Cassen (2006).

## 5.1.2 The Formation of Molecular Clouds

Molecular clouds form within the interstellar medium through gravitational collapse. Let us discuss the mechanism of this collapse in broad outline. We assume, to make things simpler, a spherical, uniform cloud (in reality, it may, of course, have a far more complex form and structure). This sphere of gas is subject to gravitational forces which tend to make it collapse, but also to a thermal (or kinetic) pressure that acts against the former. The kinetic energy $E_k$ of the sphere may be expressed as:

$$E_k = \frac{3}{2}kT\left(\frac{MkT}{m}\right) \tag{5.1}$$

where $k$ is the Boltzmann constant, $T$ the temperature of the cloud, $M$ its mass and $m$ the mass of an atom of hydrogen. The potential (gravitational) energy of the cloud, where we assume the density is constant as a function of distance from the centre, is

$$E_p = \frac{3}{5}\left[\frac{GM^2}{R}\right] \tag{5.2}$$

G being the gravitational constant and $R$ the radius of the cloud. According to the virial theorem, collapse occurs if $E_p > E_k$, that is if

$$R < 0.4\,\frac{GMm}{kT} \tag{5.3}$$

By expressing $M$ as a function of R and of density $\rho$ (number of atoms H cm$^{-3}$) we obtain:

$$M > 3.10^4 \left(\frac{T^3}{\rho}\right)^{1/2} = M_J \tag{5.4}$$

$\rho$ being the number density of particles per cubic metre. $M_J$, expressed in solar masses, is the Jeans mass, as defined by the British astronomer James Jeans, in 1926. When the medium is hot and low in density, the material is almost completely in the form of hydrogen atoms. From values typical of the interstellar medium, the Jeans mass is about $5 \times 10^7$ solar masses for the inter-cloud medium, a few tens of solar masses for the densest clouds, which is comparable with observations.

It should be emphasized again that the above treatment assumes a cloud with constant density and temperature, which is far from being realistic. In reality, stability/instability is determined not by a general virial theorem applied to a constant-density configuration, but by the details of the true equilibrium. There is also a great deal of evidence that the formation and structure of molecular clouds are controlled by magnetic fields. The effect of magnetic fields on the stability and collapse of a cloud are complex. Discussions on these processes may be found in Spitzer (1978) and, more recently, Lequeux (2005).

### 5.1.3 Collapse of a Molecular Cloud

Within a molecular cloud, the same gravitational-collapse mechanism leads to the formation of stars. The mass of the initial cloud may vary between a few tens and a few tens of thousands of solar masses. The rotation of this cloud, which is linked, for example, to the large-scale galactic rotation, may brake the compression, as may the action of the interstellar magnetic field, the energy of which, $E_m$, defined by

$$E_m = \frac{4\pi R^3 B^2}{24\pi} \tag{5.5}$$

opposes the cloud's potential energy. The collapse may continue if $E_p > E_m$, which defines a critical mass proportional to the product $BR^2$. The magnetic field of the protostar will have the effect of ejecting material in bipolar jets, such as those that may be observed in young objects of the Herbig-Haro type (see Sect. 5.2). In certain cases, the magnetic field helps the collapse, most notably by causing the loss of angular momentum. Lequeux (2005) gives a thorough discussion of collapse mechanisms and star formation.

### 5.1.4 Observation of Young Stars

Our knowledge of the processes of star formation rest primarily on observation of T-Tauri type stars – stars of spectral types G, K or M at an evolutionary stage prior to the Main Sequence. They were first noted because of strong emission lines (Balmer lines) of atomic hydrogen, but also have strong emission in the UV and IR regions. These features are now understood and explained by an accretion disk surrounding the protostar (Lynden-Bell and Pringle, 1974). The ultraviolet emission and certain millimetre and sub-millimetre signatures may be attributed to material falling onto the protostar along the magnetic-field lines. The material falling onto the centre exhibits a specific spectral signature with a double emission peak (which is, however, very difficult to observe), but which may be explained by a spherically symmetrical model of dynamical collapse (Choi et al., 1995; Fig. 5.2).

Since the 1970s, astronomers have detected the ejection of material around young stars (Fig. 5.3) by observing molecular jets delineated by mapping the millimetric transitions of CO. Such images might show a characteristic bipolar structure, with opposing lobes, one shifted towards the blue and the other towards the red, indicating massive ejection of material, the emission from which is focussed along a particular axis, and with terminal velocities of several hundred kilometres per second. These jets have been identified by line profiles in the visible and near infrared (Acker, 2005).

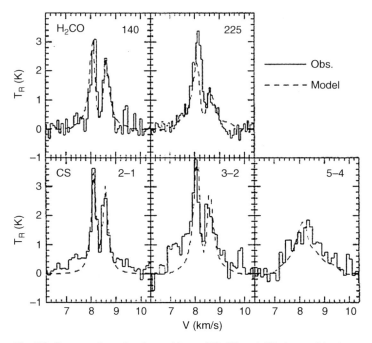

**Fig. 5.2** Spectra of rotational transitions of $H_2CO$ and CS observed in the young star B335, compared with the spherically symmetrical model of dynamical collapse (After Choi et al., 1995)

**Fig. 5.3** A diagram
illustrating star formation.
The star forms following the
collapse of a rotating cloud of
interstellar material, which
flattens into a disk,
perpendicular to its rotation
axis. The stellar wind, which
is confined by the magnetic
field, escapes in two lobes
that are aligned with the
rotation axis (After
Acker, 2005)

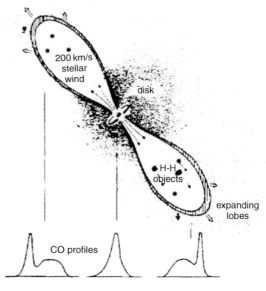

Snell 1980

Several mechanisms may be responsible for the mass loss of a single young star: stellar winds; magnetic fields; or turbulent viscosity within the protoplanetary disk. In the presence of a magnetic field, ionized particles spiralling along field lines will co-rotate with the rotating star, contributing to angular-momentum transfer outwards. As a result, the spin rate of the star will slow down (a more complete discussion may be found in Cole and Woolfson, 2002). Their analysis shows that for a rotating system of constant angular momentum, the inner material will move inwards and thus will be accreted by the star, while the outer material will orbit at greater and greater distances. This mechanism also leads to a transfer of the angular momentum from the inner part to the outer. Finally, as discussed below (Sect. 5.2.4), the formation of multiple systems is another efficient way of dissipating the angular momentum of a protostellar object.

Figure 5.4 illustrates the different components of a typical protostar, the young object HH 211 (a Herbig-Haro object, named after the two astronomers who first discovered the associated nebulae). The two elongated lobes correspond to a molecular flow, defined by contours of the emission from CO. The dark patches at the end of the lobes correspond to emission from molecular hydrogen. The length of each lobe is approximately 10 000 AU. The lobes are aligned along the protostar's rotation axis. The latter is surrounded by an accretion disk, lying in a plane perpendicular to the rotation axis, which extends out for a few hundred AU. At this stage, the object does not emit any visible radiation.

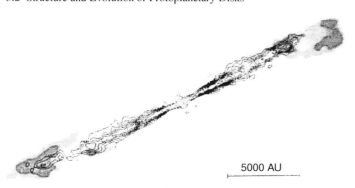

5000 AU

**Fig. 5.4** A diagrammatic representation of the object HH 211, showing the bipolar jet identified by CO emission in the millimetre band. A shock wave forms at the boundary between the jet and the interstellar medium. It is identified by the infrared emission from molecular hydrogen. The central disk, perpendicular to the axis of the bipolar flow, has been identified by its thermal emission, detectable in the millimetric continuum (After Bertout, 2003)

## 5.2 Structure and Evolution of Protoplanetary Disks

### 5.2.1 Observation of Protoplanetary Disks

Until the end of the 1970s, the existence of circumstellar disks around young stars was not commonly accepted (Meyer et al., 2006), with most astronomers preferring the idea of a spherical envelope surrounding the young object. Circumstellar disks were first revealed by their infrared emission. The infrared excess measured in the spectra of certain stars, including Vega, by the IRAS satellite (Aumann et al., 1984), proved the existence of a disk around the star, and where subsequent estimates were made of its average temperature and dimensions. Another of the discoveries by IRAS, the disk surrounding the evolved star Beta Pictoris, was subsequently mapped in the visible region, using coronagraphy, first by a ground-based instrument (Smith and Terrile, 1984), and then, highly accurately, by the HST. It should be noted that in this last case, the circumstellar disk that was observed was not a protoplanetary disk, but the evolved disk around a Main-Sequence star. Such disks have been called debris disks (see Sect. 5.3). Other protoplanetary disks have also been observed by the HST (Fig. 5.5).

A circumstellar disk becomes detectable (first in infrared and then at optical wavelengths) at the stage at which the initial collapse of the molecular cloud takes place. Observation of protoplanetary disks that are about 100 AU across requires a high angular resolution. To take an example, a disk lying at a distance of 50 pc has an angular size of 2 s of arc if its diameter is 100 AU. To resolve structure within the disk, an angular resolution of less than one arcsec is required, which is possible in the visible with the HST, or in the radio region by interferometry (for example via millimetric interferometry with the IRAM array on the Plateau de Bure, or with the VLA in the centimetric region, Fig. 5.6).

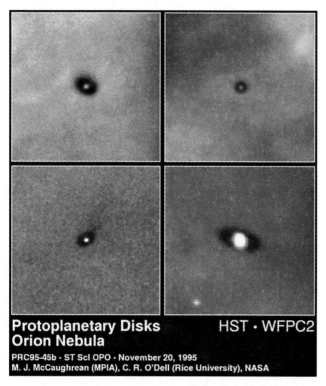

**Fig. 5.5**  Protoplanetary disks observed by the HST (image credit: courtesy NASA, ESA)

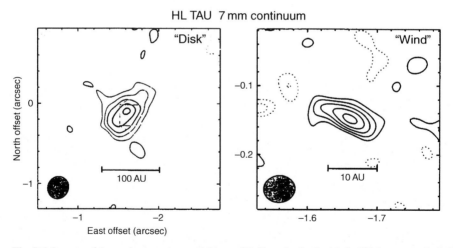

**Fig. 5.6**  Images of the environment around the star HL Tau, obtained with the VLA at a wavelength of 7 mm, and with a resolution of 40 mas. Left: the overall structure of the disk, determined from continuum emission; right (enlarged): image of the inner disk obtained from measurements made as a very high resolution, showing the collimated jets, perpendicular to the plane of the disk (After S.P. Wilmer et al., 2000, as cited by Najita, 2001)

The kinematics of disks is studied by high-resolution ($R = 10^6$) heterodyne spectroscopy in the millimetric interferometry domain – by IRAM, for example – based on the Doppler effect measured from molecular transitions (that of CO in particular). It is these observations that reveal the streams of material that are being accreted and ejected, and which play a part in star formation. Closer to the heart of the protostar, it is also possible to determine the kinematics of the hotter and denser, inner disk, by examining the vibration-rotation molecular transitions (for example the (2–0) band of CO at 2.35 μm; Fig. 5.7).

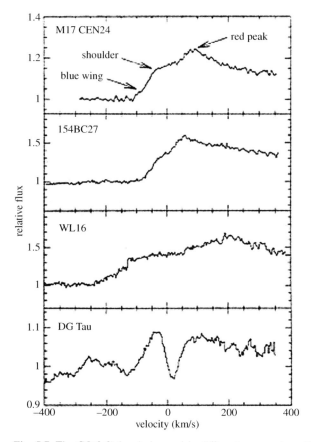

**Fig. 5.7** The CO(2-0) band observed in different young stars: M17 CEN 24, 1548C27, WL 16, DG Tau. (After Najita, 2001). The spectra show a wing in the *blue*, a shoulder, and a peak in the *red*, which are characteristic of a rotating disk

## 5.2.2 Stellar Accretion Flux

The protoplanetary disk's matter accretion rate may be estimated from the excess UV or IR radiation from the protostar. The infrared excess comes partly from the dissipation of the potential energy released when the matter spirals towards the interior of the disk (Lynden-Bell and Pringle, 1974). However, other processes contribute to the infrared emission.

The UV radiation, by contrast, comes directly from the dissipation of the gravitational energy from the transfer of material from the inner edge of the disk onto the star. It is therefore a more reliable indicator of the accretion rate. According to the most recent models, the star accretes the material along the protostar's magnetic-field lines, and not through a continuous boundary region (Fig. 5.8). This theory is based on the presence of P-Cygni profiles (which are characterized by an asymmetrical line, revealing the presence of a stellar wind, like that found in the star P Cygni, where the phenomenon was discovered) that are observed in certain Balmer lines, and which may be interpreted as the signature of material falling onto the star. Transfer models that simulate magnetic accretion manage to account for all the observed spectral characteristics. However, these observations may only be carried out on stars that have become optically visible, i.e., T-Tauri type stars, which have freed themselves from the primordial cloud.

The stellar object HH 30 (Fig. 5.9), which has been observed several times by the HST, reveals a remarkable example of a protoplanetary disk following dispersal of the primordial cloud. The opaque disk, the dark outline of which is silhouetted against the light from the star, is seen edge-on. Its diameter is several hundred AU. One jet, perpendicular to the disk, is clearly visible. It should be noted, however, that it is far less intense than in the collapse phase, an example of which we have seen in the object HH 211 (Fig. 5.4). Observations of the $^{12}CO(2-1)$ and $^{13}CO(2-1)$ transitions have enabled the velocity of the gas in the jet to be determined.

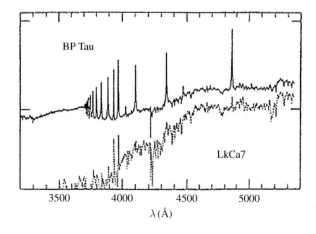

**Fig. 5.8** The UV and optical spectrum of the star BP Tau (*solid line*) compared with that of another T-Tauri star, LkCa7 (*dashed line*). The spectrum of BP Tau shows a UV excess, the sign of material falling onto the protostar. The other T-Tauri star, LkCa7, does not exhibit this sign of accretion (After Hartmann, 1998)

**Fig. 5.9** (a) Image of the object HH 30 taken by the HST; (b) A schematic interpretation of the disk and of the jet (After Bertout, 2003)

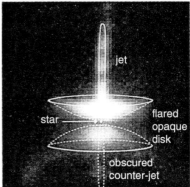

### 5.2.3 The Rotation of T-Tauri Stars

The rotation periods of T-Tauri stars may be measured by spectroscopy from the broadening of the line profiles, or by photometry by estimates from their light-curves.

The general conclusion that arises from this study is that T-Tauri stars are slow rotators: their periods (which are generally several tens of days) are about ten times the periods that would be expected as a result of the collapse of a cloud, if there had not been any loss of angular momentum. They are also significantly higher than those of stars of comparable age and spectral type that do not have a disk (where the periods are about 8 days).

This results strongly confirms the theory of the loss of angular momentum to the disk through magnetic coupling (see Sect. 5.1.4). Coupling between stellar accretion and magnetic activity is equally revealed by other observations: intense X-ray emission, non-thermal radio emission, and photometric observation of large star-spots. Direct detection of a magnetic field in a T-Tauri star has been obtained by measurement of the Zeeman broadening observed in an atomic transition (Fig. 5.10).

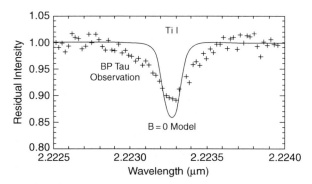

**Fig. 5.10** Measurement of the Zeeman broadening of the Ti I transition at 2.2 μm, obtained for the T-Tauri type star BP Tau. Crosses: observations; *solid line*: model in the absence of a magnetic field (After Johns-Krull et al., 1999)

How do low-mass stars acquire their mass and angular momentum? Both originate in the initial molecular cloud, whose collapse phase ends with the formation of a disk. In broad outline, the process is understood: transfer of the disk's angular momentum to the star implies the transfer of mass, which falls from the disk to the star with a free-fall time that is shorter than the rotation period. The star thus acquires a rapid rotation which results in the generation of a strong magnetic field within its convective interior. The star's magnetic field couples with the disk, at a distance of several stellar radii. This slows down the star's rotation to that of the disk at that distance. Strong stellar winds result, the terminal velocities of which and the associated mass-losses may be measured, with good agreement with predictions.

## 5.2.4 The Formation of Binary Systems

The existence of a stellar magnetic field partially explains the transfer of angular momentum, but is not always sufficient to account for the observed slow rotation rate. The mechanism is, in fact, effective during the early phases of the evolution of the cloud, when the magnetic field is strongly coupled to the gas. Ambipolar diffusion should, however, cause the gas to be decoupled from the magnetic field, which should lead to the conservation of the angular momentum during the later stages of collapse (Larson, 2003). Other mechanisms are thus necessary to account for the transfer of the angular momentum.

A very efficient mechanism is the collapse of the cloud in a binary, or even multiple, system. Most of the angular momentum is then transformed into the orbital motions of the different stars. The angular momentum of a typical molecular core is comparable with that of a binary system. Numerous theoretical studies suggest that the collapse of a molecular cloud frequently leads to the formation of binary and multiple systems (see Matsumoto and Hanawa, 2003, in particular).

Observations show that about two-thirds of all Main-Sequence stars are binary or multiple systems. Distant binaries may form directly from the collapse of molecular core. Close binaries are probably formed by more complex mechanisms, including

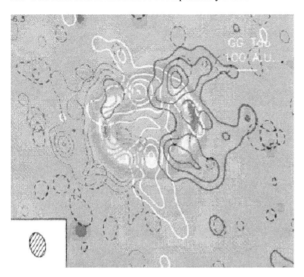

**Fig. 5.11** The binary system GG Tau, observed by millimetric interferometry by IRAM (Plateau de Bure). The two pre-Main-Sequence stars are separated by 38 AU (distance as projected on the sky). The observations were made using the $^{13}CO(2-1)$ line at 220 GHz. The emission shown in *yellow*, observed in the continuum at 1.3 mm, corresponds to the thermal emission from dust and indicates a disk-like structure surrounding the binary system. Material has been lost from the central region of the disk through tidal effects, caused by the interaction of the two components of the binary (© IRAM)

dynamical interactions with the disk and through tidal effects (*see* Chap. 6). As for isolated stars, they may result from the evolution of unstable triple systems, that subsequently divide into an isolated star and a binary system (Larson, 2003). Figure 5.11 shows the young binary system GG Tau as observed by millimetric interferometry.

## 5.2.5 The Principal Stages of Star Formation

The major stages in the formation of stars may be summarized as follows (Larson, 2003; While et al., 2006):

- collapse of the cold cloud (known as a 'class 0' object)
- protostar, an object ('class I') that is still buried within the collapsing cloud
- T-Tauri phase, a pre-Main-Sequence object ('class II'), optically visible, surrounded by a thick protoplanetary disk and with a characteristic, intense stellar wind
- young Main-Sequence object ('class III'), surrounded by a thin disk (a debris disk).

**Fig. 5.12** A time/radius
diagram of the contraction
phases of a molecular cloud.
The abscissa shows the
density in g/cm$^3$, and the
ordinate log T (K) (After
Acker, 2005)

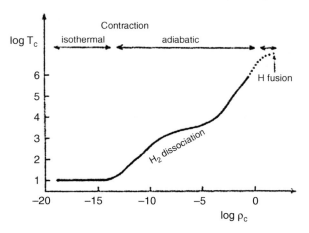

The collapse of a molecular cloud may be described in terms of several phases
(Fig. 5.12):

- Isothermal phase: This is the first step of gravitational contraction. The temperature is about a few tens of Kelvin. The gas is tenuous enough for the gravitational
  energy to dissipate through the radiation coming from the thermal excitation of
  the atoms. The temperature thus remains low and constant. The object, with a
  size of a few hundred AU, is observable through its infrared thermal emission.
- Adiabatic phase: as the density increases, the opacity of the clouds increases,
  and the energy released by the contraction of the cloud cannot escape by radiation, and the temperature rises. The contraction takes place adiabatically, without
  any exchange of heat with the exterior. The increase in temperature leads to the
  dissociation and ionization of the hydrogen molecules. The ionization phase is
  accompanied by a reduction in the rate of heating and an acceleration of the contraction phase.
- Appearance of the star: at the end of the hydrogen-ionization phase, the temperature again rises rapidly, until it reaches several million degrees. This temperature
  is sufficient to initiate the first thermonuclear reactions. Henceforward, the star
  shines in the visible region.

The evolution of the protoplanetary disk may be followed by the evolution of its
electromagnetic spectrum, from the UV to the radio region (Fig. 5.13):

- In the collapse phase (less than 10000 years), only the cold cloud is detectable
  from radiation by the dust as a black body that peaks in the far infrared; the
  spectrum is said to be of 'Class 0'.
- In spectra of Class I ($t = 10000$ years), the increase in the temperature at the
  centre of the cloud produces a shift in the black-body spectrum towards the
  near infrared. Cold material continues to fall onto the protostar, producing another component that peaks in the far infrared. This cold material accumulates
  in an equatorial disk. A UV component, very close to the protostar, may also be
  present, and is the signature of matter falling onto the central object along its
  lines of magnetic force (see Sect. 5.2.2).

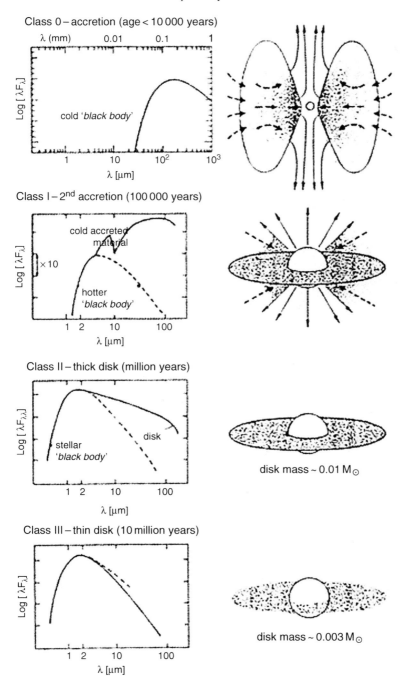

**Fig. 5.13** Evolution of the spectrum of a protostar during its four evolutionary phases (Classes 0 to III) (After André and Montmerle, 1994)

- Objects of Class II (t = about $10^6$ years) are characterized by a thick disk, which absorbs part of the light from the protostar and re-emits it in the infrared. The spectrum of the protostar is usually maximum in the near infrared, whereas the disk component, which is colder, provides a contribution at longer wavelengths. Some T-Tauri stars show a flat spectrum over the infrared range.
- Finally, in objects of Class III (t = about $10^7$ years), the disk has lost most of its mass because of the violent stellar wind, which has swept the material out into space. The spectrum of the star predominates. It peaks in the near infrared (which corresponds to a temperature of 4000 K). The peak shifts towards visible wavelengths as the star's temperature increases.

Figure 5.14 similarly shows how the infrared excess of a young star decreases with age, as the quantity of dust contained within the disk decreases.

It should be noted that the spectra of young stars may exhibit more complex forms than the simple combination of black-body emissions described above. This is particularly the case for the T-Tauri star, GM Aurigae, which exhibits, in common with a number of T-Tauri stars, a spectrum that is relatively flat over a range of frequencies (Fig. 5.15). To explain this type of spectrum, a model with a double-layer disk has been devised by Chiang and Goldreich (1997). According to this model, at a certain distance from the star, the stellar radiation is absorbed by the dust in the disk's outer layer, converted into thermal energy, and re-emitted in the form of infrared radiation both to space and towards the equatorial plane of the disk (Fig. 5.16). The dust grains, which are too small to convert all the energy that they receive, heat up to a temperature that is greater than that of the gas, producing a spectral component that is shifted towards the blue, relative to the emission from the disk.

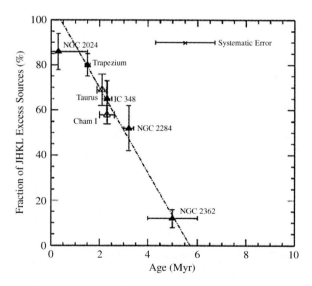

**Fig. 5.14** The quantity of dust present around the star is revealed by an infrared emission excess, which decreases as the age of the star increases (After Haisch et al., 2001)

**Fig. 5.15** The spectrum of
the T-Tauri star GM Aur. It
exhibits two components:
cold emission from the disk at
low frequencies; stellar
emission at high frequencies;
and an intermediate emission
caused by radiation from the
surface of the disk (After
Chiang and Goldreich, 1997)

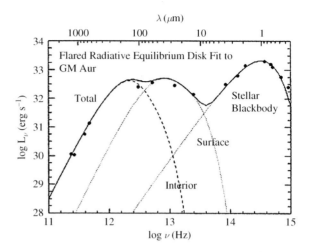

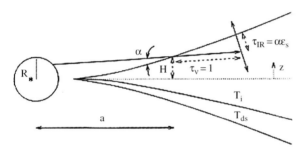

**Fig. 5.16** A model of a two-layer disk that accounts for the spectrum observed in Fig. 5.15. The thickness of the disk increases with distance from the star. The stellar radiation is absorbed by the dust in the disk's outer layer, whose temperature rises higher than that of the gas (After Chiang and Goldreich, 1997)

## 5.2.6 Later Stages of Stellar Evolution: Evolution Towards the Main Sequence

In a first approach, the evolution of the internal structure of a star beyond the T-Tauri phase may be described by means of four fundamental equations (Acker, 2005):

(1)  the perfect gas equation:

$$P = \frac{R \rho T}{\mu} \tag{5.6}$$

where $P$ is the pressure, $R$ the perfect gas constant, $\rho$ the density, $T$ the temperature, and $\mu$ the mean molecular weight;

(2)  conservation of mass:

$$\frac{dM(r)}{dr} = 4 \pi r^2 \rho \tag{5.7}$$

(3) hydrostatic equilibrium:

$$\frac{dP}{dr} = \frac{\rho \, GM(r)}{r^2} \qquad (5.8)$$

where G is the gravitational constant;

(4) conservation of energy:

$$\frac{dL(r)}{dr} = 4\pi r^2 \rho \, \varepsilon \qquad (5.9)$$

where $\varepsilon$ is the energy (of nuclear origin).

The energy is transported towards the outer layers by convection or by radiation. In the case of radiative transfer, the energy-transport depends on the opacity of the medium, which, in turn, depends on its chemical composition (gas and dust). Under the condition of LTE (local thermodynamic equilibrium, see Sect. 4.4.2.2.) the radiation, assumed to be isotropic, is given by Planck's Law. The luminosity, L, as a function of radius is given by:

$$L(r) = \frac{4\pi r^2}{3\rho \, \kappa} \cdot 16\sigma T^3 \cdot \frac{dT}{dr} \qquad (5.10)$$

an expression in which $\kappa$ is the opacity and $\sigma$ Stefan's constant. Stability of the star implies a balance between the gravitational force that induces collapse, and the radiation pressure that tends to make the star expand. The radiation pressure $P_{rad}$ is expressed as:

$$P_{rad} = \frac{4\sigma}{3c} \cdot T^4 \qquad (5.11)$$

Equilibrium cannot be maintained if the radiation pressure is less than the total pressure $P = P_{gas} + P_{rad}$. By using the hydrostatic equilibrium law, it may be shown that this condition implies that the star's luminosity remains less than a value known as the Eddington luminosity:

$$L < L_E = \frac{4\pi c GM}{\kappa} \qquad (5.12)$$

This limiting luminosity is thus proportional to the mass of the star. Beyond this value, the radiation pressure predominates, and the star becomes unstable. It should be mentioned that the Eddington limit applies to massive objects, and not to solar-type stars.

During the course of its evolution towards the Main Sequence, the star maintains an essentially constant luminosity and surface temperature. From the equations just given, it may be deduced that the luminosity is proportional to the cube of the mass, and the temperature is proportional to the ratio of the mass of the star to its radius. Observations have allowed the value of the exponent to be refined:

$$\log\left(\frac{L}{L_\odot}\right) = 3.45 \log\left(\frac{M}{M_\odot}\right) \qquad (5.13)$$

where $L_\odot$ and $M_\odot$ are the luminosity and mass of the Sun.

**Fig. 5.17** Evolution of
protostars on the
Hertzsprung-Russell
Diagram, from the collapse of
the molecular cloud to the
Main Sequence. The Hayashi
Limit, at a constant
temperature of about 4000 K,
corresponds to an instability
region. The star cannot cross
this limit, and loses
luminosity, at a constant
temperature, before joining
the Main Sequence (After
Acker, 2005)

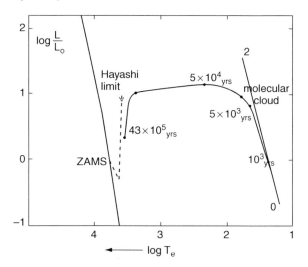

The method of energy transport depends on the mass of the star. For stars with masses comparable to that of the Sun, radiative equilibrium predominates, whereas convective equilibrium prevails in low-mass stars ($< 0.3\ M_s$). The outer convective zones reduce as the mass of the star increases.

Let us return to consideration of the evolution of stars after their T-Tauri phase. We have seen that this phase is marked by an extremely high luminosity and a relatively thick disk (a Class II object, see Fig. 5.13). One star, FU Orionis, observed by Herbig, allows us to describe the stage following the T-Tauri stage, and which is the last stage of star formation (a Class III object, see Fig. 5.13). The FU-Ori phase is characterized by an even more violent eruptive phase, during which the star ejects a large fraction (or even almost all) of the mass of the disk. During phases II and III, the surface temperatures of these objects increases to about 4000 K, then their luminosity decreases before they reach the Main Sequence (Fig. 5.17).

The lifetimes of protoplanetary disks is estimated to be some ten million years (see Fig. 5.13), as derived from observations: young stars surrounded by disks all have ages less than ten million years. This observational fact translates into an extremely powerful constraint on models for planetary formation within these disks.

## 5.2.7 The Structure of Protoplanetary Disks

As a working hypothesis, we assume a viscous disk, externally isolated (this is obviously an approximation, because accretion is also generally present). The evolution of this disk may be described by assuming Keplerian motions around the central star, also assuming that radial pressure gradients are absent. We may then write:

$$\omega(r) = (GM^*)^{1/2} . r^{3/2} \tag{5.14}$$

$$j(r) = r^2\,\omega \tag{5.15}$$

where $\omega$ is the angular velocity and $j$ the kinetic moment. $M^*$ is the mass of the central star, and $r$ the distance from the star. The vertical structure of the disk may be calculated (where the $z$-axis is perpendicular to the plane of the disk) by assuming that the gas is isothermal and in hydrostatic equilibrium along that axis. We then find the following relationship between the scale height, $h$, and the distance from the star:

$$h^2 = \frac{v_s^2}{v_k^2}\cdot r^2 = \frac{k\,T\,r^3}{G M^*\,m} \tag{5.16}$$

where $v_s$ and $v_k$ are, respectively, the velocity of sound and the Keplerian velocity at distance $r$ from the star. It will be seen that the thickness, $h$, of the disk increases with distance from the star, which agrees with observations (Figs. 5.9 and 5.16). Because the disk is thin, the Keplerian velocity is far greater than the velocity of sound:

$$V_k = \left(\frac{G M^*}{r}\right)^{1/2} >> V_s \tag{5.17}$$

Dynamical equations describing the evolution of a thin disk must take viscosity into account. This may be expressed as a couple $\tau$; the latter is proportional to the coefficient of viscosity, to the density, and to the distance from the star. By taking conservation of angular momentum into account, it may be shown that, in the inner part of the disk, the radial velocity, $v_r$, of the gas is negative (i.e., directed towards the centre). There is therefore a transfer of material inwards. The mass-loss rate may be expressed as follows:

$$\frac{dM(r)}{dt} = -2\,\pi r \rho\, h v_r \tag{5.18}$$

The equations just discussed describe the case of a disk with no gain of material from the outside. In the case of a stationary accretion disk, we make the assumption of a constant rate of accretion of interstellar material falling onto the disk, independent of time and of distance from the star, and expressed as a fraction of a stellar mass. The surface density is a function of the accretion rate and the viscosity, $v$.

The viscosity in a flow of gas with a velocity gradient is caused by an exchange of kinetic moment along the axis that is perpendicular to the motion (i.e., the effect of friction). If one only takes internal friction into account in estimating the viscosity of the disk, it seems that the dynamical effect on the disk is small. Nevertheless, a realistic calculation requires that turbulence in the disk must be taken into account. Turbulence is revealed by the presence of violent, chaotic motion, active a various spatial scales, and capable of causing significant exchange of kinetic moment. Large-scale eddies may be described as having a diameter similar to the thickness, $h$, of the disk, and a velocity equal to the speed of sound, $v_s$. In the absence of turbulence, the coefficient of viscosity $v$ may be written:

$$v = \frac{1}{3} u\, l \tag{5.19}$$

where $u$ is the average kinetic moment, and $l$ the mean free path. When turbulence is present, the viscosity coefficient needs to be modified to take account of the effects of the eddies. Empirically, we may write:

$$v = \alpha v_s h \tag{5.20}$$

where the coefficient $\alpha$ is less than 1. On a similar empirical basis, one generally use a value of $\alpha$ that is close to 0.01. Disks modelled in this manner are known as '$\alpha$-disks'. Attempts have been made to estimate the density of the surface of the disk as a function of distance $r$. By assuming that $V_s^2$ is proportional to the temperature $T$ and to $r^{3/4}$, it is found that the surface density also decreases as $r^{3/4}$.

In the case of the primordial solar nebula, the evolution of temperature and pressure as a function of distance from the Sun has been calculated, based on the $\alpha$-disk theory, as a function of $\alpha$ and of the accretion rate. Wood and Morfill (1988) carried out this modelling, expressing the opacity of the disk $\kappa$ as

$$\kappa = \kappa_0 T^2 \tag{5.21}$$

where $\kappa = 10^{-6}\,\mathrm{cm.g^{-1}K^{-2}}$ for temperatures between 160 and 1600 K (corresponding to absorption by metals and silicates) and $\kappa = 2 \times 10^{-4}\,\mathrm{cm.g^{-1}K^{-2}}$ for $T < 160\,\mathrm{K}$ (absorption by ices). From the equations for the conservation of mass, of momentum, and energy, the authors obtained the following expressions for grains predominantly consisting of water ice:

$$T(K) = 54600\,\alpha^{-1/3}.M^{1/2}.\mu^{2/3}.\kappa_0^{1/3}.r^{-3/2} \tag{5.22}$$

$$P(bar) = 1.77\,10^{-7}\alpha^{-5/6}.M^{3/4}.\mu^{2/3}.\kappa_0^{2/3}.r^{-9/4} \tag{5.23}$$

In fact, the relationship to $r$ varies as a function of the composition of the grains. It is close to $r^{-1}$ for material that is predominantly silicates or metals, which is in agreement with the relationship that has been measured for the inner Solar System (Dubrulle, 1993).

Figure 5.18 shows the distribution of $T$ and $P$ as a function of distance from the centre $r$, for $\alpha = 10^{-2}$ and an accretion rate that varies between $10^{-6}$ and $10^{-9}$ solar masses per year. The total mass of the disk (in solar masses $M_\odot$) may be expressed (Wood and Morfill, 1988) as:

$$M_d = 5.2\,10^{-5}.\alpha^{-3/2}.\mu^{1/3}.R^2 \tag{5.24}$$

where R is the maximum size of the disk, in AU. For $\alpha = 10^{-3}$, $M_d$ varies from 0.03 to 0.003 M$_s$ for an accretion rate that varies between $10^{-6}$ and $10^{-9}$ $M_\odot$/yr. For $\alpha = 10^{-3}$, the value of $M_d$ varies between 0.13 ($10^{-6}$ $M_\odot$/yr) and $10^{-9}$($10^{-9}$ $M_\odot$/yr).

More recently, more elaborate models have been developed, notably by Papaloizou and Terquem (1999), and which have used a more accurate expression for the opacity, taking account of grains, molecules, atoms, and ions. The pressure curves are in good agreement. The temperatures are, overall, in good agreement

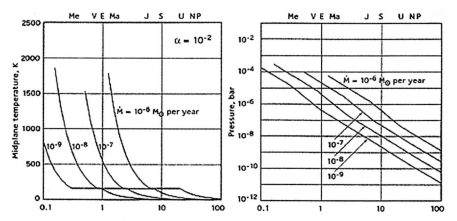

**Fig. 5.18** Evolution of temperature and pressure in the equatorial plane of a disk, according to the model by Wood and Morfill (1988). The horizontal plateau in the temperature curve corresponds to a transitional region where the opacity has values intermediate between those for ices and that for refractory materials (After Wood and Morfill, 1988)

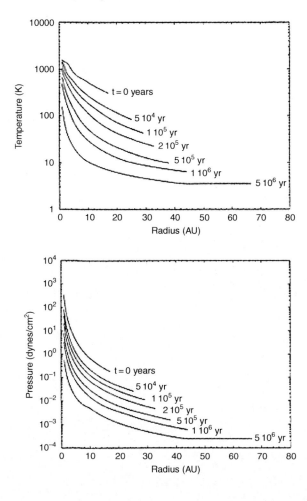

**Fig. 5.19** Evolution of the temperature and pressure profiles as a function of time within the primitive solar nebula (After Hersant et al., 2001)

with the results shown in Fig. 5.19 for $T < 160$ K, but differ significantly at higher temperatures (Wood, 2000).

In the case of the solar nebula, the evolution of the pressure and temperature curves as a function of time has been modelled by Hersant et al. (2001), assuming a monotone decrease in the accretion rate with time. Figure 5.19 shows the evolution of $T$ and $P$ over time, and where the origin for time corresponding to the moment when the Sun attained its current mass. Knowledge of the radial distribution of temperature and pressure are crucial factors in the construction of models of planetary formation, because they determine the sequence in which solids condense.

## 5.2.8 Composition of the Gas and Dust

The gas content of protoplanetary disks is dominated by molecular hydrogen. $H_2$ is detected from its quadripolar transitions, particularly in the near infrared (at 2.1 µm), and in the intermediate infrared (S(1) at 17 µm, and S(0) at 27 µm). It may be noted that hydrogen's presence is only detectable at temperatures of several hundred K. The tracer that is universally used is CO, which is observed at its millimetric and submillimetric rotational transitions. The very high spectral resolutions ($R > 10^6$) and spatial resolutions ($< 1$ arcsecond) that are reached by millimetric interferometers, in conjunction with the possibility of observing several transitions, enable the gas to be traced in the disk with spatial resolutions of several tens of AU, and also allow its velocity and temperature fields to be determined. The $^{13}CO(2\text{—}1)$ line at 220 GHz, which has relatively low opacity, enables the median plane of the disk to be probed, while the $^{12}CO(2\text{—}1)$ line at 230 GHz probes the outer surface of the disk, at about three scale heights. It appears that the temperature in the median plane is less than that at the outer surface (d'Alessio 1999), as a result of the mechanism described earlier (5.2.5, Fig. 5.16). The temperature in the median plane (about 15 K) may lead to the partial condensation of CO.

Numerous other ions and molecules (or both) have been detected in protostellar disks: CS and $H_2CO$ (see Fig. 5.2), HCN, HNC, $C_2H$, $HCO^+$, $DCO^+$, etc. (Dutrey et al., 2006).

The dust in protostellar disk exhibits the infrared spectroscopic signature of the silicates (amorphous or crystalline), carbon (either amorphous or in the form of PAH), oxides, or ices ($H_2O$) that form the mantle of the grains. The SWS spectrometer on the ISO satellite enabled these components to be positively identified. In particular, it revealed the similarity between the spectrum of Comet Hale-Bopp and that of certain disks surrounding young objects (such as HD 100546; see Fig. 5.20).

It should be noted that the typical size of the grains in protostellar disks is larger than one micrometre, which is comparable with the size of interplanetary dust, and significantly larger than interstellar grains. This property may be the sign of an accretion process in protoplanetary disks.

The slope of the spectrum of disks in the millimetre region may be used to determine the structure of the grains. The coefficient of $\kappa_0$ may be described by

**Fig. 5.20** Comparison of the
SWS-ISO spectra of Comet
Hale-Bopp (*dashed curve*)
with that of the young object
HD 100546. The similarities
between the spectral
signatures is notable. They
are attributed to forsterite (the
crystalline silicate $Mg_2SiO_4$),
to amorphous olivine, to FeO,
and to various forms of PAH
(After Malfait et al., 1998)

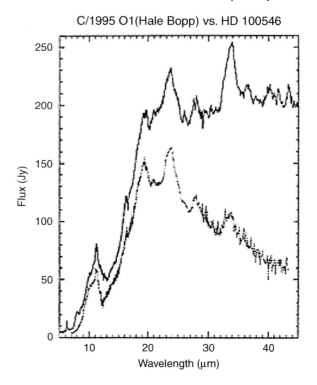

a power law:

$$\kappa_0(\nu) = C.\nu^{-\beta} \tag{5.25}$$

where $\beta$ lies between 0.5 and 0.65 (Beckwith et al., 1990). This value, which is less than that of molecular clouds, seems to favour a clumpy, and perhaps fractal, structure.

## 5.3 Planetary Disks and Debris Disks

With the protoplanetary disks just described, the mass of the disk is typically one-hundredth of the stellar mass (see Sect. 5.2.5). This mass is often too low for direct observation of the disk by imagery, but it may be deduced from measurements of its infrared flux. Hundreds of protoplanetary disks have been discovered in this way. Over the course of time, the disks will still lose a large fraction of their mass – they are objects of Class III, see Fig. 5.13 – through the effects of the violent stellar winds that accompany the T-Tauri and FU Orionis phases (or both) in the young star. The typical duration of this transition phase is shorter than ten million years. It appears to be shorter, the more massive the disk is initially.

The observation of thin disks is more difficult than for disks of Class II. We do, however, know of a few example of evolved disks, and these enable us to study their properties. The first of these is that of Beta Pictoris, first detected by IRAS in 1983 from its infrared excess, and was subsequently imaged with a coronagraph in 1984. β Pictoris is a star on the Main Sequence, and the disk surrounding it is relatively thin, when compared with the disks surrounding younger objects. Another interesting example observed recently is that of HR 4796A, an evolved, pre-Main-Sequence star, which exhibits a denser disk (less evolved) than that of β Pic. Some ten debris disks have been observed in scattered light. Some disks have also been mapped by submillimetre interferometry at 850 μm, in particular Fomalhaut, Vega, β Pic, and ε Eri. These observations enable the structure of the cold dust to be mapped and to reveal density fluctuations or gaps, which are possible signatures of the presence of a planet (Meyer et al., 2006). In the case of ε Eri, an exoplanet has, in fact, been detected by velocimetry, and the presence of a second exoplanet is suspected. Table 5.1 lists the properties of several debris disks observed by visible, infrared, or millimetric imagery.

Most of the debris disks have a very tenuous, or even non-existent, gaseous component. The latter may, however, be observed in certain cases. Studies undertaken with the Spitzer infrared telescope have shown that these exhibit major variations in the gas:dust ratio, and certain disks, such as that of HD 105, being devoid of gas (Meyer et al. 2006).

**Table 5.1** Characteristics of some resolved debris disks (adapted from Meyer et al., 2006)

| Star | Spectral type | Age (Myr) | Size of the disk (AU) | Albedo |
|------|---------------|-----------|------------------------|--------|
| HR 4796A | A0 | 8 | 70 | 0.1–0.3 |
| HD 32297 | A0 | 10? | 400 | 0.5 |
| β Pictoris | A5 | 12 | 10–1000 | >0.4 |
| AU Mic | M1 | 12 | 12–200 | 0.3 |
| HD 181327 | F5 | 12 | 60–86 | 0.5 |
| HD 92945 | K1 | 20–150 | 120–146 | – |
| HD 107146 | G2 | 30–250 | 130 | 0.1 |
| Fomalhaut | A3 | 200 | 140 | 0.05 |
| HD 139664 | F5 | 300 | 110 | 0.1 |
| HD 53143 | K1 | 1000 | 110 | 0.06 |
| Vega | A0 | 200 | > 90 | |
| ε Eridani | K2 | < 1000 | 60 | |
| η Corvi | F2 | 1000 | 100 | |
| τ Ceti | G8 | 5000 | 55 | |

## 5.3.1 Observation of the Disk of HR 4796A

HR 4796A is a star of type A0V, lying at the end of the pre-Main-Sequence stage (or at the very beginning of the Main Sequence phase), and thus well past the T-Tauri

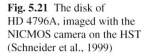

**Fig. 5.21** The disk of
HD 4796A, imaged with the
NICMOS camera on the HST
(Schneider et al., 1999)

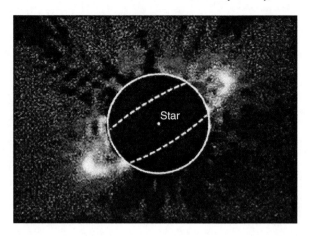

or FU-Orionis phases, and far younger than β Pic. Its disk (Fig. 5.21) is denser
than that of β Pic, and its luminosity, when compared with that of its star, is the
highest among stars of this age class ($L_{IR}/L_S = 10^{-3}$; Artymowicz, 2000). The
disk has been observed by imagery at 18 μm and then in the near infrared with
the HST (Schneider et al., 1999). The observations are in good agreement with a
radial distribution of dust according to a power law, out to distances from the star
between 55 and 120 AU. The disk is thin, with an abrupt limit at 120 AU, which
might be caused by perturbations by the other star, HR 4796B, in the binary.

By comparison of images a different infrared wavelengths, the size of the grains
has been determined to be 2–3 μm (Telesco et al., 2000). If the grains primarily
consist of silicates – material commonly observed in the interstellar medium – then
we would expect that these grains would be swept outwards by radiation pressure.
The presence of small grains in the disk of HR 4796A therefore presents a problem
that has yet to be resolved (Artymowicz, 2000).

### 5.3.2 Observation of the Disk of β Pic

The first problem posed by the existence of the disk around β Pic is the age of the
star: β Pic is a star that is several hundred million years old, or more than ten times
the lifetime of the disks that we have described. In the absence of any mechanism of
renewal, the disk should have dissipated; its presence therefore implies the existence
of a process for the renewal of the disk from the initial reservoir. The process by
which the material of the disk is accreted into planetoids and small bodies, just
before the T-Tauri phase, provides a natural solution to this problem. The material
accreted into these objects may subsequently re-supply the disk, after the T-Tauri
phase, by evaporation close to the star, or through the effects of collisions.

Figure 5.22 shows the disk of β Pic, as observed by the HST. It has been detected
with a diameter of nearly one thousand AU, and the enlargement of the central

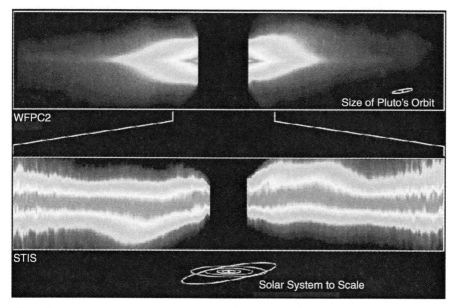

**Fig. 5.22** Coronographic images of the disk of β Pic obtained with the STIS instrument on the HST. *Top*: overall image; *bottom*: enlargement of the central portion, with a vertical exaggeration of 4 times, to reveal the warping of the disk. The intensity (and thus the quantity of dust) is a maximum near the equatorial plane (After Heap et al., 2000)

portion shows that the disk there is warped over a distance of about 100 AU. It appears to be devoid of material in the very centre out to a radius of about 20 AU.

The images of the disk obtained in the visible and infrared have enabled constraints to be set of the size distribution of the grains and their spatial distribution. The dimensions range form the sub-micron to the millimetre, with a maximum between 1 and 20 μm. The scattering properties of the grains (Artymowicz, 2000) have been obtained from visible-light coronographic images of the central region. They suggest a composition that is basically crystalline silicates (olivine and pyroxene). The high albedo (>0.4) differentiates the material from that in the rings or the small bodies in the Solar System, which is darker.

Spectroscopic observation of the disk in the visible and UV regions has revealed an unexpected phenomenon. Certain narrow lines of neutral atoms (Na, C) or ionized atoms ($Ca^+$, $Mn^+$, $Fe^+$, $Zn^+$) are observed repeatedly. They have allowed the abundance of the gas to be determined: the gas/dust ratio thus obtained is very low ($< 3 \times 10^{-3}$), whereas it is estimated to have been 100 in the early Solar System. So the disk of β Pic is therefore devoid of gas, which suggests that the low fraction that is observed to be present derives from evaporation or collisions between the small bodies.

This theory is reinforced by the discovery in visible and UV spectra of variable lines, that are strongly shifted towards the red (Vidal-Madjar et al., 1994; Lecavelier des Etangs et al., 2000; Fig. 5.23). These have been interpreted as the signature of

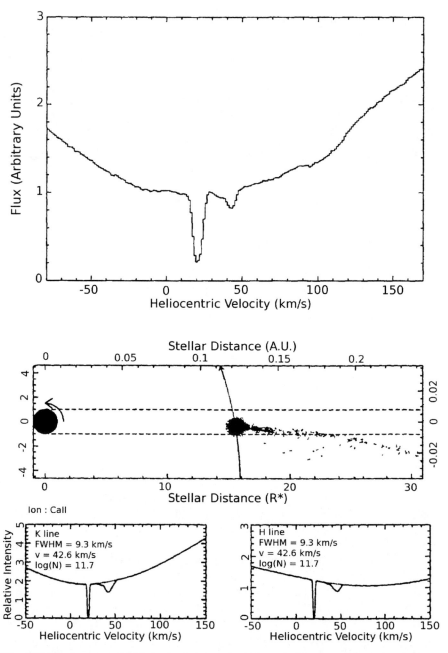

**Fig. 5.23** The spectrum of β Pic in the Ca⁺ line. *Top:* The scale on the abscissa is the velocity in km/s. The stable component corresponds to a velocity of 20 km/s relative to the star. Two components shifted towards the red are also detected, one narrow one at 50 km/s and the other, wider one at 100 km/s. *Centre:* Geometrical model of the phenomenon. *Bottom:* Numerical simulation based on the model (After Lecavelier des Etangs et al., 2000)

planetesimals, typically 1 km in diameter, that are evaporating and falling onto the central star with velocities that may reach 300 km/s. The frequency of these events, which varies from one year to the next, may be as many at 200 per year.

Could this mechanism explain the longevity of the disk around β Pic? Taking account of the thickness and size of the disk, the evaporation mechanism, which is perforce limited to the central region, is undoubtedly inadequate. We therefore need to imagine that multiple collisions occur at large distances from the star.

The frequency at which planetesimals decay onto β Pic is several orders of magnitude greater than the fall of comets into the Sun, which implies the existence of a different mechanism. Perturbations by planets could explain the phenomenon. Beust and Morbidelli (1996) have studied the effect of a massive planet with an eccentricity of 0.05; they have shown that its perturbations would lead to decay of particles in a 4:1 resonance after 10 000 revolutions.

Other factors also favour the theory that at least one planet exists within the disk around β Pic (Artymowicz, 2000): the dispersion in the orbital inclinations of the dust; certain photometric variations; the central gap at distances less than 40 AU; the asymmetries in the disk and the fact that it is warped. The last could be explained by the presence of a planet following an inclined orbit within the central gap, the existence of which it would also explain.

## 5.4 The Formation of Planetesimals and Planetary Embryos

How do planet form within the protoplanetary disk? They are the results of the accretion of dust particles, which have agglomerated into larger and larger bodies. For collisions between particles to lead to the formation of a larger body, and not simply result in their being dispersed in smaller fragments, there are specific limitations regarding the collision velocities, and the physical characteristics of the grains (internal cohesive force, porosity, etc. ...).

### 5.4.1 From Microscopic Particles to Centimetre-Sized Grains

The mechanism by which particles that are microscopic in size aggregate within a disk has been studied, particularly by Weidenschilling and Cuzzi (1993), and by Weidenschilling (1997). These particles stick together as a result of inelastic collisions and van der Waals forces acting on the surface of the grains. The rate and velocity of the collisions primarily depends on the properties of the gas, which is by far the largest component of the disk, in which the particles are suspended. The solid particles, being denser than the gas, tend to migrate towards the median plane, which has the effect of increasing the collision rate. The velocities of the small particles (which are less than 10 μm across) are linked to that of the gas surrounding them, and depend on its temperature, whereas the large bodies (with sizes greater than one kilometre) are subject to gravitational perturbations. At intermediate sizes, the particles are affected by the motion of the gas.

The first stage in accretion, caused by thermal motion, is possible thanks to the low relative velocities of the particles, which come to collide and then adhere to one another thanks to the van der Waals forces at their surface. The structures that result, which are both predicted by modelling and actually observed in the laboratory, are porous in nature, and fractal, up to sizes of about one centimetre (Weidenschilling, 2000).

## 5.4.2  From Centimetre-Sized Grains to Kilometre-Sized Bodies

The growth of the grains between sizes of a few centimetres to a few kilometres is not as well understood. The surface adhesion forces are inadequate, and gravity is still ineffective, with the drag force exerted by the gas still being far greater than the escape velocity from the surface of a small body. Safronov (1969) and Goldreich and Ward (1973) proposed a mechanism of gravitational instability causing a mass of material to collapse into a body about a kilometre across. This mechanism, however, cannot be involved, because of effects linked to turbulence, which keep the density below the critical value needed for collapse (Cuzzi et al., 1993). Although the clumping mechanism remains poorly understood, it may be noted that the drag exerted by the gas favours collisions between particles of different sizes, which have different velocities relative to one another.

Modelling the formation of planetesimals has been carried out for small icy bodies in the Solar System (Weidenschilling, 2000). The simulation takes as its starting point, grains 1 μm across uniformly mixed into a disk of given thickness. The results are shown in Fig. 5.24. After a time, $t$, equal to 1000 years, the peak of the distribution corresponds to grains that are centimetre-sized and which formed through thermal accretion, and which rapidly fall to the median plane and sweep up any small particles surrounding them. The mean plane becomes more and more dense, which favours the formation of a few larger bodies. As soon as the latter reach about a kilometre in size (at $t = 2000$ years), the escape velocities from their surface become greater than the radial velocities, and gravitational perturbations become significant.

Similar models have been constructed for other protoplanetary disks. The overall evolution of the size distributions is, on the whole, similar, and only the time scale changes. The time required to form planetesimals is proportional to the local orbital period. From 2000 years at 1 AU, it becomes $3 \times 10^5$ years at 30 AU, a time that is far less than the lifetime of protoplanetary disks. The presence of planets that have actually formed or are in the process of formation within these disks is therefore completely plausible.

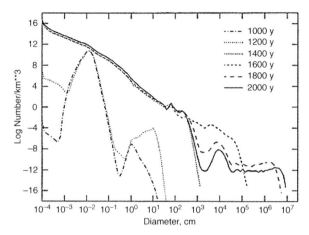

**Fig. 5.24** Evolution of the size distribution of particles with a disk as a function of time. Initially, the grains have a diameter of 1 μm. The conditions in the protoplanetary disk are those in the primordial solar nebula, for a distance of 1 AU from the Sun (After Weidenschilling, 2000)

## 5.4.3 From Protoplanets to Planets

Kilometre-sized objects move in Keplerian orbits and are no longer susceptible to drag forces exerted by the surrounding gas. The orbits are almost circular and coplanar, but the differential between the Keplerian velocities creates collisions and gravitational interactions. These interactions have the effect of increasing the eccentricity and inclination of the bodies, whereas collisions and friction with the gas tends to circularize the orbits. Because the relative velocities of the bodies are related to their velocities and their masses, the size distributions and the velocities evolve together and in a non-linear fashion.

The following section summarizes the discussion by de Pater and Lissauer (2001) on the growth rate of planetesimals.

### 5.4.3.1 The Effect of Collisions

Collisions between solid bodies may result in accretion, fragmentation or inelastic rebound. The impact velocity at which two solid bodies collide is:

$$v_i = \sqrt{v^2 + v_e^2} \tag{5.26}$$

where $v$ is the relative velocity of one body with respect to the other far from encounter, and $v_e$ is the escape velocity at the point of impact:

$$v_e = \left( \frac{2G\,(m_1 + m_2)}{R_1 + R_2} \right)^{1/2} \tag{5.27}$$

where $m_1$ and $m_2$ are the respective masses of the two bodies, and $R_1$ and $R_2$ are their respective radii. The rebound velocity is $\varepsilon v_i$, with $\varepsilon \leq 1$. Accretion may occur if $\varepsilon v_i$ is smaller than $v_e$. For a 10-km rocky object, the escape velocity is about 6 m/s. This is larger than the typical relative velocities of planetesimals, so that a 10-km body is likely to accrete the surrounding planetesimals, whereas fragmentation will preferentially occur for very small planetesimals.

The mean growth of a planetary embryo of mass $M$ is

$$\frac{dM}{dt} = \rho_s v \pi R^2 F \tag{5.28}$$

where $\rho_s$ is the density of the planetesimals, and $v$ is the average relative velocity between the embryo and the small bodies (assumed to be much smaller than the embryo). $F$ is the gravitational enhancement factor, given by

$$F = 1 + (v_e/v)^2 \tag{5.29}$$

in the 2-body approximation.

It is possible to express the growth rate of the embryo's radius as a function of the surface mass density $\sigma_p$, the embryo density $\rho_P$ and the Keplerian orbital angular velocity $n$:

$$\frac{dR}{dt} = \sqrt{\frac{3}{\pi} \frac{\sigma_p n}{4\rho_P}} \tag{5.30}$$

This equation leads to a growth time of about $2 \times 10^7$ y for the Earth, and more than $10^8$ yr for Jupiter. In the latter case, we know that other more efficient factors have been involved, because Jupiter and Saturn must have formed within $10^7$ years, before the T-Tauri phase and the dissipation of the gas.

### 5.4.3.2  The Runaway Growth of Planetary Embryos

As the embryo grows, its escape velocity increases. If the relative velocity of the embryo versus the swarm of planetesimals remains small, the $F$ factor may increase by large factors, leading to the runaway growth of the embryo. The embryo's feeding zone is limited to the annulus of planetesimals which the embryo may perturb gravitationally. Thus, rapid runaway will stop when the embryo has consumed the matter contained in the annulus. This mechanism thus leads to the formation of gaps inside protoplanetary disks. The size of the object that is formed depends on the material available within the gap that it creates.

The size of the gap depends primarily on the mass of the object that is forming. For a disk with no gas, the size of the gap is equal to a few times $r_H$, where $r_H$ is the Hill radius, the latter being defined as the distance beyond which the gravitational force exerted by the star exceeds that of the protoplanet:

$$r_H = a \left( \frac{m}{M} \right)^{1/3} \tag{5.31}$$

where $a$ is the semi-major axis of the protoplanet's orbit, $m$ is the mass of the protoplanet, and $M$ the mass of the star. In the case of a disk with a high gas content that is strongly viscous, the width of the gap is a function of the Reynolds number $\mathfrak{R}$:

$$\Delta = 0.29a(m/M)^{2/3}.\mathfrak{R}^{1/3} \tag{5.32}$$

where $\Delta$ is the width of the gap (Varnière et al., 2004; Beust, 2006).

Another situation occurs when the relative velocity of the swarm of planetesimals is comparable with or higher than the escape velocity of the embryo. In this case, the $F$ factor remains close to unity and the evolutionary path of the planetesimals exhibits an orderly growth in the entire size distribution. This is why numerical simulations of planetary formation typically lead to two types of solutions:

- a burst of growth, which results in the rapid growth of one body, which sweeps up material from the surrounding space;
- orderly growth, resulting in several large bodies of similar mass, with a power distribution of smaller bodies.

In the latter case, the small bodies could have a very long lifetime. Such evolution could possibly explain the existence of debris disks around evolved stars (Table 5.1).

# Bibliography

Acker, A., *Astronomie-Astrophysique, Introduction*, Dunod, Paris (2005)

André, J.-P. and Montmerle, T., 'From T-Tauri stars to protostars: circumstellar material and young stellar objects in the Rho-Ophiuchi cloud', *Astrophys. J.*, **420**, 837–862 (1994)

Artymowicz, P., 'Beta Pictoris and other solar systems', in *From Dust to Terrestrial Planets*, (eds) Benz, W., Kallenbach, R. and Lugmair, G.W., 69–86, Kluwer, Doordrecht (2000)

Aumann, H.H., Beichmann, C.A., Gillett, F.C., 'Discovery of a shell around Alpha Lyrae', *Astrophys. J.*, **278**, L23–L27 (1984)

Beckwith, S.V., Sargent, A.I., Chini, R.S. and Guesten, R., 'A survey for circumstellar disks around young stellar objects', *Astron. J.* **99**, 924–945 (1990)

Bertout, C., *Mondes lointains*, Flammarion, Paris (2003)

Beust, H., 'Modélisation des disques de débris, in, Formation planétaire et exoplanètes', in *Comptes-rendus de l'école thématique de Goutelas 2005*, (eds) Halbwachs, J.-L., Egret, D. and Hameury, J.-M., 155–190, Observatoire de Strasbourg/SF2A (2006).

Beust, H. and Morbidelli, A., 'Mean motion resonances as a source of infalling comets toward Beta Pictoris', *Icarus*, **120**, 358–370 (1996)

Cassen, P., 'Protostellar disks and planet formation', in *extrasolar planets*, Saas-Fee Advanced Course 31, (eds) Cassen, P., Guillot, T. and Quirrenbach, A., 369–444, Springer-Verlag, Heidelberg (2006)

Chiang, E.I. and Goldreich, P., 'Spectral energy distributions of T Tauri stars with passive circumstellar disks', *Astrophys. J.*, **490**, 368–376 (1997)

Choi, M., Evans, N.J. II, Gergersen, E.M. and Wang, Y., 'Modeling line profiles of protostellar collapse in B335 with the Monte Carlo method', *Astrophys. J.* **448**, 742–747 (1995)

Cole, G.H.A. and Woolfson, M.W., *Planetary Science*, Institute of Physics Publishing, Bristol and Philadelphia (2002)

Cuzzi, J.N., Dobrovolski, A.R. and Champney, J.M., 'Particle-gas dynamics in the mid-plane of the solar nebula', *Icarus*, **106**, 102–134 (1993)

Dutrey, A., Lecavelier des Etangs, A. and Augereau, J.-C., 'The observation of circumstellar disks: dust and gas components', in *Comets II*, (eds) Festou, M.C., Keller, U. and Weaver, H.A., 81–95, University of Arizona Press, Tucson (2004)

Elmegreen, B.G., Efremov, Y., Pudritz, R.E and Zinnecker, H., 'Observation and theory of star cluster formation' in *Protostars and Planets IV*, (eds) Mannings, V. et al., 179–215, University of Arizona Press, Tucson (2000)

Goldreich, P. and Ward, W.R., 'The formation of planetesimals', *Astrophys. J.*, **183**, 1051–1061 (1973)

Hartmann, L., *Accretion Processes in Star Formation*, Cambridge University Press, Cambridge (1998)

Heap, S.R., Lindler, D.J., Lanz, T.M. and Cornett, R.H. et al., 'Space telescope imaging spectrograph coronographic observations of Beta Pictoris', *Astrophys. J.*, **539**, 435–444 (2000)

Hersant, F., Gautier, D. and Huré, J., 'A two-dimensional model for the primordial nebula constrained by D/H measurements in the solar system: implications for the formation of giant planets', *Astrophys. J.*, **554**, 391 (2001)

Johns-Krull, C.M., Valenti, J.A. and Koresko, C., 'Measuring the magnetic field of the classical T-Tauri star BP Tau', *Astrophys. J.*, **510**, L41–44 (1999)

Larson, R.B., 'Star formation in groups', *Mon. Not. Roy. Ast. Soc*, **272**, 213–220 (1995)

Larson, R.B., 'The physics of star formation', *Reports on Prog. in Physics*, **66**, 1651–1697 (2003)

Lecavelier des Etangs, A., Hobbs, L.M. and Vidal-Madjar, A. et al., 'Possible emission lines from the gaseous Beta Pictoris disk', *Astron. Astrophys.*, **356**, 691–694 (2000)

Lequeux, J., *The Interstellar Medium*, Springer-Verlag, Heidelberg (2005)

Lynden-Bell, D. and Pringle, J.E., 'The evolution of viscous disks and the origin of the nebular variables', *Mon. Not. R. Astron. Soc.*, **168**, 603–637 (1974)

Malfait, K., Waelkens, C. and Waters, L. et al., 'The spectrum of the young star HD100546 observed with the Infrared Space Observatory', *Astron. Astrophys.*, **332**, L25–L28 (1998)

Matsumoto, T. and Hanawa, T., 'Fragmentation of a molecular cloud core versus fragmentation of the massive protoplanetary disk in the main accretion phase', *Astrophys. J.*, **595**, 913–934 (2003)

Meyer, M.R., Hillenbrand, L.A. and Backman, D.E., et al., 'The formation and evolution of planetary systems: placing our solar system in context with Spitzer', *Pub. Astron. Soc. Of the Pacific*, **118**, 1690–1710(2006)

Najita, J., 'Star formation', in *Encyclopedia of Astronomy and Astrophysics*, 3016–3027, Nature/IoP Publishing, Bristol (2001)

Papaloizou, J.C. and Terquem, C., 'Critical protoplanetary core masses in protoplanetary disks and the formation of short-period giant planets', *Astrophys. J.*, **521**, 823–838 (1999)

Pfenniger, D. and Combes, F., 'Is dark matter in spiral galaxies cold gas? II. Fractal models and star non-formation', *Astron. Astrophys.*, **285**, 94–118 (1994)

Pfenniger, D., Combes, F. and Martinet, L. 'Is dark matter in spiral galaxies cold gas? I. Observational constraints and dynamical clues about galaxy evolution', Astron. Astrophys., **285**, 79–93 (1994)

Safronov, V.S., *Evolution of the Protoplanetary Cloud and Formation of the Earth and Planets* (Nauka, Moscow, 1969; English translation: NASA TTF-677 (1972)

Schneider, G., Smith, B.A. and Becklin, E. et al., 'NICMOS imaging of the HR 4796A circumstellar disk', *Astrophys. J.*, **513**, L127–L130 (1999)

Smith, B.A. and Terrile, R.J., 'A circumstellar disk around Beta Pictoris', *Science*, **226**, 1421–1424 (1984)

Spitzer, L., *Physical Processes in the Interstellar Medium*, Wiley & Sons, New York (1978)

Telesco C.M., Fisher R.S., Pina R.K., et al., 'Deep 10 and 18 micron imaging of the HR 4796A circumstellar disk: Transient dust particles and tentative evidence for a brightness asymetry', *Astrophys. J.*, **530**, 329–341 (2000)

Varnière, P., Quillen, A.C. and Frank, A., 'The evolution of protoplanetary disk edges', *Astrophys. J.*, **612**, 1152–1162 (2004)

Vidal-Madjar, A., Lagrange-Henri, A.-M. and Feldman, P.D., et al., 'HST-GHRS observations of Beta Pictoris: additional evidence for infalling comets', *Astron. Astrophys.*, **290**, 245–258 (1994)

Weidenschilling, S.J., 'The origin of comets in the solar nebula: a unified model', *Icarus*, **127**, 290–306 (1997)

Weidenschilling, S.J., 'Formation of planetesimals and accretion of the terrestrial planets', in *From Dust to Terrestrial Planets*, (eds) Benz, W., Kallenbach, R. and Lugmair, G.W., 295–310, Kluwer, Doordrecht (2000)

Weidenschilling, S.J. and Cuzzi, J.N., 'Formation of planetesimals in the solar nebula', in *The Formation and Evolution of Planetary Systems*, (eds) Weaver, H. and Danly, L., 1031–1060, Cambridge University Press, Cambridge (1993)

White, R.J., Greene, T.P. and Doppmann, G.W. et al. (2007), 'Stellar properties of embedded protostars', in *Protostars and Planets V*, (eds) Reipurth, V.B., Jewitt, D. and Keil, K., 117–132, University of Arizona Press, Tucson (2007)

Wood, J.A. 'Pressure and temperature profiles in the solar nebula', in *From Dust to Terrestrial Planets*, (eds) Benz, W., Kallenbach, R. and Lugmair, G.W., 87–96, Kluwer, Doordrecht (2000)

Wood, J.A. and Morfill, G., 'A review of solar nebular models', in *Meteorites and the Early Solar System*, (eds) Kerridge, J.F. and Mathhews, M.S., 329–347, University of Arizona Press, Tucson (1988)

# Chapter 6
# The Dynamics of Planetary Systems

The structure of the Sun's planetary system has been the subject of numerous studies ever since it was discovered that the orbits of the planets were governed by an extremely simple law: the law of gravitation. Celestial mechanics has allowed the positions of the planets and satellites to be predicted with great precision. Moreover, the study of the stability of orbits has revealed the fundamental role of resonant interactions that govern complex configurations that are sometimes stable, and sometimes chaotic. Exosystems offer dynamicists a new field to explore or to use to test the mechanisms worked out in the Solar System.

## 6.1 Characteristics of the Orbits

### 6.1.1 Calculation of Radial Velocities

Up to now, exosystems have not been observed directly. The characteristics of the orbits of the planets are primarily deduced from analysis of the motion of the central star. The analysis techniques are the same as those used for binary-star systems. These techniques were established at the time of the first spectroscopic detections of binary stars, which date back to 1890.

The radial-velocity technique measures the velocity of the star in a reference frame containing the observer. Reduction of the observations therefore requires that the motion of the observer should also be taken into account. This has several components. Rotation of the Earth itself translates, as far as the observer is concerned, into a motion along the direction of the geographical parallel with a velocity of $460$ m/s. $\cos(\varphi)$, where $\varphi$ is the latitude of the observing site. The motion of the Earth every 28 days around the Earth—Moon barycentre translates into a motion at a rate of 13 m/s. The Earth—Moon barycentre itself orbits around the Sun, with an average velocity of 29.8 km/s and an oscillation (the orbit is an ellipse) with an amplitude of 0.5 km/s, and the Sun itself orbits the Solar System's barycentre with an average velocity of 13 m/s.

M. Ollivier et al., *Planetary Systems*. Astronomy and Astrophysics Library,
DOI 978-3-540-75748-1_6, © Springer-Verlag Berlin Heidelberg 2009

Measurements of radial velocities are obtained with accuracies of one metre per second. Allowance for the different components of the observer's motion should, therefore, be made with a higher degree of accuracy.

There is an online tool that enables one to calculate these motions with a nominal accuracy of 2 cm/s. This is VSOP87E, accessible on the Centre de Données Stellaires site at Strasbourg (http://cdsweb.u-strasbg.fr/ftp/cats/VI/81/).

Another effect that must be taken into account is the fact that light takes about 1000 s to go from a point on Earth to the same point when the Earth is at the opposite point of its orbit. The time of an event will not be the same depending on the position of the Earth at the time of observation. It is therefore necessary to relate the observations to a virtual clock, located at the centre of the Sun or at the Solar System's barycentre.

When the time to be measured concerns the arrival of pulses from a pulsar, the accuracy that needs to be attained also requires relativistic effects to be taken into account.

### 6.1.2 Orbital Characteristics from Radial-Velocity Curves

If the star is isolated, its radial velocity relative to the Sun is constant. If it has a companion (planet or star), the star revolves around the barycentre of the two bodies. In a reference frame centred on the star, the position of the planet is:

$$R = r\cos(f)\vec{i} + r\sin(f)\vec{j} \tag{6.1}$$

where $f$ is the true anomaly, i.e., the angle between periapsis and the position of the companion (planet or star). $(\vec{i}, \vec{j})$ are the orthogonal unit vectors in the plane of motion, $\vec{i}$ being directed towards periapsis. If $r = |R|$,

$$r = \frac{a(1 - e^2)}{1 + e\cos f} \tag{6.2}$$

$$\frac{df}{dt} = \frac{2\pi a^2}{Tr^2}\sqrt{1 - e^2} \tag{6.3}$$

where $T$ is the orbital period, $e$ the eccentricity, and $a$ the semi-major axis. The velocity of the planet is therefore:

$$V_p = -\frac{2\pi a}{T\sqrt{1 - e^2}}\left[\sin f.\vec{i} - (e + \cos f).\vec{j}\right] \tag{6.4}$$

From this equation we can deduce the velocity of the star relative to the system's barycentre:

$$V_p = \frac{m}{(m + M)}\frac{2\pi a}{T\sqrt{1 - e^2}}\left[\sin f.\vec{i} - (e + \cos f).\vec{j}\right] \tag{6.5}$$

where $m$ is the mass of the companion, and $M$ the mass of the star.

**Fig. 6.1** The reference
system, showing the plane of
the sky and the plane of the
planet's orbit. Angles are
measured from point ♈(After
Ferraz-Mello et al., 2006)

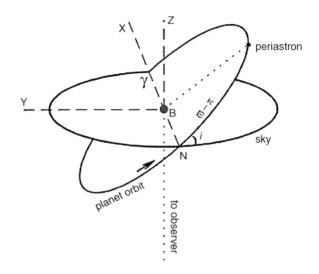

This velocity must be projected along the line of sight. The system of axes chosen
for this purpose is defined by a z-axis along the line of sight, increasing away from
the observer. The x-axis is defined by the intersection of the plane of the sky and the
orbital plane of the planet, increasing towards the point crossed by the planet when
it approaches the observer (point ♈in Fig. 6.1). The y-axis is defined such that the
system of axes is direct.

The measured velocity is the component of the star's velocity along the z-axis
added to the velocity of the system's barycentre, $V_G$:

$$V_z = K.\left[\cos(f + \omega) + e\cos\omega\right] + V_G \tag{6.6}$$

where $K$ is the semi-amplitude of the variation in the radial velocity, and $\omega$ is the
longitude of periapsis of the planet's orbit. Then:

$$K = \frac{m}{m+M}\frac{2\pi a}{T}\frac{\sin i}{\sqrt{1-e^2}} \tag{6.7}$$

For a Jupiter-type planet orbiting a solar-type star, and if $i = 90°$, $K$ is equal to
12.7 m/s.

The radial velocity is considered to be positive when the star is receding from
the observer. For a non-circular orbit, the variation in $f$ is not uniform, and thus the
curve for $V_z$ is not sinusoidal (Fig. 6.2). Modelling the curve enables us to determine
the eccentricity, the position of periapsis, and the 'mass function' $f(M,m,i)$ of the
system:

$$f(M,m,i) = \left(\frac{m}{m+M}.\sin i\right)^3.(m+M) \tag{6.8}$$

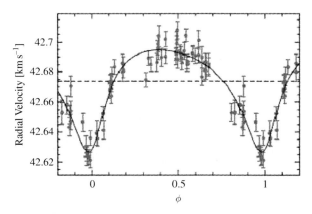

**Fig. 6.2** The radial-velocity curve of the star HD 65216 superimposed on a Keplerian orbital model. The parameters of the planet deduced from this model are $a = 1.37$ AU, $e = 0.41$, and $\omega = 198°$ (After Mayor et al., 2004)

The mass of the star is measured independently by spectroscopic observations and modelling. However, the best models do not enable us to determine this mass with an accuracy better than 8 per cent, which limits the accuracy of the planet's elements.

Knowledge of the mass of the star enables $m.\sin i$ to be calculated, which gives a lower limit for the mass of the planet (see Chap. 2).

### 6.1.3 Multiple Systems Case

If the system includes two (or more) planets, the radial velocity, $V_z$, becomes:

$$V_z = \sum_k K_k \left[ \cos(f_k + \omega_k) + e_k \cos \omega_k \right] + V_r \tag{6.9}$$

where $K_k$ is defined for each planet by applying Eq. (6.3). The mass function is:

$$f(M, m_k, i_k) = \frac{m_k^3 (M + m_k)}{\left( M + \sum_k m_k \right)^3} \cdot \sin^3 i_k \tag{6.10}$$

It may be noted that if the two planets are coplanar, observations enable the ratio of the masses of the two planets to be calculated.

These equations assume that the planets are in Keplerian orbits around the star. This approximation is valid provided mutual perturbations between the planets are not too large. If the planets are close to one another, the observations cannot be

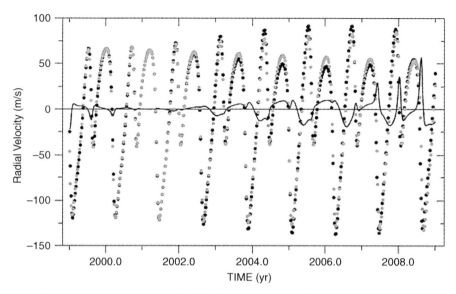

**Fig. 6.3** Radial velocities of the star HD 82943, calculated with a Keplerian model (*grey*), and with a 3-body model (*black*), coinciding with observations made in 2001. The line shows the difference between the models, assuming sin $i = 1$ (After Ferraz-Mello et al., 2005)

analyzed using this model. Comparison with a model where the planets interact would be required to calculate the system's parameters. This effect may be seen from analysis of the curve of HD 82943 (Fig. 6.3).

The observations made in 1999—2003 are in good agreement with the Keplerian model, but longer-term observations require the three-body model. It should be noted that the model is calculated with the planetary masses equal to the minimum masses, that is, assuming sin $i = 1$. It is not possible to determine the angle of inclination, except in the case where the planet transits the star. Some information may be gained for other systems from comparison of observations with simulations (see later). So, in the case of the system of HD 82943, for example, if the observations diverge from the model, this would indicate that the orbit of the planets is inclined with respect to the line of sight (sin $i \neq 1$). Modelling the observations would allow us to arrive at the inclination of the system and thus at the masses of the planets. These models all assume that the orbits are coplanar.

### 6.1.4 Exoplanets and Known Multiple Systems

Table 6.1 summarizes the properties of planets in multiple systems. The name of a planet is defined by the name of the star, followed by a letter: $b, c, d$, etc. The letter $a$ corresponds to the star itself. It should be noted that the alphabetical order corresponds to the chronological order of the discoveries. That order corresponds

**Table 6.1** The characteristics of 20 systems known around Main-Sequence stars at the beginning of 2007. [From the database at http://exoplanet.eu]

| Name | Mass ($M_j$) | Period (days) | Ratio of periods** | a (AU) | Ecc. | Incl. (°) |
|------|------|------|------|------|------|------|
| 47 UMa b | 2.54 | 1089 | - | 2.09 | 0.061 | |
| 47 UMa c | 0.79 | 2594 | 2.38 | 3.79 | 0 | |
| | | | | | | |
| 55 Cnc b | 0.784 | 14.67 | 5.22 | 0.115 | 0.0197 | |
| 55 Cnc c | 0.217 | 43.93 | 2.99 | 0.24 | 0.44 | |
| 55 Cnc d | 3.92 | 4517.4 | 102.83 | 5.257 | 0.327 | |
| 55 Cnc e | 0.045 | 2.81 | - | 0.038 | 0.174 | |
| | | | | | | |
| Gliese 876 b | 1.935 | 60.94 | 2.02 | 0.20783 | 0.0249 | 84 |
| Gliese 876 c | 0.56 | 30.1 | 15.53 | 0.13 | 0.27 | 84 |
| Gliese 876 d | 0.023 | 1.93776 | - | 0.0208067 | 0 | |
| | | | | | | |
| HD 108874 b | 1.36 | 395.4 | - | 1.051 | 0.07 | |
| HD 108874 c | 1.018 | 1605.8 | 4.06 | 2.68 | 0.25 | |
| | | | | | | |
| HD 12661 b | 2.3 | 263.6 | - | 0.83 | 0.35 | |
| HD 12661 c | 1.57 | 1444.5 | 5.48 | 2.56 | 0.2 | |
| | | | | | | |
| HD 128311 b | 2.18 | 448.6 | - | 1.099 | 0.25 | |
| HD 128311 c | 3.21 | 919 | 2.05 | 1.76 | 0.17 | |
| | | | | | | |
| HD 160691 b | 1.67 | 654.5 | 68.53 | 1.5 | 0.31 | |
| HD 160691 c | 3.1 | 2986 | 4.56 | 4.17 | 0.57 | |
| HD 160691 d | 0.044 | 9.55 | - | 0.09 | 0 | |
| HD 160691 e | 0.522 | 310.55 | 32.52 | 0.921 | 0.0666 | |
| | | | | | | |
| HD 168443 b | 7.2 | 58.116 | - | 0.29 | 0.529 | |
| HD 168443 c | 17.1 | 1739.5 | 29.93 | 2.87 | 0.228 | |
| | | | | | | |
| HD 169830 b | 2.88 | 225.62 | - | 0.81 | 0.31 | |
| HD 169830 c | 4.04 | 2102 | 9.32 | 3.6 | 0.33 | |
| | | | | | | |
| HD 190360 b | 1.502 | 2891 | 169.06 | 3.92 | 0.36 | |
| HD 190360 c | 0.057 | 17.1 | - | 0.128 | 0.01 | |
| | | | | | | |
| HD 202206 b | 17.4 | 255.87 | - | 0.83 | 0.435 | |
| HD 202206 c | 2.44 | 1383.4 | 5.41 | 2.55 | 0.267 | |
| | | | | | | |
| HD 217107 b | 1.37 | 7.1269 | - | 0.074 | 0.13 | |
| HD 217107 c | 2.1 | 3150 | 441.99 | 4.3 | 0.55 | |
| | | | | | | |
| HD 37124 b | 0.61 | 154.46 | - | 0.53 | 0.055 | |
| HD 37124 c | 0.683 | 2295 | 2.72 | 3.19 | 0.2 | |
| HD 37124 d | 0.6 | 843.6 | 5.46 | 1.64 | 0.14 | |

**Table 6.1** (continued)

| Name | Mass ($M_j$) | Period (days) | Ratio of periods** | a (AU) | Ecc. | Incl. (°) |
|------|--------------|---------------|---------------------|--------|------|-----------|
| HD 38529 b | 0.78 | 14.309 | - | 0.129 | 0.29 | |
| HD 38529 c | 12.7 | 2174.3 | 151.95 | 3.68 | 0.36 | |
| HD 69830 b | 0.033 | 8.667 | - | 0.0785 | 0.1 | |
| HD 69830 c | 0.038 | 31.56 | 95.94 | 0.186 | 0.13 | |
| HD 69830 d | 0.058 | 197 | 6.24 | 0.63 | 0.07 | |
| HD 73526 b | 2.9 | 188.3 | - | 0.66 | 0.19 | |
| HD 73526 c | 2.5 | 377.8 | 2.01 | 1.05 | 0.14 | |
| HD 74156 b | 1.86 | 51.643 | - | 0.294 | 0.636 | |
| HD 74156 c | 6.17 | 2025 | 39.21 | 3.4 | 0.583 | |
| HD 82943 b | 1.75 | 441.2 | 2.01 | 1.19 | 0.219 | |
| HD 82943 c | 2.01 | 219 | - | 0.746 | 0.359 | |
| HIP 14810 b | 3.84 | 6.674 | - | 0.0692 | 0.148 | |
| HIP 14810 c | 0.951 | 113.8267 | 16.97 | 0.458 | 0.2806 | |
| υ And b | 0.69 | 4.617 | - | 0.059 | 0.012 | |
| υ And c | 1.89 | 241.5 | 52.31 | 0.829 | 0.28 | |
| υ And d | 3.75 | 1284 | 5.32 | 2.53 | 0.27 | |

** Ratio of the period of the planet with respect to the period of the planet closest to the star

frequently (but not always) to increasing distance from the star, which may cause some confusion.

Planets in multiple systems have similar orbital characteristics to those of isolated planets close to their star, and with very eccentric orbits (Table 6.1). But apart from the planets themselves, the systems, taken overall, have orbital properties that reveal a turbulent past and major interactions between the planets. In particular, several systems exhibit mean-motion resonances, i.e., the periods of the planets are rational ratios (Fig. 6.4). The periods of planets in a 2:1 resonance have a ratio of 2. These resonances are believed to be the result of a single phenomenon: the migration of planets (see Sect. 6.2). The observations do not allow us to know if these planets are still in the process of evolution. But because theoretical models frequently associate migration with the stage at which planets form within a disk, and most of the systems observed do not have a disk, we may assume that the orbits of the planets have stabilized.

The migration process leads to very stable configurations where the periapses are aligned (or anti-aligned). This configuration, which is observed in several systems, whether resonant or not, may be explained only if the planets have migrated and have passed through configurations in which they were in resonance. A system may then remain frozen, or may continue to evolve, but the orbits will retain this property. The system of υ Andromedae is the best example of planets whose orbits are aligned without being in resonance.

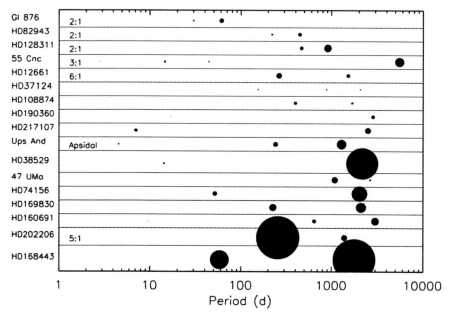

**Fig. 6.4** Multiple systems. The diameter of the points is proportional to $m.\sin i$ of the planet. Systems in resonance are indicated at left by the type of resonance (After Marcy et al., 2005)

Our knowledge of multiple systems is still very incomplete, with some twenty known systems. The orbital parameters are still subject to large error bars. However, exploration of this field is full of promise: A full knowledge of the orbits in a multiple system, together with studies of the dynamics of the system is very rewarding in imposing constraints on the formation, or at least the evolution, of the planets.

## 6.1.5 Rotation of the Planets

An important dynamical characteristic of planets is their rotation. It is not possible, at present, to know the rotation periods of exoplanets. However, it is possible to predict them for planets whose orbits have evolved through tidal effects.

Tidal effects are a well-known phenomenon in the Solar System, in particular those between a planet and a satellite. Gravitational attraction, coupled with deformation of the two bodies, results in circularization of the orbit of the satellite and evolution towards a rate of rotation that is synchronized with the rate of revolution around the planet. In the Solar System, the Moon and all the regular satellites are in synchronous rotation about their respective planets. Over a longer timescale, the system evolves towards synchronizing the revolution period of the satellite with the rotation period of the planet. This tidal effect on the rotation of bodies is also known to occur between the two stars in a close binary system.

When an exoplanet approaches the star, the star's tidal effect will perturb its orbit. The most important consequence is circularization of the orbit. This is why the 'hot Jupiters', lying at distances of less than 0.05 AU, have circular orbits. It is often assumed that these planets have reached a state of equilibrium, co-rotating with their stars (Fig. 6.5). Models show, however, that the effects of atmospheric tides may lead to other limiting solutions, and even to retrograde rotation like that which may have occurred with Venus.

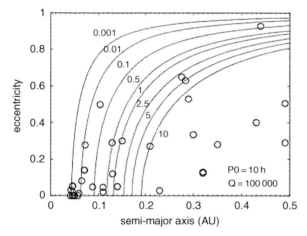

**Fig. 6.5** Dissipation through tidal effects. The curves represent the time ($10^9$ years) required to reach a state of equilibrium, as a function of the characteristics of the orbit: semi-major axis and eccentricity. The circles are exoplanets. This study has been made, assuming a dissipation coefficient and an initial rotation period similar to those of Jupiter. Exoplanets to the right of the curve for $10 (\times 10^9)$ years may be considered not to have undergone tidal effects. The planets to the left of this curve, and particularly the planets at 0.05 AU, have reached a state of equilibrium determined by the star's tidal effects (After Laskar and Correia, 2008)

## 6.2 Migration

### 6.2.1 Migration in the Solar System

Observations of extrasolar planetary systems suggest that the phenomenon of migration has played a significant role in their evolution. What is involved? The idea is that once a giant planet has formed outside the ice limit, it may change its position within the system, generally moving inwards, through the effects of interactions with the disk of gas, the disk of planetesimals, other planets, or a companion to the central star. This mechanism was known to planetologists who conceived it to explain the existence of resonant satellite systems. It came late to the history of scenarios for the formation of the planets, because the planets in the Solar System have not undergone extensive migration.

The key stage in the formation of a planet is the formation of a solid core. The time formation takes and the mass of this core depend on the density of solid material in the protoplanetary disk. This density is more significant outside the ice line, the minimum distance for the condensation of water ice, rather than inside, where only rocky material and metals condense. This limit lay at about 5 AU under the conditions in the primitive Solar System. Within the ice limit, the protoplanets were not sufficiently massive to accrete a gaseous envelope. Outside, the greater density of solid material favoured the rapid formation of massive cores, which attracted the surrounding gas (see Chap. 4). The natural location for the formation of giant planets is therefore outside the ice line. In the Solar System, however, the terrestrial planets and the giant planets are, in effect, on opposite sides of this boundary. In addition, Jupiter lies at just the distance from the Sun where the density was greatest. In this scheme, migration did not appear to be required.

It was only with the discovery of trans-Neptunian objects in orbits in resonance with Neptune, and time-scales that were rather too long for the formation of Uranus and Neptune, that the migration mechanism found a place in the story of the Solar System's formation.

The mechanism of migration within the Solar System has been the subject of various models. One example is shown in Fig. 6.6. Although all the models agree that Jupiter's migration has been of very little extent, some predict migration outwards, whereas others suggest migration inwards. All the models predict a migration of several astronomical units outwards for Uranus and Neptune, which therefore formed closer to the Sun. Certain models predict that Neptune was initially

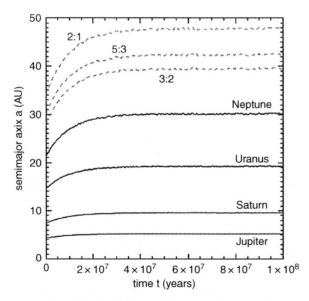

**Fig. 6.6** Numerical modelling of the migration of the Solar System's giant planets. The dashed lines represent the positions of resonances with Neptune (After Hahn and Malhotra, 2005)

inside Uranus, and that their orbits crossed. Their formation in a denser zone of the protoplanetary disk would therefore have been faster than originally expected. This resolves a problem, because the formation of the cores of these two planets seemed to require longer than the lifetime of the disk of gas. Even if Uranus and Neptune had significant cores of ice, their gaseous envelopes prove that their cores formed before the gaseous disk dissipated.

During this migration, the planets drove the Kuiper-Belt objects outwards, forcing them into resonances with Neptune. The migration was sufficiently slow for the Kuiper-Belt objects to remain trapped in the resonances and for their eccentricities to increase, eventually leading to the current configuration.

This phenomenon of migration was undoubtedly operative in the circumplanetary disks around the giant planets. This is suggested by the existence of several systems of satellite that are in resonance, in particular, the Galilean satellites of Jupiter. The satellites migrated until they became trapped in what is known as a mean-motion resonance. Once trapped, the orbits could continue to migrate, preserving the resonant relationship.

## 6.2.2 Migration in Exosystems

From the time of the first discoveries, the properties of exoplanets contradicted the scheme that had been developed of the formation of the Solar System. For example, several planets were discovered 0.05 AU from their star, where the temperature of the protoplanetary disk would not allow ices to condense, nor even refractory materials such as silicates. So the formation of a solid core was not possible. Yet a number of extrasolar planets were found at this distance from their stars (see Chap. 4).

Some authors have suggested other mechanisms for forming a planet very close to a star. Even if these models are dynamically possible, they show that a planet would not be able to cool down, and would lose its atmosphere by evaporation. By contrast, studies show that a planet that formed and had time to cool, and thus contract, far from the star, could then migrate down to 0.05 AU and remain stable. This stability only relates to the core of a planet. The orbital stability of these planets, in contrast, is far less understood (see below).

It is, therefore, generally assumed that most of the giant planets observed close to their parent stars formed like the planets in the Solar System, i.e., outside the ice line, and then migrated to their current positions. To confirm this theory, we still need to find the one or more mechanisms responsible for this migration, and find indications proving that this hypothesis is relevant.

It is important to note that the observations are explained by migrations towards the interior of planetary systems, but several of the mechanisms envisaged by theorists lead to migrations that are either towards the interior of a system or toward the exterior, depending on the initial parameters.

It should also be noted that the discovery of numerous exoplanets with short periods arises from an observational bias: planets with periods greater than the age of the earliest observations have yet to be discovered. In 2007, such observations are 13 years old, in other words the period of a planet close to the ice line, in a system similar to the Solar System. So observations are only just starting to probe the natural domain of giant planets.

So it is possible that the mechanism of migration towards the interior of a planetary system is not a universal phenomenon, and that the exoplanets discovered in the near future will correspond to systems without migration, or even migration towards the outside of planetary systems.

### 6.2.3 The Different Migration Mechanisms

The mechanism most frequently advanced to explain the migration of exoplanets proposes an interaction between the planet and a disk of gas. This mechanism, which was devised to account for planetary rings, occurs within the protoplanetary disk before the disk of gas disperses. The interaction takes the form of a couple between the disk and the planet. This transfer may result, under certain conditions, in a decrease in the semi-major axis of the planet's orbit.

For a planet of low mass with respect to the mass of the disk, the migration is known as Type I migration. The timescale of the migration is:

$$\tau_I(years) \approx 10^5 \left(\frac{M_p}{M_{Earth}}\right)^{-1} \left(\frac{\sigma}{600 g.cm^{-2}}\right)^{-1} \left(\frac{a}{AU}\right)^{-1/2} \left(\frac{H}{a}\right)^2 .10^2 \quad (6.11)$$

where $a$ is the semi-major axis and $H$ the thickness of the disk. $M_p$ is the mass of the planet and $\sigma$ is the surface density of the disk. For a planet of one Earth mass at one AU, $\tau_I \approx 10^5$ yr. This time is shorter than the lifetime of disks of gas or the time for the formation of planets.

When the planet is more massive (Type II migration), it creates an empty zone in the disk of gas, the width of which is about twice the Hill radius (see Sect. 5.4.3). In this case, the timescale of the migration depends on the viscosity of the disk, but no longer depends on the mass of the planet:

$$\tau_{II}(years) \approx 0.05 \frac{1}{\alpha} \left(\frac{a}{H}\right)^2 \left(\frac{a}{AU}\right)^{3/2} \quad (6.12)$$

Unlike Type I migration, here the planet migrates at the same time as the disk, and not relative to the disk. If $a/H = 10$ and $1/\alpha = 1000$, $\tau_{II}$ amounts to about 5000 years at a distance of one AU, which is, like Type I migration, far shorter than the time for the formation of the planets and the lifetime of gaseous disks.

The transition between the two types of migration occurs when the Hill's radius of the planet exceeds the scale height of the disk, i.e., for a planetary mass $M_p$ such that:

$$\frac{M_P}{M_*} > 3 \left(\frac{H}{a}\right)^3 \quad (6.13)$$

For a/H = 10 and a solar-mass star, the transition occurs at 3 $M_J$. Numerous numerical studies have shown that this mechanism can account for the observations. Several problems persist, however. For certain configurations, migration takes place towards the exterior of the system. In some cases, the evolution of orbital momentum may also be a chance affair, which does not follow a systematic course.

Interaction between protoplanets and a disk of planetesimals has been suggested to explain the outward migration of Uranus and Neptune. However, to explain the decrease of several AU that is required by the orbits of observed exoplanets, a very massive disk would be required, amounting to about 10 per cent of the stellar mass. In addition, this mechanism may also result in an increase in the semi-major axes, as in the case of the Solar System.

Another mechanism is interaction with a stellar companion (the Kozaï mechanism). If the system initially includes a planet in an inclined orbit, perturbations by the stellar companion may cause the planet's orbit to evolve towards a smaller, less inclined orbit with high eccentricity. If the planet's periapsis comes close enough to the star, its orbit is re-circularized through tidal effects. This theory requires the discovery of a mechanism to explain a planet that is initially situated in an extremely inclined orbit.

Interaction between planets may also lead to migration. If a system contains two planets whose separation is less than:

$$\frac{|a_1 - a_2|}{a_1} < 2.4 \left( \frac{M_1}{M_*} + \frac{M_2}{M_*} \right)^{1/3} \tag{6.14}$$

where $a_i$ is the orbital radius of the planet of mass $M_i$, one possible evolution is an inwards migration of the inner planet, with an increase in its eccentricity. The

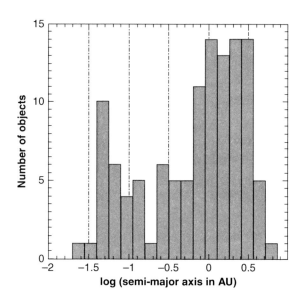

**Fig. 6.7** Distribution of semi-major axes of exoplanets

outer planet migrates outwards. If the inner planet stabilizes, the outer planet also stabilizes, and its presence may be sought to test this mechanism.

Yet another proposed mechanism is the capture of an 'orphan planet', i.e., one that is not bound to a star, by a protostar. The planet, in an almost parabolic orbit is subsequently braked by tidal effects.

Whatever the migration mechanism, one question that still remains largely in suspense is the end of this migration. What stops the planet's decay and prevents it from falling onto the star? In fact, the distribution of orbits (Fig. 6.7) clearly shows an accumulation of planets around 0.05 AU, and a gap at about 0.3 AU. This distribution, which is not caused by any observational bias, is statistically significant. It cannot be explained if planets end their orbital contraction by falling into the star. Some braking mechanisms have been proposed, based on interactions with the star or properties of the disk, but they require specific configurations (see Sect. 6.2.5).

## 6.2.4 Observational Indications

The properties of the orbits (Fig. 6.8) may be examined for observational proof of orbital migration. These properties also serve as constraints in choosing the migration model. There is not a regular distribution of orbits, but rather an accumulation at short distances from the star (Fig. 6.7).

Planets with very short periods (4 days) have circular orbits whereas planets with periods of 7–12 days (known as 'borderline' planets) display a dispersion in eccentricity (Fig. 6.8, *top*). The tidal mechanism suggested to circularize orbits is most effective with stars that rotate slowly. It may be noted that this mechanism may be tested by considering the borderline planets. Those associated with young, rapidly rotating stars (for example, those in the Pleiades cluster), should have greater eccentricities than those associated with older stars that are rotating slowly (for example, those in the Hyades cluster).

One of the consequences of migration is heating of the planet, which may be subject to evaporation, losing its gaseous atmosphere, and retaining only its core of ice or rock. A Jupiter-type planet may thus evolve into a planet similar to Neptune. The first example of this mechanism has perhaps been observed in the planet HD 209458b, where detection of an extended hydrogen exosphere, observed in the Lyman-$\alpha$ line, enable us to estimate an extremely high rate of evaporative loss of the atmosphere ($10^{10}$ g/sec). This rate is such that in a few million years it may lead to a giant planet without an atmosphere.

Systems with several planets are favourable places to search for signs of migration (see Sect. 7.3). Several of these systems include planets whose orbits are in mean-motion resonance. Modelling shows, however, that if the migration is sufficiently slow, the orbits of the planets that are trapped in resonance display a specific configuration of the angle between the periapses of the planetary orbits. The two systems in a 2:1 resonance and whose orbits are known with sufficient accuracy, GJ 876 and HD 82943, satisfy these conditions. The 55-Cancri system, with a 3:1 resonance, also displays a configuration of this type. All the resonant systems known so far are thus the result of a migration mechanism.

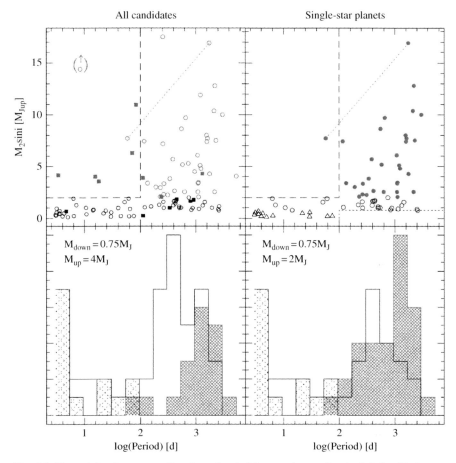

**Fig. 6.8** *Top:* minimal mass as a function of period for exoplanets. On the left the filled squares indicate the planets orbiting binary stars and the circles indicate planets orbiting single stars. The circle within parentheses indicates HD 162020, which is probably a brown dwarf. On the right, just planets of single stars are shown. In this diagram the symbols represent the mass of the planets: filled circles are the massive planets ($\geq 2\,M_J$), open circles intermediate masses (between 0.75 and $2\,M_J$) and the triangles, light planets ($< 0.75\,M_J$). The components of the HD 168443 system are joined by a dotted line. *Bottom:* distribution of periods of planets for different masses: the grey, white and red histograms are those for light, intermediate, and massive planets, respectively. The masses delimiting the three populations differ on the left (0.75 and 4 $M_J$) and on the right (0.75 and 2 $M_J$) (After Udry et al., 2003)

If we ignore multiple star systems, where the mechanisms governing formation may be different, there is no planet with a mass greater than 2 $M_J$ and a period less than 100 days (Fig. 6.8, *top right*). This is not an observational bias, because these planets are the easiest to detect.

It seems that the maximum mass increases with distance from the central star. There are no light planets ($M.\sin i < 0.75\,M_J$) with periods above 100 days.

## 6.2.5 The End of the Migration and Tidal Effects

Several of the migration mechanisms that have been suggested go through a phase where the planet has a very eccentric orbit, with periapsis very close to the star. Dynamical models show that successive passes of a gaseous planet close to a star leads to the eventual re-circularization of the orbit at an equilibrium distance that is about twice the Roche limit. This agrees with observations (Fig. 6.9). The maximum loss of energy by the planet occurs when periapsis is close to the Roche limit[1]. If the planet passes closer to the star, it loses material and moves farther away from the star. It may even be ejected from the system. It is therefore possible that the migration of a planet ends more often in its ejection from the planetary system than by falling into the star. Tidal effects also affect the rotation of a planet (see Sect. 6.1.5).

Gaseous planets that pass close to their star may lose all their hydrogen by thermal or non-thermal effects. They may create a new atmosphere by evaporation from their icy core, and thus create new terrestrial-type planets. Two possible candidates are shown in Fig. 6.9.

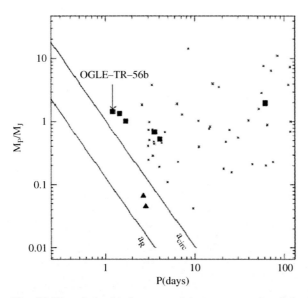

**Fig. 6.9** The relationship between minimum mass and period, of exoplanets, compared with the Roche limit calculated for a planet with a radius equal to that of Jupiter ($a_R$) and twice the Roche limit ($a_{circ}$). The rectangles are planets for which the inclination (and thus the mass) is known. The two triangles correspond to the super-Neptunes GJ 436b and 55 Cnc e, whose internal structure may be different from the other, more massive, planets. If these planets were icy, this would explain how they resisted the dismemberment mechanism described in the text (After Faber et al., 2005)

---

[1] The Roche limit is the distance from the star within which a planet is fragmented by tidal effects. This limit is equal to $2.456.R_* \left(\frac{\rho_*}{\rho_P}\right)^{1/3}$, where $R_*$ is the radius of the star, and $\rho_P$ and $\rho_*$ are the densities of the planet and the star, respectively.

## 6.3 Stability of Planetary Systems

In a sample of 179 known planetary systems, 21 have at least 2 planets, including one system around a pulsar. Of 20 multiple systems around Main-Sequence stars, 14 have 2 planets, 4 systems possess 3 planets, and 2 systems have 4 (Table 6.1). This proportion of 12 per cent multiple systems among known planetary systems is a minimum, because possible low-mass planets or ones distant from their stars cannot yet be detected. The orbital parameters of the systems are calculated from observations made over a short period of time. They correspond to elliptical orbits, which are not the planets' real orbits, particularly because the planets mutually per-turb one another. So it is not possible to know the future evolution of the planetary positions by simple extrapolation from their motions. In addition, these parameters are obtained with a certain degree of inaccuracy and using specific assumptions, es-pecially regarding the inclination of the orbits. Dynamical studies allow us to test the stability of the systems and their past and future evolution. To do so, we need to calculate the evolution of the systems taking account of the gravitational interactions between the bodies, and by exploring all the orbital parameters that are compatible with observations.

### 6.3.1 Dynamical Categories

For two planets in any given system, it is possible to look at the ratio of the orbital periods. For all the pairs of successive planets, the figures form a series running from values close to 2 to more than 300 (Fig. 6.10).

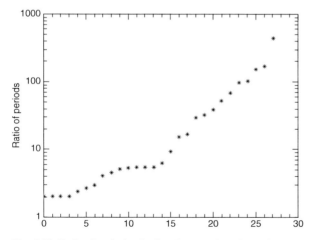

**Fig. 6.10** Ratio of periods of pairs of successive planets in exosystems (Table 6.1)

The systems, or to be more accurate, the pairs of planets, may be classified into three dynamical categories:

**Resonant, or Class I, systems** contain two planets whose orbits are very close to a mean-motion resonance of 2:1 or 3:1. These planets are strongly coupled by gravitation. Their orbits are perturbed over very short timescales and they would not be stable but for the fact that they have become trapped in a mean-motion resonance system.

**Interactive, or Class II, systems:** They contain planets whose orbits are not in resonance, but which are close enough to perturb one another. The ratio of their periods is greater than 4.6:1, which means that it is difficult for them to be captured in a resonance. The interactions between the planets may be strong, but conservation of angular momentum limits any variations in eccentricity. The Solar System is a member of this class. These systems may exhibit specific configurations, such as the alignment or anti-alignment of the periapses. The dynamical boundary between this class and the next one is not easy to define. The stability of orbits of Class II is, however, extremely sensitive to the values of the system's parameters, which is not the case with systems of Class III. It sometimes happens that the parameters that are deduced directly from the observations correspond to those of an unstable system and that the real parameters of the system are to be found by investigating stable systems that are close to the initial solutions and are also compatible with the observations.

**Separated, or 'hierarchical', Class III systems:** If the periods of the planetary orbits have a ratio greater than 30:1, any interactions between the planets are very weak. The eccentricities of the inner planets in such systems may, however, exhibit significant variations, as is the case with HD 74156 b. The evolution of these orbits is not at all sensitive to variations in the orbital elements.

HD 82943 and GJ 876 are in resonant systems. Interactive systems include 47 UMa, υ And, and the Solar System. HD 168443 and HD 83443 are separated systems.

This classification enables us to understand the behaviour of systems a bit better, but it is not enough. Dynamics experts place the two inner planets of υ And in Class II, because of their interactions, whereas the ratio of their periods, 52.2:1, would put them in Class III. A Class Ib has been created to include planets without significant period ratios, and having low eccentricities. In particular, this class contains the planets around pulsar PSR 1257+12 and probably 47 UMa as well.

For several of the systems observed, a systematic study of the stability over $10^6$ years has been made, varying the orbital parameters but remaining within the error bars obtained by an analysis of the radial velocities. Each set of parameters either led to a stable or unstable configuration. Such studies have shown that the stable zones are similar within each category:

- Resonant systems have very narrow stable zones within their phase space. The stability depends on the ratio of the periods of the two planets and, to a lesser extent, on the eccentricity of the more massive of the planets.
- Interactive systems have quite broad stable zones. The stability is linked to the eccentricities of the two planets. Even in a planetary system with low eccentricities, such as the Solar System, the neighbourhood of the planets is filled with

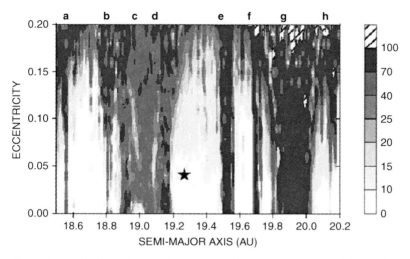

**Fig. 6.11** A plot of stability near Uranus. The white areas correspond to stable orbits. The hatched area corresponds to rapid ejection of the planet. The letters indicate resonances with Jupiter, Saturn, and Neptune (After Ferraz-Mello et al., 2005)

a dense set of resonances. Jupiter and Saturn, for example, are close to a 5:2 ratio; Uranus is between the 7:1 resonance with Jupiter and the 2:1 resonance with Neptune on the one hand, and the 3:1 resonance with Saturn on the other. Neptune is close to the 2:1 resonance with Uranus. Figure 6.11 illustrates how Uranus is isolated within the middle of a zone of unstable orbits.

- Separated systems are always stable.

Probabilities of survival to $10^6$ years have been calculated for all the different configurations. These probabilities are 20 per cent in resonant systems, 80 per cent in interactive systems, and 100 per cent in separated systems. These figures do not express the stability of the real system but correspond to the proportion of stable systems among the virtual systems that were obtained with parameters close to the real ones.

Here are some of the systems known at the beginning of 2007. It is only a small sample of the great diversity found among exoplanetary systems. In addition, these systems may contain planets that have not yet been discovered because they are too small or have periods that are too long.

### 6.3.2  The GJ 876 System

The two planets are in orbits with a 2:1 resonance. Modelling the observations needs to take account of gravitational interactions between the planets because the periods are very short, 30 and 60 days, and the masses are high relative to the mass of the star, $(m_1 + m_2)/M = 0.0074$. The two orbits are almost aligned, and the three angles

$\lambda_1 - 2\lambda_2 + \varpi_1$, $\lambda_1 - 2\lambda_2 + \varpi_2$, and $\lambda_1 - 2\lambda_2 + \varpi_1 - \varpi_2$ oscillate around 0 with a low amplitude, $\lambda_1$ being the mean longitude of the planet and $\varpi_1$ the longitude of periapsis. A third planet has subsequently been discovered. With a period of 1.9 days, it is too far from the two others to perturb the resonant system.

This resonant configuration cannot be explained except by migration of the planets: during the course of their movement, the planets crossed a resonance zone and have remained trapped in that configuration. This explanation is incomplete, however, because this mechanism should have resulted in orbits with greater eccentricity than those observed for the planets, 0.31 and 0.05.

### 6.3.3 The HD 82943 System

Another informative example is the HD 82943 system. The two planets discovered lie close to the 2:1 resonance. The planets are more distant than those in the preceding system, with periods of 220 and 440 days. The masses are smaller $(m_1 + m_2)/M = 0.003$, and a model with the planets on Keplerian orbits is in good agreement with the observations, but the corresponding system is dynamically unstable (Fig. 6.3). There are stable configurations close to this solution. Longer-term observations, together with dynamical studies should enable us to improve the model and even to resolve the uncertainty over the inclination of the system.

### 6.3.4 The υ Andromedae System

This system consists of 3 planets, of which the innermost is on a circular orbit, while the others have eccentric orbits (around 0.3). Tidal effects have circularized the innermost orbit, and the other two planets interact very strongly. The orbits are in a specific alignment: the periapses of the two planets oscillate around the same value with a low amplitude. This configuration may be caused by the migration of the planets through interaction with a gaseous disk. However, later observations favour the theory of a close encounter between υ And c and a fourth planet. The models show that the eccentricity of υ And d remains at a value close to 0.03 and that the eccentricity of υ And c is evolving with large-amplitude changes. These parameters confirm a scenario where the eccentricity of υ And d has undergone sudden variation and not a slow evolution. A possible process is a close encounter with an outer planet, which was subsequently ejected from the system (Fig. 6.12).

Modelling these observations also provides an indication of the inclination of the orbits: if the three planets are in the same plane, the system becomes unstable if sin $i$ is less than 0.5 (where sin $i = 1$ corresponds to a system seen side-on). The system also becomes unstable if the inclination of the planetary orbits is greater than $40°$.

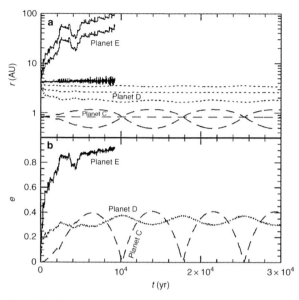

**Fig. 6.12** The dynamical evolution of a virtual planetary system similar to that of υ Andromedae: the planets C, D, and E correspond to the planets υ And c, υ And d, and an additional planet. For each planet, the three lines correspond to the semi-major axis between periapsis and apoapsis. The planets are initially on quasi-circular orbits. The planet D receives an impetus that increases its eccentricity, and then the two planets C and D enter a regular cycle of interaction. The planet E is ejected, without that ejection modifying the system of C and D. The planet υ And b is not taken into account in this model (After Ford et al., 2005)

### 6.3.5 The HD 202206 System: A Circumbinary Planet?

This is perhaps the first example of a planet around a binary system. The first companion was detected in 2002 around HD 202206. Its $m.\sin i$ is close to the limit for brown dwarfs, $16.5 M_J$. Its orbit has a semi-major axis of 0.83 AU, and an eccentricity of 0.43. A second planet was announced in 2004, far less massive than the first ($2.44 M_J$), but farther from the star (2.55 AU). Modelling the system with three bodies, and comparing this with observations, allows us to determine the orbital parameters of the system (Table 6.2). Figure 6.13 superimposes the correlation levels with the observations (contour lines), with tones of grey representing the stability of the system. It will be seen that the parameters that best fit the observations correspond to a white zone, i.e., to unstable orbits. An external planet in such a system is ejected in 40 000 years. Close to this solution we can see a dark zone that corresponds to stable orbits. It is therefore very likely that the real system lies within this zone of stability, which only represents a small increase in the residuals. This solution corresponds to masses of $17.43 M_J$ and $2.44 M_J$, respectively, and semi-major axes of 0.83 AU and 2.54 AU.

**Table 6.2** Orbital parameters of the two planets in the HD 202206 system corresponding to a stable configuration. (After Correia et al., 2005)

| Param. | S5 | Inner | Outer |
|---|---|---|---|
| $a$ | [AU] | 0.83040 | **2.54200** |
| $\lambda$ | [deg] | 266.22864 | 30.58643 |
| $e$ | | 0.43492 | 0.26692 |
| $\omega$ | [deg] | 161.18256 | **55.50000** |
| $i$ | [deg] | 90.00000 | 90.00000 |
| $m$ | [$M_{\text{Jup}}$] | 17.42774 | 2.43653 |
| Date | [JD–2 400 000] | | 52250.00 |
| rms | [m/s] | | **10.73** |
| $\sqrt{\chi^2}$ | | | **1.67** |

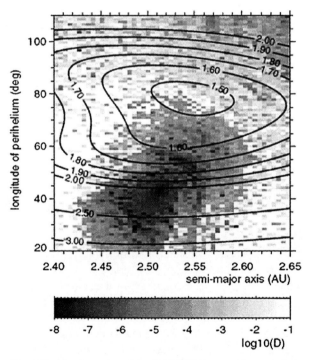

**Fig. 6.13** The dynamics of the HD 202206 system for variations of the semi-major axis and of the longitude of periapsis of the outer planet. The greyscale represents the stability index: the darkest zone is the most stable. The lines correspond to the residuals of the correlation between the system and observations (After Correia et al., 2005)

The model also shows that modifications in the orbits of the planets will be detectable in future. Measurements of them will give the inclination of the system and thus the true masses of the planets.

This system poses the question of the formation of such objects. If the outer planet formed within the circumstellar disk, it is similar to the inner planet, which is not, therefore, a brown dwarf (formed like a star), but a 'super-planet' (formed from a solid core). On the other hand, if the inner object is a brown dwarf, it formed like a star and the planet formed within a circumbinary disk. Such circumbinary disks have been observed, as, for example, around the binary system GG Tau A and B.

### 6.3.6 The HD 69830 System: Three Neptunes and a Ring of Dust

The HD 69830 system consists of three low-mass planets, between 5 and 20 times the mass of the Earth. The semi-major axes of the orbits are 0.08, 0.2, and 0.6 AU. The spectrum of HD 69830 shows an infrared emission excess, which indicates the presence of a disk of dust that is less than one micrometre in size. This thin disk lies less than 1 AU from the star. The observations exclude the presence of a planet with a mass greater than that of Saturn less than 4 AU from the star. Dynamical models of the system assume that the orbits are coplanar, and use two hypothetical values for the inclination of the system relative to the plane of the sky (and thus about the masses). These two values lead to stable systems.

Modelling of the system also allows us to say that the stability zones for the disk of dust lie either between 0.3 and 0.5 AU or beyond 0.8 AU (Fig. 6.14).

Comparison between the characteristics of the planets and formation models indicates that the inner planets should be rocky, while the outer planet probably has a gaseous envelope surrounding a rocky and icy core. It may also be noted that the outer planet lies within the habitable zone, that is, water at the surface of any solid body, a satellite of this planet, for example, is in the liquid state.

## 6.4 Planetary Systems Around Pulsars

In 1992, Wolszczan and Frail detected modulations in the arrival time of pulses from the millisecond pulsar PSR 1257+12, at a distance of 300 parsecs from the Sun. From this they deduced the presence of two planets orbiting the pulsar. Two years later, a third planet was announced, on an inner orbit. The planets remain the least massive exoplanets so far discovered, with one of them having a mass similar to that of the Moon (Table 6.3).

From the point of view of exobiology, planets around a pulsar are of little interest, because the electromagnetic environment leaves little chance of the possible development of any form of life on those planets. However, the dynamics of these systems, particularly that of the multiple system around PSR 1257+12 is extremely

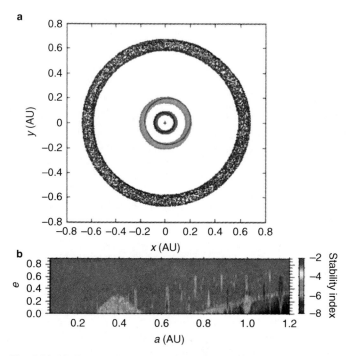

**Fig. 6.14**  (**a**): Examination of the stability of the three planets in the HD 69830 system. The points are the positions of the planets every 50 000 years. The system remains stable over a period of $10^9$ years. (**b**): The stability zone for mass-less particles. The mid-grey zones correspond to unstable orbits; the light-grey and very dark zones correspond to stable orbits where the ring of dust, detected in the infrared, may be located (After Lovis et al., 2006)

interesting (Fig. 6.15). It is one of the rare systems where the planets interact gravitationally and where the orbital eccentricities are low, as in the Solar System. The ratio of the periods of planets d and c is close to 3/2. This proximity to a mean-motion resonance produces perturbations that are observable from Earth.

**Table 6.3**  Characteristic of planets around pulsars

| Planet | $M.\sin i$ ($M_{JUP}$) | Period (days) | Semi-major axis (AU) | Ecc. | Incl. (deg) |
|---|---|---|---|---|---|
| PSR 1257+12 b | 6.29e-05 | 25.26 | 0.19 | 0 | – |
| PSR 1257+12 c | 0.0135 | 66.54 | 0.36 | 0.0186 | 53 |
| PSR 1257+12 d | 0.0122 | 98.21 | 0.46 | 0.0252 | 47 |
| PSR B1620-26 b | 2.5 | 100 years | 23 | - | 55 |

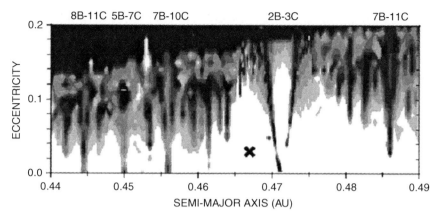

**Fig. 6.15** Plot of stability close to planet c in the PSR 1257+12 system. The dark areas correspond to unstable orbits. The resonances with the other planets in the system are indicated at the top of the diagram (After Beaugé et al., 2005)

The existence of planetary systems around pulsars is an enigma: How could the planets have survived the supernova phase that gave birth to the pulsar? Their composition must be very different from that of a planet of a Main-Sequence star.

According to one hypothesis, these objects did not form in the circumstellar disk of the pulsar's parent star. It is possible that these planets may have formed after the supernova phase, in a disk of 'circumpulsar' material. Few studies have been made of this process, but searches have been made to detect the sites at which such planets might form, using infrared observations of millisecond pulsars.

It should be noted that the low number of pulsar planetary systems detected is because of the low number of millisecond pulsars known to date (about one hundred).

## 6.5  The Dynamics of Debris Disks

The formation of stars is accompanied by the formation of circumstellar disks, known as 'primary' disks. These disks have the same composition as the local interstellar medium. The mass of these primary disks decreases with the age of the star as a result of the clumping of dust in the disk into larger bodies and through the disk being swept the very violent stellar wind from the protostar (see Chap. 6).

A significant number of Main-Sequence stars are still surrounded by disks of dust. These disks, more evolved, as known as debris disks (see Sect. 6.3). They are detected from the infrared excess in the star's spectrum. Analysis of the infrared spectrum provides information on the radial structure of the disk and on the size of the particles. The characteristics of these disks are different from those of primary disks; in particular, they do not contain any gas, or very little, and have a low optical depth.

The properties of these disks of dust give cause to suspect that they are hosts to more massive bodies, which might be planets or planetesimals. Because dust grains are rapidly ejected by radiation pressure, there must be a source that re-supplies the population. Two processes may generate the dust: collisions between planetesimals or degassing like that found in comets. The dust detected in debris disks is therefore the visible portion of disks that also contain more massive bodies. Other indications prove the existence of cometary bodies in some of these disks, and in particular in the disk of β Pictoris. The spectrum of this star shows variable absorption lines cause by the passage of cometary tails in front of the disk of the star. Models indicate that to explain these observations, these comets must be perturbed by one (or several) planets.

Another characteristic, common to several disks, may also be the signature of the presence of one or more planets. This is a gap, several astronomical units wide, near the star, which could have been cleared by planetary perturbations. Only a few of the disks detected in the infrared have been imaged (Fig. 6.16), but all these disks clearly show complex structures, rings, spirals, arcs (HD 141569), concentrations of material (β Pictoris, ε Eridani, Vega, Fomalhaut), asymmetries in azimuth (HD 141569, HR 4796), braiding, or vertical asymmetries (Table 6.4). Debris disks, particularly those of β Pictoris and HR 4796 have also been discussed in Sect. 6.3.

These structures may be compared with those observed in planetary rings, in particular with Saturn's rings: The presence of sharp boundaries or narrow rings is associated with resonances with the satellites. The satellites also generate vertical undulations, or gravity waves. The stability of arcs is also explained by specific configurations of satellites in inclined or eccentric orbits (or both). Small satellites close to the rings may be involved, or more distant and massive satellites, which create perturbations in regions where the ring particles are in resonance with them.

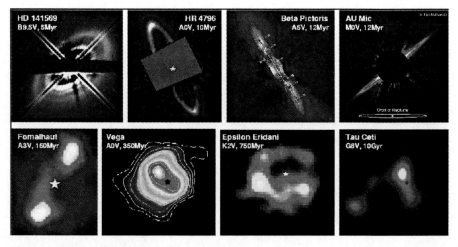

**Fig. 6.16** Resolved disks of Main-Sequence stars

**Table 6.4** Characteristics of resolved disks: age, distance, star's spectral type, distance at which the disk has a density peak, typical width of the ring, inclination of the disk, and resolution of the observations [After Augereau, 2004]

| | Age (millions of years) | Distance (parsecs) | Spectral type | ΔR (AU) | DR (AU) | i (°) | Resolution (AU) |
|---|---|---|---|---|---|---|---|
| HD 141569 | 5 | 99 | B9.5V | 200–310 | 45–120 | 55 | 5 |
| HR 4796 | 8 | 67 | A0V | 70 | 12 | 73 | 4 |
| β Pictoris | 12 | 19.2 | A5V | 90 | 80 | 90 | 1 |
| Fomalhaut | 150 | 7.7 | A3V | 155 | 60 | 70 | 58 |
| Vega | 350 | 7.8 | A0V | 100 | 40 | 5 | 110 |
| ε Eridani | 730 | 3.2 | K2V | 65 | 30 | 30 | 45 |

The gravitational perturbations between a circumstellar disk and a planet are the same as the perturbations created in a planetary ring by one or more satellites.

Modelling the structures observed in circumstellar disks allows us to work back to the characteristics of the planets that cause these structures. For example, the principal mechanism capable of generating asymmetrical brightness in a disk of dust is trapping in resonance with a planet. This mechanism has been advanced to explain the asymmetries seen in the extrasolar disks such as β Pictoris or Vega (Fig. 6.17). In the Solar System, dust has been detected, trapped in resonance with the Earth by the same mechanism.

The disk of β Pictoris (Sect. 6.3) has been better studied than any other debris disk. It shows other signs of the presence of one or more planets, in addition to the absorptions in the star's spectrum just mentioned. A short decline in the flux from the star has been interpreted as an eclipse by a planet. The disk shows a vertical asymmetry (a butterfly-like effect), which also betrays the presence of a planet in an inclined orbit.

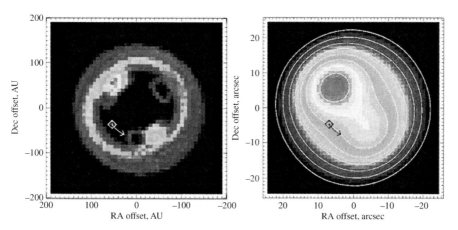

**Fig. 6.17** Numerical models of dust trapped in resonance with a planet (*left*) and comparison with Vega's disk (*right*). The square indicates the position of the planet (Wyatt, 2003)

# Bibliography

Augereau, J-C., 'Structures in dusty disks, extrasolar planets: today and tomorrow, ASP conference proceedings', Vol. 321, Edited by Jean-Philippe Beaulieu, Alain Lecavelier des Etangs and Caroline Terquem, 305 (2004)

Beaugé, C., Callegari, J. and Ferraz-Mello, S., et al., 'Resonances and stability of extra-solar planetary systems', in *Dynamics of Population of Planetary Systems*, Proceedings IAU Colloquium No 197, (eds) Knezevic, Z. and Milani, A., 3–18, Cambridge University Press, Cambridge (2005)

Correia A., Udry, S. and Mayor, M., et al., 'The CORALIE survey for southern extra-solar planets XIII. A pair of planets around HD202206 or a circumbinary planet?', *Astron. Astrophys.*, **440**, 751 (2005)

De Pater, I., Lissauer, J.J., *Planetary Sciences*, Cambridge University Press, Cambridge (2001)

Faber J., Rasio, F. and Willems, B., 'Tidal interactions and disruptions of giant planets on highly eccentric orbits', *Icarus*, **175**, 248 (2005)

Ferraz-Mello, S., Michtchenko, T.A., Beaugé, C., 'The orbits of the extrasolar planet HD 82943c and b', *Astrophys. J.*, **621**, 473 (2005)

Ferraz-Mello, S., Michtchenko, T. and Beaugé, C., et al., 'Extrasolar planetary systems' in *Chaos and Stability in Extrasolar Planetary Systems*, (eds) Dvorak, R., Freistetter, F. and Kurths, J., Lecture Notes in Physics, Springer-Verlag, Heidelberg (2006)

Ford, E., Lystad, V. and Rasio, F., 'Planet-planet scattering in the upsilon Andromedae system', *Nature*, **434**, 773 (2005)

Hahn, J. and Malhotra, R., 'Neptune's Migration into a stirred-up Kuiper belt: a detailed comparison of simulations to observations', *Astron. J.*, **130**, 2392 (2005)

Haisch, K. and Lada, E., 'Disk frequencies and lifetimes in young clusters', *ApJ*, **553**, L153 (2001)

Laskar J. and Correia, A., *The Rotation of Extra-solar Planets*, ASP Conference Series (in press, 2008)

Lovis, C., Mayor, M., Pepe, F. and Alibert, Y. et al, 'An extrasolar planetary system with three Neptune-mass planets', *Nature*, **441**, 305 (2006)

Marcy, G., Butler, R. and Fisher, D., et al., 'Observed properties of exoplanets: masses, orbits, and metallicities', in *Progress of Theoretical Physics Supplement*, **158**, 24–42 (2005)

Mayor, M., Udry, S. and Naef, D., et al., 'The Coralie survey for southern extra-solar planets XII. Orbital solutions for 16 extra-solar planets discovered with CORALIE', *Astron. Astrophys.*, **415**, 391 (2004)

Udry, S., Mayor, M. and Santos, N., 'Statistical properties of exoplanets I. The period distribution: constraint for the migration scenario', *Astron. Astrophys.*, **407**, 369 (2003)

Wyatt, M., 'Resonant trapping of planetesimals by planet migration: debris disk clump and Vega's similarity to the Solar System', *ApJ*, **598**, 1321 (2003)

# Chapter 7
# Structure and Evolution of an Exoplanet

We have seen that the process of formation (or evolution) of exoplanets seems to be very different from the planets in the Solar System, because giant exoplanets are discovered in close proximity to their stars. What about their internal structure and their atmospheres? We know the principal external factors responsible for the structure of planetary atmospheres (solar radiation, magnetic field, and interaction with the surface). In addition, the temperature and cloud structures of planetary atmospheres depend upon their composition, which determines, at each level, the opacities of the gaseous and solid phases (Sect. 4.4.2.2). We can also model a synthetic exoplanet spectrum that corresponds to each model in terms of comparing it with experimental data, when instrumental methods allow us to observe the spectra.

Modelling the internal structure of a sub-stellar object requires determining the temperature gradient and thus knowing the energy-transfer mechanism. Three mechanisms may be involved: radiation, conduction, and convection. A complete description of these processes may be found, in particular, in de Pater and Lissauer (2001) and Guillot (2006). A key parameter involved in all cases is the opacity, which depends on the atmospheric composition, pressure, and temperature, and determines how much energy is absorbed at each level. At pressures higher than a few bars, the collision-induced absorption caused by $H_2$–$H_2$ and $H_2$–He collisions must be taken into account. In addition, molecules present in planetary atmospheres ($CH_4$, $NH_3$, $H_2O$ ...) contribute to the opacities through their rotational and vibration-rotation bands. As a result, the spectrum of a planetary atmosphere is strongly dependent upon the wavelength (*see* Sect. 4.4.2.2.). Finally, condensates may play an important role by contribution to the opacity; their abundances mostly depend on the temperature (*see* e.g., Pollack et al., 1994).

Radiation transfer typically dominates in planetary stratospheres and upper tropospheres, at pressures lower than 0.1 bar. In the conduction regime, energy is transferred by collisions between particles. This mechanism is efficient, in particular, near the surface of the terrestrial planets. Finally convection is the energy transport caused by large-scale motions induced by density gradients resulting from temperature differences. It may be shown (Guillot, 2006) that convection dominates in isolated sub-stellar objects (weakly irradiated exoplanets or brown dwarfs). It is

M. Ollivier et al., *Planetary Systems*. Astronomy and Astrophysics Library,
DOI 978-3-540-75748-1_7, © Springer-Verlag Berlin Heidelberg 2009

also the case for dense atmospheres (with pressure higher than 1 bar) and molten interiors of planets.

We can also model a synthetic exoplanet spectrum that corresponds to each model in terms of comparing it with experimental data, when instrumental methods allow us to observe the spectra.

In this chapter we shall discuss the internal structure of exoplanets (giant exoplanets, ocean-type exoplanets, and terrestrial-like exoplanets), and then the atmospheric structure of giant exoplanets. The atmospheric models may be tested by comparing them with synthetic spectra that correspond to existing observations. A specific section is devoted to the case of habitable planets (these being defined as planets where the surface is at least partially covered in liquid water), and their spectral characteristics. The problem of searching for life on planets in the habitable zone is discussed in Appendix.

## 7.1 The Internal Structure of Giant Exoplanets

Giant exoplanets are considered to be analogous to the giant planets in the Solar System as far as their formation and evolution are concerned. We believe that they formed by accretion of a massive core of ices (about 10 Earth masses at least) and through the collapse of the surrounding nebula, which primarily consisted of hydrogen and helium (*see* Sect. 4.3.2.3). Subsequently they cooled and contracted gravitationally. A more comprehensive review may be found in Guillot (2006).

### 7.1.1 The Observable Features

In the case of exoplanets, we have only two observable features at our disposal: the minimum mass and, in a few cases, the exact mass and the radius. For the planets in the Solar System, we have a considerable number of far more significant data. In particular, several types of observational data allow us to put constraints on the internal structure of the giant planets in the Solar System:

a) the mass (determined very accurately thanks to the motion of satellites);
b) the equatorial radius, $R_e$, and the polar radius, $R_p$ (these being measured in the visible region);
c) the rate of rotation (the internal rotation, defined from the magnetic field; it is slightly different from the period derived from surface features);
d) the gravitational moments J2, J4 and J6, determined from the paths of space-probes; these are defined as follows:

$$V(r, \theta) = \frac{-GM}{r} \left( 1 - \sum \left( \frac{R_e}{r} \right)^i J_i P_i \cos \theta \right) \qquad (7.1)$$

where $V$ is the gravitational potential, $P_i$ Legendre polynomials, and $J_i$ the gravitational moments of order $i$. The giant planets are close to hydrostatic equilibrium and only even-ordered coefficients cannot be neglected;

e) the atmospheric composition, in particular the abundance ratios of helium and heavy elements (carbon for the four giant planets, as well as S, N, and the rare gases for Jupiter, following the measurements made by the Galileo probe);

f) the thermal profile, determined down to a pressure-level of several bars by measurements of the radio occultations of the Voyager probes and, more accurately, down to 22 bars on Jupiter, thanks to the Galileo probe;

g) the energy balance for the giant planets, determined from measurements of the overall infrared emission by the IRIS experiment on the Voyager probes. An internal source was discovered for Jupiter, Saturn and Neptune. These three planets emit significantly more energy than they receive from the Sun. the intrinsic fluxes are 5500, 2000, and 430 ergs$^{-1}$cm$^{-2}$, respectively, for Jupiter, Saturn, and Neptune.

## 7.1.2 The Equations of Internal Structure

Four fundamental equations govern the internal structure of fluid spheres, and reveal the balance between internal pressure and gravity (*see* Sect. 6.1.2): the hydrostatic equation; the perfect gas equation; the conservation of mass; and the conservation of energy.

### 7.1.2.1 Low-Mass Objects

From the hydrostatic equation, we may obtain the magnitude of the pressure at the centre of a sphere $P_c$, assuming constant density. This approximation leads to a simple equation:

$$P_c = \frac{3}{8\pi} \frac{GM}{R^4} \tag{7.2}$$

where G is the gravitational constant, $M$ the mass of the object, and $R$ its radius. For objects of low mass relative to the Earth, where the effects of compression are low, this equation provides a satisfactory approximation. The internal pressure may then be written as:

$$P(r) = \frac{4\pi}{6} G R^2 \rho^2 \left[1 - (\frac{r}{R})^2\right] \tag{7.3}$$

$r$ being the distance from the centre, and $\rho$ the mean density.

The temperature at the centre may be estimated by assuming convection, as a function of the temperature gradient $dLnT/dLnP$, empirically determined as close to 0.3, and by choosing conditions at the appropriate limits (*see* Sect. 4.4.2.2).

## 7.1.2.2  An Object That is a Compressible Sphere

In the case of massive objects, compression becomes important and the density becomes a function of pressure. The precise determination of the pressure, density and temperature profiles would require a full integration of the set of differential equations (hydrostatic equation, transfer equation, mass conservation, and energy conservation). An approximate analytical solution may be obtained using the polytropic relation between pressure and density:

$$P = K\rho^{(1+1/n)} \qquad (7.4)$$

where $K$ is constant. This relation may be used when the density does not depend on temperature. The index $n$ (the polytropic index) takes the value of 0 for noncompressible objects (the density being constant as a function of $P$); the value $n = 3/2$ corresponds to a perfect, monatomic gas (a pure hydrogen plasma). The mass-radius relationship may be written as:

$$R \propto M^{\left(\frac{1-n}{3-n}\right)} \qquad (7.5)$$

For an object with the mass of Jupiter, the value of $n$ is close to 1. For an object of 10 Jupiter masses, we find that $n = 1.3$: The effect of compression becomes such that the radius decreases as the mass increases. This effect is shown in Fig. 7.1, which shows the mass-radius relationship for different compositions. The curve

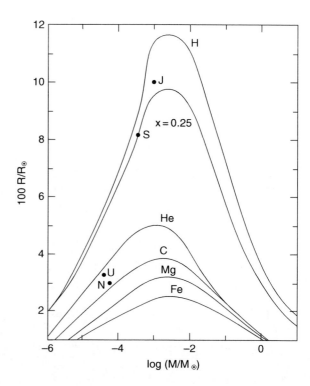

**Fig. 7.1** The mass-radius relationship for planets of different chemical compositions. The position of the four planets is indicated by their respective initials. The curve marked $x = 0.25$ represents a mixture of hydrogen and helium of solar composition ($Y = 0.25$) (After Marley, 1999)

marked $x = 0.25$ represents a mixture of hydrogen and helium of solar composition ($Y = 0.25$). It will be seen that Jupiter and Saturn consist mainly of hydrogen and helium, while Uranus and Neptune primarily consist of ices.

### 7.1.3 Rotation Effects

The giant planets in the Solar System are rotating rapidly, which results in flattening at the poles. This characteristic can provide constraints on their internal structure. Here we follow the discussion by Guillot (2006), based on a work initiated by Lagrange, Clairaut, Darwin and Poincaré, and detailed by Zharkov and Trubitsyn (1978).

We have seen that the gravitational potential $V(r,q)$ may be expressed at a function of the gravitational moments and Legendre polynomials (*see* 7.1.1). Similarly, the centrifugal potential $W(r,q)$ may be expressed as:

$$W(r,\theta) = \frac{1}{3}\omega^2 r^2 [1 - P_2 \cos\theta] \tag{7.6}$$

where $P_2$ is the second-degree Legendre polynomial. The equipotentials for a planet are defined as the surfaces at which the total potential $U = V + W$ is constant. These are also pressure surfaces where pressure, density, and temperature are constant.

We introduce the factor $q$, defined as the ratio between the centrifugal acceleration at the equator and the gravitational acceleration's principal term:

$$q = \frac{\omega^2 R_e^3}{GM} \tag{7.7}$$

Re is the equatorial radius of the object. From the parameter $q$, which has been measured for the four giant planets, it is possible to calculate the axial moment of symmetry from the following equation:

$$\frac{C}{MR_e^2} = \frac{2}{3}\left(1 - \frac{2}{5}\left[\frac{5}{\frac{3J_2}{q}+1} - 1\right]^{1/2}\right) \tag{7.8}$$

We find a value for $C/[MR_e^2]$ that lies between 0.22 and 0.26, significantly less than the corresponding value for a sphere of homogeneous density (0.40), which indicates that the giant planets are differentiated, with a central region that is more dense.

### 7.1.4 Equations of State

Construction of a realistic model for internal structure requires a knowledge of the equations of state, derived from experimental and theoretical studies, linking the pressure of a group of components to its temperature, density and composition. At the very high pressures and temperatures encountered in planetary interiors (and *a fortiori* in brown dwarfs), the perfect gas law, valid at the surface, no longer applies.

For pressures less than or equal to approximately 2 Mbar, it is possible to carry out experimental simulations by research with shock waves. A projectile impacts the sample being studied at high velocity, and by compression creates, for a very short time ($< 100\,\text{ns}$), temperature and pressure conditions comparable with those being studied.

### 7.1.4.1 Hydrogen

For pressures below 1 Mbar, the behaviour of molecular hydrogen is well understood, both theoretically and experimentally. At higher pressures, hydrogen passes into a phase known as 'metallic hydrogen' in which the atoms are ionized by pressure. According to experiment, the transition occurs around 1.4 Mbar, at a temperature of about 3000 K. At still higher pressures, liquid metallic hydrogen consists of an ionized mixture of protons and electrons at temperatures above 10 000 K. The phase diagram of hydrogen is shown in Fig. 7.2.

At the boundary between molecular hydrogen and metallic hydrogen, some authors have suggested the presence of a discontinuous phase transition called the 'plasma phase transition', based on the different nature of their action potentials: the potential is weakly repulsive in metals and strongly repulsive in insulators. It has not been possible to detect this discontinuous transition in the laboratory, but it should be noted that if it exists, it would have important consequences for the internal structure of planets, because it would create an impenetrable barrier to convection. In

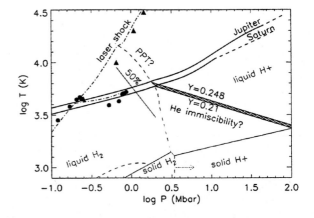

**Fig. 7.2** The phase diagram for hydrogen at high temperatures and pressures. The actual $T(P)$ profiles for Jupiter and Saturn are shown by the solid lines. The filled triangles and circles correspond to experimental results. The hatched region indicates the limit at which helium separates from liquid hydrogen (*see* Sect. 7.1.4.1) (After Guillot, 2006)

such a case, the atmospheric abundances of the elements, measured in the outer layers would not be representative of the deeper layers.

### 7.1.4.2 Helium and Other Elements

Laboratory experiments provide data up to a pressure of several hundred kilobars. The equation of state for a mixture of hydrogen and helium is calculated from those of H and of He, using the following equations:

$$\rho^{-1} = (1-Y)\rho_H^{-1} + Y\rho_{He}^{-1} \tag{7.9}$$

$$U = (1-Y)U_H + YU_{He} \tag{7.10}$$

$$S = (1-Y)S_H + YS_{He} + S_{mix}(Y) \tag{7.11}$$

in which Y is the helium fraction by mass and $S_{mix}$ the mixing entropy. Under certain temperature conditions and mixing ratios, helium is no longer soluble in hydrogen (see Fig. 7.2). The helium then separates out and falls as droplets down into the interior, as far as regions that are still hotter at which mixing becomes possible again. The effect of this mechanism is to deplete the outer mixture of helium and enrich the interior. It is accompanied by an exchange of energy towards the outside. This phenomenon may currently be occurring in the atmosphere of Saturn, and possibly also, to a lesser extent, in Jupiter. It may be partially responsible (together with the gravitational energy associated with the contraction) for the internal energy observed with those two planets.

The other elements that must be taken into account are ices ($H_2O, CH_4, NH_3$) and rocky materials (silicates, magnesium oxides, and metals). In the case of the ices, laboratory experiments have been conducted at pressures up to about 2 Mbar and at a temperature of 4000 K. The elements are in liquid form up to a pressure of approximately 300 kbar, but are then ionized, forming an electrically conduction fluid. Empirical $P(\rho)$ relationships have been established for the ices and rocky material.

## 7.1.5 Construction of Models of Internal Structure

Static models of internal structure may be built from the following equations:

- the equation of hydrostatic equilibrium in the presence of rotation:

$$\frac{dP}{dr} = -\rho(r)g(r) + \frac{2}{3}R\omega^2\rho(r) \tag{7.12}$$

where $g(r)$ is the gravitational acceleration at distance $r$, and $\omega$ is the angular velocity;

- the equations of state $P(r, T, x_i)$;
- the temperature profile $T = T(P)$; and
- the abundance profiles of the different components $x_i = x_i(P)$.

It should be mentioned that these models sacrifice reliance on an energy equation in favour of matching the effects of rotation through the gravitational moments and adopting a $T(P)$ relation.

The model, based on initial assumptions about the density and temperature profiles, should satisfy the observational constraints (*see* Sect. 7.1.1): i.e., the equatorial radius, the gravitational moments, and the abundances of the elements. The best fit is obtained by successive iterations. The final model does not, however, give a unique solution, mainly because of the uncertainties in the gravitational moments.

### 7.1.5.1 Jupiter and Saturn

Most of the models of Jupiter and Saturn assume that the interior consists of three regions: molecular hydrogen (a zone depleted in helium, *see* 7.1.4.2), metallic hydrogen (a zone enriched in helium), and a central core. Figure 7.3 shows an example of the density profile as a function of the normalized radius for Jupiter and Saturn (Marley, 1999). In this particular case, Jupiter's central core is approximately 5–10 Earth masses, while that of Saturn is significantly smaller. Other models (Guillot, 2006) imply that Saturn has a core with a mass comparable with that of Jupiter. With Marley's model (Fig. 7.3), the transition from the core to metallic hydrogen occurs at 39 Mbar; that for Saturn is at 13 Mbar and 11 900 K.

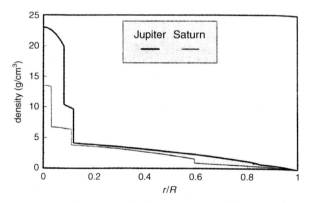

**Fig. 7.3** Examples of models of the internal structure of Jupiter and Saturn. The inner core consists of two parts, a rocky inner component surrounded by an outer component, rich in ices. In both cases, the radius of the overall central core is about one tenth of the planetary radius. The transition between metallic hydrogen and molecular hydrogen occurs at 85 per cent of the radius for Jupiter and 60 per cent of the radius for Saturn (After Marley, 1999)

### 7.1.5.2 Uranus and Neptune

The models assume a three-layer structure: rocky central core, a layer of ices, and an envelope of molecular hydrogen and helium.

Uranus and Neptune have comparable masses and sizes, so one might expect that their internal structure would be similar. However, the measurements of the gravitational moments obtained by Voyager 2 have shown that the internal structure of the two planets is different, with heavy elements less concentrated towards the centre in Neptune (Fig. 7.4). In both cases, the mass of hydrogen and helium is just a few Earth masses, compared with about 300 and 80 Earth masses for Jupiter and Saturn, respectively.

In the case of Uranus, it is not possible to interpret the measurements of the gravitational moments by a model with three homogeneous layers, which suggests that a significant fraction of the planet's interior is not mixed in a homogeneous manner. This peculiarity may explain why the internal energy of Uranus is very weak, because the process whereby heat would escape by convection would be inhibited (Guillot, 2006).

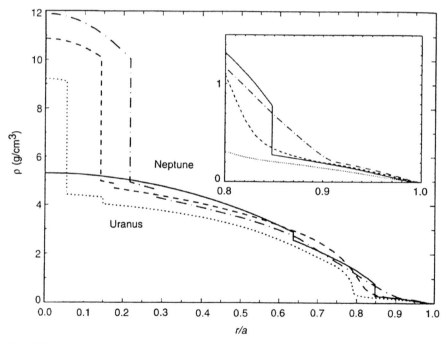

**Fig. 7.4** Examples of models of the internal structure of Uranus (*dotted line*) and Neptune (3 models, continuous, *chain-dotted* and *dashed lines*). In the case of Neptune, the size of the central core is poorly constrained (After Marley, 1999)

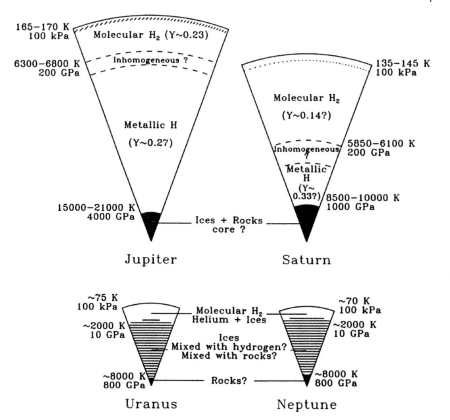

**Fig. 7.5** Schematic representation of the interiors of the giant planets with the uncertain ranges of various parameters. The size of the central core is very uncertain in all the cases (After Guillot, 2006)

Figure 7.5 shows the typical characteristics of models of the interiors of the giant planets, and their uncertainties. [After Guillot, 2006]

## 7.1.6 Evolutionary Models

We have seen that the giant planets, except Uranus, have a source of internal energy (*see* Sect. 7.1.1). The most plausible explanation is the gravitational energy accumulated during the formation of the planet, converted into heat in the interior and released outwards in the cooling phase. During a initial cooling phase, the planets also contracted, liberating an even more significant amount of gravitational energy. In the current phase, when the internal pressure hardly depends on temperature, the cooling takes place with an essentially constant radius. This mechanism is known as Kelvin-Helmholtz cooling (Marley, 1999).

Knowing the energy excess L for each planet, we can obtain an estimate of the planet's duration of cooling, assuming homogeneous cooling over the course of time, by the use of the following equation:

$$L = 4\pi R^2 \sigma (T_e^4 - T_0^4) = \frac{-d(MC_vT_i)}{dt} \qquad (7.13)$$

where $T_e$ is the actual effective temperature, $T_0$ the effective temperature in the absence of an internal source (and thus linked to the amount of solar energy that is received), and $T_i$ is an average internal temperature. $C_v$ is the mean specific heat per gramme. The radius of the planet, $R$, is assumed to be constant throughout the contraction, and $T_i$ is assumed to be similar to $T_e$.

The time required to cool from the initial temperature $T_{ei}$ to the current temperature $T_e$ is designated $\tau$. Because $T_{ei} \gg T_e$, the choice of $T_{ei}$ is not critical in calculating the integral. We obtain the following result:

$$\tau = \frac{MC_v}{4L} \qquad (7.14)$$

For the calculation to make sense, it is essential that the value obtained for $\tau$ should be similar to the age of the Solar System, i.e., 5000 million years.

For Jupiter the result is conclusive: To obtain $\tau = 4.5 \times 10^9$ years, we must start with an initial temperature $T_{ei}$ above 340 K, which is a reasonable assumption. For Saturn, in contrast, the value obtained for $\tau$ is only half, which means that the excess internal energy in Saturn is too great to be explained by Kelvin-Helmholtz cooling. Here, the mechanism by which helium separates out from the metallic hydrogen becomes involved, and this could be responsible for some of Saturn's internal energy. Because Saturn, less massive than Jupiter, is always colder, the condensation of helium occurs earlier in the cooling phase, and the effect is therefore more obvious (*see* Fig. 7.2).

More elaborate models have been able to simulate the evolution of Jupiter and Saturn, assuming a completely convective transfer and also the existence of a radiative zone (Guillot, 2006). Figure 7.6 illustrates the evolution of the temperature and radius as a function of time for Jupiter.

In the cases of Uranus and Neptune, the problem is the opposite of that with Saturn: The calculated times are greater than the age of the Solar System. One would therefore expect to observe an excess internal energy greater than what actually measured for these two planets. The problem is particularly severe for Uranus, because no internal-energy excess whatsoever has been detected. One possible cause could

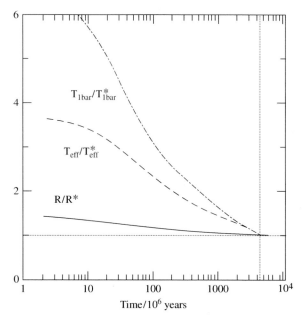

**Fig. 7.6** Contraction and cooling in Jupiter. The temperature at 1 bar and the effective temperature are shown as a function of time. $T_{1bar}^*$, $T_e^*$ and $R^*$ are the current values of these parameters. The vertical dotted line indicates the age of the Solar System (After Guillot, 2006)

be the inhibition of convection, mentioned earlier, which could prevent energy from being released towards the exterior.

## 7.2 The Internal Structure of Terrestrial-Type Exoplanets and Ocean Planets

For a start, the methods by which exoplanets are detected (by velocimetry and transits), for reasons of sensitivity, favour the discovery of massive and large planets. With advances in techniques, the limit of detectability for exoplanets has decreased below 10 Earth masses, and new discoveries are expected in the coming years, notably with the COROT mission. These results and these prospects have given rise to new theoretical studies aimed at modelling the interior of terrestrial-type exoplanets and ocean planets.

As examples of these studies, we describe below two sets of synthetic models (Valencia et al., 2006; Sotin et al., 2007). By analogy with the Earth, these models assume convection within each layer and conduction at the surface and across the boundary zones.

## 7.2.1 Terrestrial-Type Exoplanets

Valencia et al. (2006) have calculated the internal structure of exoplanets of terrestrial type (i.e., rocky) with radii appropriate for a range of one and ten Earth masses, taking as their point of departure the Earth's internal structure. The density, gravity, mass, and pressure are expressed as a function of the distance from the centre of the planet. By analogy with the Earth, various regions are specified (upper mantle, transition zone, lower mantle, outer core, and inner core). The mineral phases considered are olivine for the mantle and iron (with 8 per cent silicon) for the core. Still by analogy with the Earth, the model takes account of convection in the upper mantle, lower mantle, and the core, and conduction at the surface and across the boundary zones.

First of all, a simple model without any phase transition (i.e., with just a mantle and a core) enables us to establish an initial temperature profile. This profiles enables an initial determination of the location of the various phase transitions that are present in the Earth's interior. Subsequently, the thickness of each transition zone, the temperature profile and the locations of the phase transitions are adjusted by successive iterations. Figure 7.7 shows the density and temperature obtained for exoplanets of 1–10 Earth masses ('Super-Earths'), assuming terrestrial composition. Different compositions are also considered. In the case of the Super-Earths, the authors obtain the following power law (Valencia et al., 2006):

$$R \propto M^{0.267-0.272} \tag{7.15}$$

Similar modelling has been carried out by Valencia et al. (2006) for exoplanets less massive than the Earth ('Super-Mercurys', Fig. 7.8). Their composition is assumed similar to that of the Earth, but the mass fraction in the core is higher (60 per cent, corresponding to the internal structure of Mercury). In this case, the scale law becomes:

$$R \propto M^{0.3} \tag{7.16}$$

In parallel with these studies, Sotin et al. (2007) carried out modelling of exoplanets of the terrestrial type, basing this on the terrestrial planets and outer-planet satellites in the Solar System. The parameters entered into the models were the composition of the star, the magnesium content of the mantle, the mass fraction of water, and the overall mass of the planet.

The models by Sotin et al. (2007) are based on 5 layers: at the centre, a core rich in iron, assumed to be fluid and consisting of an Fe-FeS mixture with 80 per cent pure iron; a silicate lower mantle, consisting of mixed silicates at high pressure, rich in iron and magnesium; an upper mantle consisting of olivine and enstatite; a layer of water ice at high pressure; and a hydrosphere, where water might be in for form of a liquid or ice.

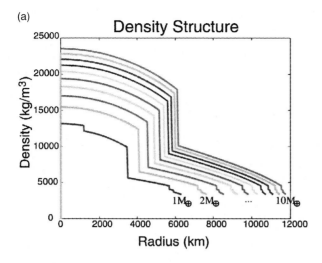

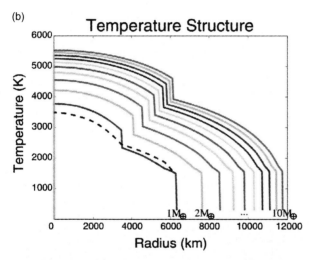

**Fig. 7.7** Density (**a**) and temperature (**b**) profiles obtained for exoplanets of 1–10 Earth masses (Super-Earths). The composition, the mass fraction of the core (32.59 per cent, by analogy with the Earth), and the surface temperature are fixed (After Valencia et al., 2006)

The mass-radius relationship deduced from these calculation is in good agreement with the values found by Valencia et al. (2006). It is therefore reasonable to use these models to try to predict the structure of exoplanets with masses comparable with that of the Earth.

(a)

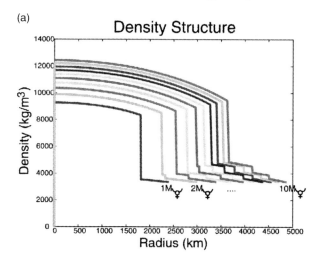

(b)

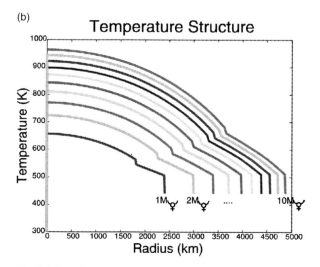

**Fig. 7.8** Density (**a**) and temperature (**b**) profiles obtained for exoplanets with masses 1–10 times the mass of Mercury, that is, 0.05–0.5 Earth masses (Super-Mercurys). The composition, the mass fraction in the core (60 per cent) and the surface temperature are fixed (After Valencia et al., 2006)

## 7.2.2 Ocean Planets

The models by Sotin et al., (2007) allow the simulation of the internal structure of ocean planets, whose existence was first suggested by Léger et al. (2004). These planets would have formed in the outer regions of the stellar system, where condensation of water took place, and subsequently migrated inwards. The discovery of an exoplanet of 5.5 Earth masses (Beaulieu et al., 2006), reinforces this assumption:

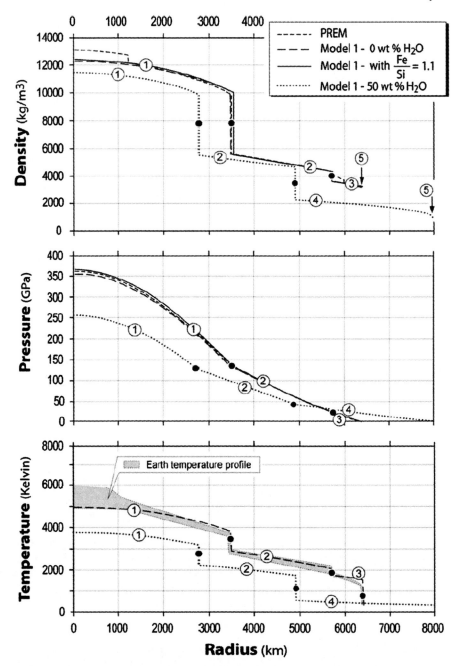

**Fig. 7.9** Density, pressure, and temperature profiles calculated for an Earth-like exoplanet and for an ocean planet. In both cases, the mass of the exoplanet is 1 Earth mass. The models are compared with the PREM model used for the Earth, and incorporate seismic data (After Sotin et al., 2006)

The planet lies 5 AU from its star, in a region where the expected temperature is around 50 K; so water should be trapped in the form of ice in the planetesimals, and where it forms a significant fraction of them.

The calculations considered a mass fraction of $H_2O$ of 50 per cent. They showed that an ocean planet of one Earth mass would have a radius of 8000 km rather than 6400 km; if its mass is 10 Earth masses, its radius is 15 000 km instead of 12 000 km for a Super-Earth. The depth of the liquid ocean varies from 3000 km to 6000 km for ocean planets whose mass varies between 1 and 10 Earth masses.

Figure 7.9 shows density, pressure and temperature profiles calculated for terrestrial-type planet, and for an ocean planet.

Figure 7.10 shows the mass-radius relationship for terrestrial-type exoplanets and for ocean planets. It will be seen that regardless of the mass of the object, the radius of an ocean planet of equal mass, is 26 per cent greater than that of a terrestrial-type exoplanet. It should, therefore, be possible, in principle, to distinguish between exoplanets of the terrestrial and ocean types, from simultaneous velocimetric and transit observations.

It will also be seen that the evolution of the two curves in Fig. 7.9 is very similar. The figure shows that Ganymede, Callisto, and Titan should have about 50 per cent water by mass. So there are 'ocean satellites' in the Solar System, but their method of formation is different from that envisaged for ocean-type exoplanets: they formed directly when the local nebula that accompanied the formation of their giant planet collapsed, and have not undergone any – or at least very little – migration. Because

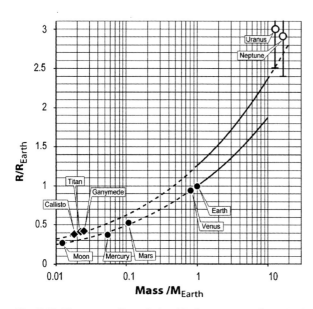

**Fig. 7.10** The mass-radius relationship for an ocean planet (*top*) and a terrestrial-type planet (*bottom*). The values of $M$ and $R$ are normalized with respect to those for the Earth (After Sotin et al., 2006)

of the temperature conditions in their environment, water is not liquid at their surfaces, but is in the form of ice.

## 7.3 The Atmospheres of Exoplanets: Their Structure, Evolution and Spectral Characteristics

The observational study of exoplanets is a young science, but, even so, is one that has passed the simple detection stage. The transits of Pegasids – giant exoplanets – when associated with measurement of their mass by velocimetry, will henceforth allow direct comparison of masses and radii of these planets with the models. Thanks to the Hubble Space Telescope, the components of the upper atmospheres of Pegasids could be identified during their transits. More recently, another space observatory, Spitzer, has been able to detect the thermal emission from Pegasids, and thus established extremely tight constraints on their atmospheric composition and circulation. These spectroscopic observations allow us to imagine the potential of future missions, such as the JWST, which will allow us to obtain the infrared spectra of short-period, giant exoplanets. In the longer term, smaller and smaller, and colder and colder planets will be able to be studied through low- and medium-resolution spectra with instruments such as Darwin/TPF-I or TPF-C (Terrestrial Planet Finder), space telescopes dedicated to the detection and description of habitable terrestrial-type planets.

### 7.3.1 Giant Exoplanets

From a knowledge of the equations of state and opacities of hydrogen and helium at sufficiently high temperature and pressure states (20 000 K, 50 Mbar), and assuming a structure in hydrostatic equilibrium as well as a given initial state at $t = 0$, we can calculate the evolution of a planet of mass M, and especially predict observable parameters such as the radius or the luminosity. From these observables, the theoretical model thus enable us to deduce the age of the system or set constraints on the composition of the object. We can, in fact, refine this model by adding a solid core of mass $M_c$, consisting of rocky material or ices, and an enrichment ($M_{Z, env}/M_{env}$) of the envelope of mass $M_{env}$. This results, for the same overall mass of the planet, in a more compact object. The nature of the solid portion by itself has a limited effect: A core of ice ($\sim 3$ g/cm$^3$) and a core of rock ($\sim 6$ g/cm$^3$) with the same mass give a similar evolution. Here we assume that the equations of state for the hydrogen/helium mixture, and for the enriched hydrogen/helium mixture are known. In reality, however, this is a vast field of both experimental and theoretical research which is in constant development (*see* for example Guillot, 2005).

In the earliest stages, the theoretical evolution of a planet of mass $M = M_{env} + M_c$ strongly depends on the state (radius and temperature) at $t = 0$, that is, at the moment when accretion ceases, which corresponds to the disappearance of the circumstellar disk of gas. This state at $t = 0$ may be obtained from formation models (Pollack et al., 1996, Alibert et al., 2005), but the complexity of the accretion phenomenon does not enable us to determine the structure of the planet at the end of accretion with sufficient accuracy. The radius and the temperature at $t = 0$ are thus very poorly constrained, and it is important to note that for younger objects (less than 10 million years old), which are also the most luminous and the most likely to be detected directly, we do not have a reliable age-mass-luminosity relationship (Chabrier et al., 2006). For these young objects, knowledge of the mass and the luminosity, for example, do not allow us to deduce the age of the system accurately. Similarly, knowledge of the age and the luminosity do not allow the mass to be inferred.

Nevertheless, as Fig. 7.11 shows, this uncertainty no longer affects the evolution after the first 100 million years, and we may predict the evolution of the radius and luminosity of an older planet of given mass. These models have given excellent results for low-mass stars and brown dwarfs. However, strongly irradiated giant planets cannot be modelled as if they were isolated objects, such as brown dwarfs, because the incident solar flux affects both the outer structure of the atmosphere and the evolution of the planet.

Including the effect of irradiation in the calculation of the structure and evolution of a giant planet is not a trivial exercise. In fact, modelling the atmosphere – here, we arbitrarily describe an atmosphere as the outermost portion of the envelope, where

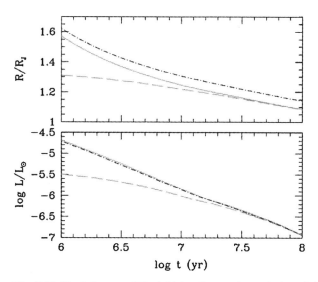

**Fig. 7.11** The influence of the initial radius on the evolution of the radius and luminosity of an non-irradiated exoplanet. The continuous-line and dashed curves show the evolution of a planet of 1 $M_{\mathrm{Jup}}$, where the initial radius is set at 3 and 1.3 $R_{\mathrm{Jup}}$. In both cases, the planet has a rocky core of 6 $M_{\mathrm{Earth}}$ and an envelope that is enriched with respect to the solar composition ($M_{Z,\,\mathrm{env}}/M_{\mathrm{env}} = 0.1$). The chain-dotted curve shows the evolution of a planet without a solid core that is of solar composition and with an initial radius of 3 $R_{\mathrm{Jup}}$ (After Chabrier et al., 2006)

the pressure is less than $10^3$ bar – requires an internal heat flux as a condition for the lower limit. Yet this heat flux depends on the age and should be given by an evolutionary model, because at present there is no complete model that simultaneously treats evolution and detailed radiative transfer in the irradiated zone. Let us consider a non-irradiated planet as the initial evolutionary state: The heat flux at the planet's surface (and thus its luminosity) is calculated by a model of the internal structure, and may be written $4\pi R^2 \sigma T_{eff}^4$, where $T_{eff}$ is the temperature of a body equivalent to the object (*see* Sect. 4.4.2.2). Now let us consider the same planet, but irradiated. Determination of the atmospheric structure requires us to know the incident stellar flux (as an upper limit) and also the internal heat (as a lower limit). The latter is then given by the non-irradiated model for the internal structure. The atmospheric profile obtained may then be reintroduced as an upper limiting condition to calculate the evolution over an interval of time $\Delta t$. The evolution may thus be described in a quasi-static manner. In practice, instead of recalculating the atmospheric structure at each time-step, the atmospheric profile is interpolated on the basis of pre-calculated models with a grid of values for $T_{eff}$ and $\log g$ (gravity).

In Fig. 7.12, we may compare a non-irradiated atmospheric profile (night on the figure) with an irradiated profile (day), obtained for the same values of $T_{eff}$ and

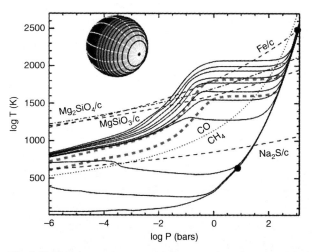

**Fig. 7.12** Temperature/pressure profiles calculated for a Pegasid. Each profile shown as a continuous line corresponds to a different irradiation geometry, from the substellar point (the point directly facing the star), *top*, to the non-irradiated profile (terminator and night side), *bottom*. In this model, no redistribution of the incident stellar flux ($f = 1$) and no transport are included. The profiles in broken grey lines represent a 'mean dayside' profile with redistribution of the incident flux over one hemisphere ($f = 1/2$, *top*) and a mean profile with global redistribution ($f = 1/4$, *bottom*). The black dots indicate the separation between the convective and radiative zones for the substellar and night-time profiles. The composition of elements is assumed to be solar, and the equilibrium chemical composition is given by the temperature. Dashed lines indicate the regions in which the principal grains condense. The dotted curve separates the domains where carbon is in the form of CO (hot) or $CH_4$ (cold). The parameters used (mass and irradiation) are those for the planet TrES-1. Modelling carried out by Barman et al. (2005), using the PHOENIX model

log $g$. It will be seen that the strong irradiation produces a hot and quasi-isothermal radiative zone that forces the convective zone to higher pressures and temperatures. Because of this radiative heating from above, the heat flux from the interior towards outer regions becomes less effective, slowing down cooling and the contraction of the planet. At any given age and for the same mass, a non-irradiated planet is thus more compact than an irradiated planet with the same mass and composition.

Once irradiation is included in the modelling, the mass/radius relationship may be compared with observations of Pegasids that transit their parent star, and where we can measure the radius (by the transit) and the mass (from the radial velocity). The theory reproduces the radius for most of the Pegasids that transit in a satisfactory manner (Chabrier et al., 2004), but underestimates the radius of some of them, in particular that of HD 209458b, where the radius is inexplicably large. Figure 7.13 indicates the masses and radii, with their associated errors, of a few Pegasids, known to transit, as well as the theoretical evolution $R(t)$ obtained for the range of masses and irradiation found for these planets. The composition is assumed to be solar (no enrichment of the envelope, and no solid core). Some of these planets have a radius

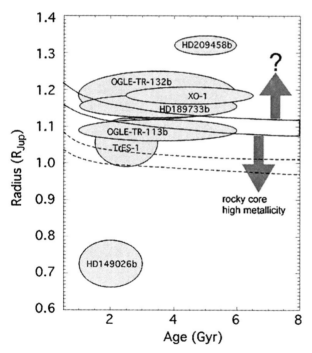

**Fig. 7.13** Comparison between the theoretical and observed radius for Pegasids that transit their star. Each ellipse corresponds to the radius and age of one Pegasid and the associated uncertainties. The dotted curves are the boundaries of the theoretical evolution (using an isolated-planet model) of the radius for the range of masses of the planets shown here. The solid lines are the boundaries for evolution of the radius when irradiation by the central star is included. The theoretical evolutionary curves are those for a planet without a solid core and of solar composition

less than that predicted for a planet of solar composition. This may be completely explained by an enrichment in heavy elements or by the presence of a rocky core (or both). For example, the mass of the planet HD 149026b is dominated by heavy elements other than H and He, and it undoubtedly possesses a significant rocky core, which may amount to as much as $80\,M_{\mathrm{Earth}}$ (Fortney et al., 2006). Although the enrichment of this planet seems to be in qualitative agreement with the star's greater than solar metallicity, it is not easy to explain quantitatively in terms of current planetary formation scenarios. The loss of part of the hydrogen-helium envelope, either by XUV irradiation (Baraffe et al., 2004), or by the collision of two giant planets during the course of their migration (Ikoma et al., 2006) could be the origin of such a 'metallic' planet.

Although planets that are more compact than their theoretical 'solar' model may be explained by significant enrichments, planets that have an even larger radius ($\sim$20 per cent in the case of HD 209458b) still defy understanding. This difficulty is all the greater, because Pegasids orbit stars with high metallicity and we would therefore expect them to possess a rocky core and an enriched envelope. The departure from theory may therefore be even more tangible (Guillot et al., 2006). Although the models of the internal structure and the atmosphere suffer from uncertainties linked to the equations of state and opacities (particularly in the case of an envelope with high metallicity), the discrepancy is too great to be ascribed to them. To maintain the radius of HD 209458b, of XO-1, or of HD 189733b at their observed values, it would be necessary to dissipate $10^{20}$ Watts (about 1 per cent of the stellar flux intercepted by the planet) in the planet's convective layers, and to maintain this contribution of energy throughout the whole of the planet's evolution. Showman and Guillot (2002) have suggested dynamical mechanisms created by the major asymmetry in the way in which the stellar energy is deposited, which could convert the radiative stellar energy into kinetic energy (winds), transport it, and dissipate it within zones in the interior. Tidal interactions are another means of dissipating the energy within the interior layers but they assume that orbital parameters such as the discrepancy between the orbital period and the period of the rotation, eccentricity or obliquity are not zero. But the orbital evolution of Pegasids should quickly synchronize and circularize the planet's orbit and reduce the obliquity to 0, unless another planet in the system perturbs such evolution. In the case of HD 209458b, the eccentricity could be measured, and has been found to be essentially zero, thus excluding the influence of another planetary body. Another possibility is that the planet might be trapped in a Cassini resonance with non-zero obliquity: the energy dissipated would then be sufficient, but the probability of such a configuration is low and it would be difficult to explain a high proportion of 'inflated' planets (Levrard et al., 2007). A third effect that could affect the radius is atmospheric loss created by the XUV irradiation and interaction with the stellar wind. If the resulting mass loss is very rapid it affects the evolution of the planet (Baraffe et al., 2004). This occurs when the characteristic escape time becomes less than the Kelvin-Helmholtz time, which may be expressed as:

$$\frac{m}{\mathrm{d}m/\mathrm{d}t} < \frac{Gm^2}{RL} \tag{7.17}$$

where $L$ is the luminosity of the planet. This condition assumes that a considerable amount of energy is deposited in the outer layers of the atmosphere, which does not seem realistic for a giant planet of $\sim 1\ M_{Jup}$. In addition, an increase in the radius requires an even greater amount of energy to be deposited and a more effective escape mechanism. This positive feedback would produce runaway effect leading to the rapid total loss of the hydrogen envelope, and so it is unlikely that we would observe a planet in this phase. Such a dramatic evolution could nevertheless have occurred with less massive planets, such as the hot Neptunes (Baraffe et al., 2005).

So the radius of HD 209458b still remains mysterious. Other exoplanets that transit also seem to have radii greater than predicted by theory, but the error bars accompanying their measurements are very large. In contrast, the radius of HD 209458b has been measured with a far higher accuracy than those of other planets. We must await new measurements with the HST to know the exact number of 'inflated' planets in our sample of transiting exoplanets.

Apart from determination of the radius and the inclination, transits allow us to obtain information about the atmospheres of exoplanets. For a planet with an atmosphere, the planetary radius measured by the decrease in the stellar flux during a transit depends on the wavelength at which the flux is measured. Indeed, if a component present very high in the atmosphere absorbs radiation at a wavelength $\lambda_1$ when the atmosphere is transparent at $\lambda_2$ down to very dense layers, we will measure $R(\lambda_2) < R(\lambda_1)$. So if we are able measure accurately the difference in depth of a transit in two spectral bands, we will be able to detect the components in the atmosphere (Ehrenreich et al., 2006). This is how sodium has been identified in the upper atmosphere of HD 209458b (Charbonneau et al., 2002), as well as a cloud of neutral gas around the planet, consisting of hydrogen (Vidal-Madjar et al., 2003), and possibly carbon and oxygen (Vidal-Madjar et al., 2004). The interpretation currently accepted for this cloud of neutral atoms is that it represents gas escaping from the atmosphere.

From now on, at last, we are able to detect the infrared thermal emission from Pegasids thanks to the Spitzer space observatory. Such detection was first made for planetary transits (HD 208458b, TrES-1, HD 149026b, and HD 189733b) by measuring the variation in luminosity of the star + planet system during the secondary transit, in the infrared bands with the IRAC, MIPS and IRS instruments (Deming et al., 2005; Charbonneau et al., 2005; and Deming et al., 2006). This variation gives the luminosity of the planet, observed just before and just after the secondary transit (i.e., when the planet passes behind the star), when its day side is turned towards the observer (*see* Fig. 7.14).

Recently, Spitzer has been able to measure the light-curve at 16 µm, that is the variation in luminosity as a function of orbital phase, for a Pegasid that does not exhibit transits: υ Andromedae (Harrington et al., 2006 – Fig. 7.15). The amplitude of the variations is determined by the inclination of the system and by the day-night contrast. This therefore allows us to set constraints on the flux-redistribution parameter ($f = 1$, 0.5 or 0.25) in theoretical models. In fact, the atmospheric models used to calculate the structure and composition, to construct synthetic spectra, and to calculate the evolution of the radius and the luminosity of exoplanets are, for the most

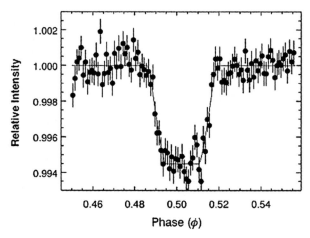

**Fig. 7.14** Secondary transit of HD 189733b (the passage of the planet behind the star), observed by Spitzer/IRS in a band at 16 μm. The Spitzer space observatory has been able to detect the thermal emission from Pegasids during their secondary transits. The flux of the system star + planet decreases when the planet passes behind the star. The difference in the flux before and after, and during the transit therefore gives the thermal emission from the day side of the planet (After Deming et al., 2006)

part, 1-dimensional models (plane-parallel or spherical) in which the irradiation flux at the top of the atmosphere is fixed. Several methods of tackling this problem exist: We may assume that atmospheric dynamics plays a negligible role and that the planet's rotation is very slow, and calculate a 1-dimensional atmospheric profile for a ring around the sub-steller point, defined by its zenith angle (*see* Fig. 7.12). The overall planetary spectrum is then reconstructed from the individual contributions from each ring. We can calculate a night-time, non-irradiated profile in this way, as well as a day-time profile where the flux at the top of the atmosphere is the stellar flux intercepted by the planetary disk of surface area $\pi R^2$ and redistributed solely on the day hemisphere, of surface area $2\pi R^2$. The redistribution factor $f$ is then equal to 0.5. Finally, we can represent the planet by a single profile, assuming that the flux it intercepts is spread over the whole of the planetary surface ($f = 0.25$), which may be an acceptable approximation when the planet's rotation is sufficiently rapid.

Because of its significant amplitude, which bears witness to a strong difference in temperature between the day and night sides of the planet, the light-curve of $\upsilon$ Andromedae allows us to eliminate the model with global redistribution of the flux ($f = 0.25$), where atmospheric dynamics and the rotation would create uniform temperature conditions at the levels probed by the thermal emission.

The Spitzer observations of secondary transits have been carried out in different spectral bands, which allows comparison of the measurements with theoretical spectra calculated for the Pegasids. Figure 7.16 shows just such a comparison. The Spitzer measurements provide information at very low resolution and affected by significant noise, which only allows very limited information about the nature of the Pegasids to be obtained. Nevertheless, similar observations, but of better quality

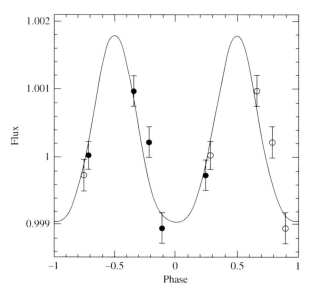

**Fig. 7.15** Infrared radiation curve for a Pegasid as measured by Spitzer. Observed by the MIPS instrument in a band at 24 μm, the flux received from υ Andromedae (star + planet) is modulated by the planet's thermal emission, which varies as a function of its orbital phase. The variations observed are the sign of a significant temperature contrast between the day and night sides. The black points are the observations, and the white points correspond to the same observations but offset by one period. The curve is obtained from a model without redistribution of the incident flux ($f = 1$), similar to that shown in Fig. 7.13 (After Harrington et al., 2006)

should be obtained in the future with the JWST (NASA, ESA), and should allow us to set better constraints on the numerous uncertainties that still affect theoretical models (including heavy-element composition, clouds, atmospheric dynamics, and photochemistry).

Currently calculation of synthetic spectra for Pegasids, and for exoplanets in general, rest on the following assumptions:

1.  solar abundance of elements,
2.  absence of vertical or horizontal transport,
3.  molecular composition in chemical equilibrium (determined by temperature),
4.  absence of photochemistry (the dissociation of chemical species by incident UV radiation is neglected).

These approximations, although by no means justified, are linked to practical limitations. Regarding point 1, and as may be seen in Figs. 7.16 and 7.18, it should be mentioned that the models will, henceforth, take account of a possible enrichment in heavy elements as characterized by the metallicity of the envelope, $M_{Z,env}/M_{env}$. This enrichment has a significant effect on the abundances of molecules and grains formed by condensation.

Atmospheric circulation and the transport of chemical species, and of the heat that is associated with it, are likely to strongly affect thermal atmospheric profiles and their distribution in longitude and latitude, the chemical composition, and thus

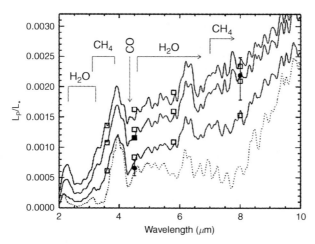

**Fig. 7.16** Theoretical spectra for TrES-1 and comparison with Spitzer observations. The curves shown with a continuous line show theoretical spectra of the day hemisphere of TrES-1, the hemisphere that is visible during observations close to secondary transit. These spectra have been obtained by Barman et al. (2005) and correspond to the profiles in Fig. 7.12: without redistribution of the incident stellar flux ($f = 1$, *top*), with redistribution over the whole planetary surface ($f = 0.25$, *bottom*), and redistribution over the day hemisphere ($f = 0.5$, *centre*). The spectra are given as the ratio of planetary to stellar flux. The white squares indicate measurements made by the IRAC instrument on Spitzer in the 3.6, 4.5, 5.8, and 8.0 μm bands for each of the models. The black square in the 4.5 μm band corresponds to a model with $f = 0.5$ and an enrichment in heavy elements of 10 times the solar value. The observations made with IRAC at 4.5 and 8.0 μm (Charbonneau et al., 2005) are given by the black circles and the associated error bars. The spectrum shown by the dotted line is that of an isolated brown dwarf ($T_{\text{eff}} = 1150$ K)

the appearance of the planet's spectrum. In fact, if the molecules are transported in times shorter than the characteristic chemical-reaction times, the composition is no longer a reflection of the local temperature. For example, in a model of partial redistribution ($f = 1$, or $f = 0.5$) where the night side is very cold and the day side very hot (at P = 1 bar, $T_{day} = 2000$ K and $T_{\text{night}} = 300$ K, *see* Fig. 7.16), the CO/CH$_4$ ratio varies strongly from one hemisphere to the other, while such a thermal gradient, even assuming synchronous rotation, should produce violent winds which would tend to equalize the differences. In this case, the CO produced at high temperatures in the day hemisphere is transported to, and survives in, the night hemisphere (Cooper and Showman, 2006 – Fig. 7.20). This phenomenon of quenching also occurs vertically, as observed on Jupiter: the CO formed at high pressures and high temperatures rises towards levels at lower pressures and temperatures, where it is detected, despite predictions made on the basis of equilibrium models. In addition, the temperature itself is evened out horizontally by winds. These dynamical considerations tend to favour models with global redistribution ($f = 0.25$), which seems, however, to contradict the amplitude of the infrared radiation curve for υ Andromedae measured by Spitzer. At any rate, this shows that coupling between radiative transfer, chemistry and dynamics is essential for realistic modelling of the atmospheres of exoplanets and of their spectra.

The dissociation of molecules by UV radiation is an essential element in the chemistry of planetary atmospheres in the Solar System. In the giant planets and on Titan, it initiates (*inter alia*) the formation of hydrocarbons and clouds of photochemical origin which modify the scattering and absorption properties of the atmosphere. The closer a planet is to its parent star and the hotter the atmosphere, the more justified it is to assume chemical equilibrium. The UV irradiation also increases, however, when the orbital distance decreases, which tends to increase the photochemical effects. Including photochemistry causes the modelling to become very complex, because it presupposes that the chemical composition is calculated from a set of several hundred reactions, the rates of which are often uncertain, in particular when we are dealing with conditions that are far outside the normal range of temperatures. Despite preliminary studies which tend to minimize its role in Pegasids (Liang et al., 2004), photochemistry must eventually be introduced into exoplanet models.

Modelling the atmospheres of exoplanets and their spectra is also rendered uncertain by the tricky treatment of condensation and of cloud formation. The range of temperatures and pressures encountered in the atmospheres of giant exoplanets is vast, and numerous phase transitions are able to occur, causing the formation of mineral grains (silicates and iron), droplets, or ice crystals. The microphysical processes responsible for the nucleation of particles combine with horizontal and vertical motions in the atmosphere (winds, convection, and advection) to form cloud layers. In the atmospheres found in the Solar System, these clouds are generally of finite horizontal extent (they cover only part of the surface), and of variable vertical extent. There properties are only included in the models *a posteriori*, and in an empirical fashion. Under these conditions, it is clear that we should not expect these one-dimensional models to reproduce the properties of extrasolar clouds in a realistic fashion. The observed spectra will probably reveal a structure intermediate between cloud-free models (where whatever condenses falls down to layers with both elevated pressure and high opacity) and models with clouds (where the particles remain in the layers in which they form). It is, however, important to note that these two types of model give very different results (*see* Fig. 7.17 for a model with clouds, Fig. 7.18 for spectra with and without clouds, and Fig. 7.19 for spectra without clouds). In Fig. 7.17 it is possible to see the possible effect of clouds as a function of orbital distance. In the case of a Pegasid ($a = 0.04$ AU, $T_{eff} = 1440$ K), the clouds of silicates and iron at high altitude mask most of absorbing material in the atmosphere, and the spectrum is dominated by the signatures of Na and K. At 0.1 AU ($T_{eff} = 870$ K), the clouds form at a lower level and allow the most prominent absorptions by gaseous components to be seen. At 0.04 and 0.1 AU, the thermal emission from the planet makes a strong contribution to the spectrum, whereas for more distant and cooler objects it is negligible relative to the reflected starlight. In the hottest objects, carbon is primarily in the form of CO, whereas methane becomes dominant farther out.

Apart from the effect of the clouds themselves, condensation depletes the atmosphere of certain potential absorbents, which may strongly affect the atmospheric structure and the spectrum. This effect is shown in Fig. 7.21: VO and TiO absorb

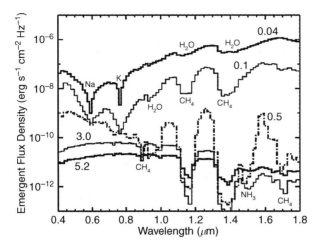

**Fig. 7.17** Synthetic spectra of giant planets as a function of orbital distance. This diagram shows the visible and near-infrared spectra of exoplanets calculated by J. Fortney and M. Marley, for a solar composition and a mass of 1 $M_{Jup}$. The orbital distance in AU is indicated on each of the spectra. The top spectrum is that of a Pegasid (a = 0.04 AU, $T_{eff}$ = 1440 K). The spectrum at 5.2 AU corresponds with that of Jupiter (After Marley et al., 2007)

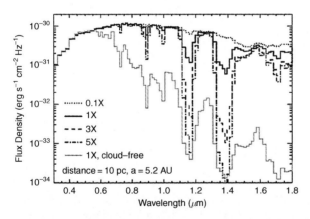

**Fig. 7.18** Sensitivity of synthetic spectra to metallicity and to clouds. Shown here are theoretical spectra for an exo-Jupiter (at 5.2 AU) with different enrichments or depletions in heavy elements, relative to a solar-type composition. (The atmosphere of Jupiter is about 3× the solar value.) The absorption bands are those of methane ($CH_4$). The bottom curve represents a spectrum without clouds (After Marley et al., 2007)

in the UV and visible regions, where the incident stellar energy is significant. The addition of these components thus modifies the UV and visible albedo and produces stratospheric heating, which affects the atmospheric structure and the spectrum as far as the thermal infrared. We should not expect these two specific components to occur at high altitudes in sufficient quantities to create such an effect, because

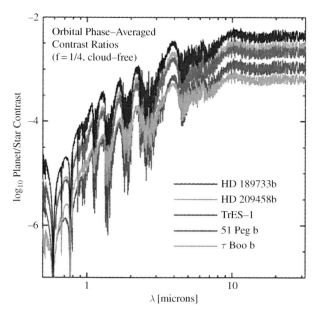

**Fig. 7.19** Comparative synthetic spectra for different Pegasids, using a cloud-free model and with complete redistribution of the incident stellar flux ($f = 0.25$) (After Burrows et al., 2006)

they condense at lower levels, but the diagram illustrates the great theoretical (and perhaps even real) diversity in possible spectra.

## 7.3.2 Terrestrial Planets and Habitable Planets

The atmospheres of the terrestrial planets primarily arise from the degassing of volatile components that were trapped in the planetesimals that were involved in the accretion of the planet (Fig. 7.22). This atmosphere, degassed either by impacts during the accretion phase, or by volcanism, was subsequently partially lost and fractionated by gravitational escape (Selsis, 2006) under the influence of stellar radiation. The importance of the escape varies from one planet to another as a function of its mass, its magnetic field (itself a function of the mass), and its orbital distance. The D:H ratio measured in the atmospheres of Venus and Mars shows that they have undergone significant escape (*see* Sect. 4.4.3.3). As for the role of escape in the evolution of the Earth's atmosphere, it is poorly constrained. The depletion of the Earth's atmosphere in certain volatile elements, as well as the fractionation of rare gases certainly bears witness to escape phenomena (Zahnle et al., 2007) but these probably took place before the formation of the Earth proper, in the fragile atmospheres of the planetary embryos, which were eroded by X-ray and EUV radiation from the young Sun, and by the bombardment by planetesimals. The atmospheres

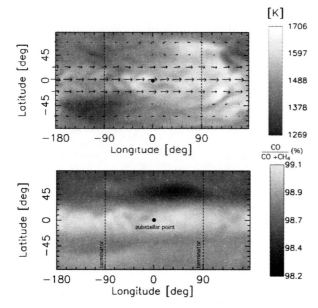

**Fig. 7.20** The effect of atmospheric circulation on the horizontal distribution of temperatures and on the CO:CH₄ ratio. This simulation, carried out by Cooper and Showman (2006) shows how the atmospheric circulation evens out temperature (*top*) and the abundance of CO (*bottom*). The atmospheric level represented is at a pressure of 1 bar and the planet's parameters are those of HD 209458b. The initial state in this simulation has a day-night contrast of 1000 K at the top of the atmosphere, and an equilibrium chemical composition (enhanced CO:CH₄ ratio on the day side and negligible on the night side). The planet is assumed to be in synchronous rotation. After simulating a period of 1000 days, the atmosphere is in the state shown here, where the violent winds (of several kilometres per second) have evened out the temperature and the CO abundance

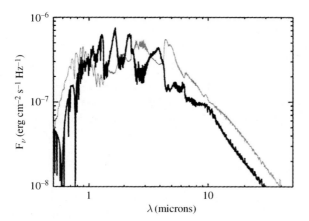

**Fig. 7.21** Sensitivity of the synthetic spectrum of a Pegasid to the presence of TiO and VO. These two theoretical spectra of a Pegasid, calculated by Hubeny et al. (2003), show the profound alterations caused by the presence of visible and UV absorbers (here TiO and VO) in the profile of the overall spectrum, from the UV to the thermal infrared (Burrows et al., 2006)

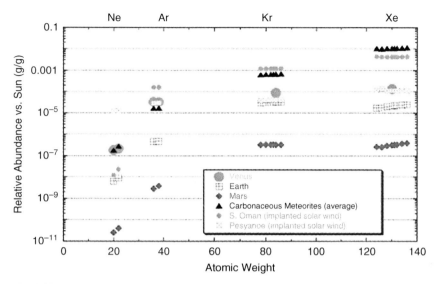

**Fig. 7.22** Abundances of rare gases and their isotopes in the atmospheres of the terrestrial planets and in meteorites. A solar composition would be represented here by a straight, horizontal line. Except for xenon, the rare-gas composition of Earth and Mars resembles that of carbonaceous chondrites. That of Venus is closer to the solar wind implanted in meteorites (but it should be noted that the measurements of the atmosphere of Venus are very uncertain). This diagram shows that the atmospheres of the terrestrial planets are the result of degassing of the volatile component of solid materials accreted during their formation. This volatile component is depleted and fractionated by escape processes (Zahnle et al., 2007)

of the terrestrial planets is therefore, by its very nature, extremely different from the gas in the protosolar nebula, and the composition of rare gases is similar (in relative abundances) to that of carbonaceous meteorites (*see* Fig. 7.22). So starting with cosmic abundances of elements, as is done to model the giant planets, does not make any sense when it comes to the terrestrial planets. One possible approach would be to take the mass abundances of volatile components (H, C, O, and N) trapped in meteorites and from those deduce the mass and composition of elements in the atmosphere. This would, however, neglect the escape phenomena that took place during the formation and primordial evolution of the atmosphere, as well as the division of the gases between the planetary interior and the atmosphere. The Earth's interior (mantle and crust) contains approximately $5 \times 10^5$ times more $CO_2$ than the atmosphere, most being held within the mantle or stored in sedimentary rocks in the form of carbonates. In the case of Venus, a significant fraction, if not almost all of the $CO_2$ is found in the atmosphere (90 bars) and corresponds more or less to the whole of the terrestrial reservoir, estimated at about 150 bars. This difference in distribution is explained by the absence of liquid water on Venus, because this allows the formation of carbonates at low $CO_2$ pressures, and also the functioning of the terrestrial carbon cycle.

In addition, within a single protoplanetary system, the planetesimals taking part in the formation of the terrestrial planets do not all have the same content of volatile components. The abundances of water, carbon, and nitrogen increase significantly with orbital distance. In the Solar System, meteorites originating in asteroids orbiting at less than 2.6 AU are water- and gas-poor. Those originating in more distant zones contain up to 15 per cent of water in the form of hydrated minerals. Bodies formed outside the ice line (4–5 AU, *see* Sect. 4.3.2.2) consist, on average, of 50 per cent water ice. From which of these populations of planetesimals did Earth's atmosphere originate? This question is still subject to considerable debate, but it seems that the majority of the population was poor in volatiles (ordinary chondrites or enstatites), and the water and gas primarily came from a lesser contribution of carbonaceous-chondritic material originating in the outer region of the asteroid belt (2.6–3.5 AU) (Raymond et al., 2007 and Sect. 4.4.6.1). The cometary contribution (ice) is thought to be very minor for the Earth. So it seems that the existence of oceans and a dense atmosphere on Earth depends in a crucial manner on a small contribution by mass of planetesimals formed relatively far away from the Earth's orbit. This contribution of material rich in volatiles during the accretion phase is likely to vary quantitatively from one terrestrial planet to another inside a single planetary system, and also from one planetary system to another. This contribution is indeed particularly sensitive to the location and properties of the giant planets, which formed before the terrestrial planets; the diversity of which has already been shown by recently detected exoplanets. To sum up, two terrestrial planets of the same mass, forming at the same orbital distance from two stars of the same type, could well draw on different reservoirs (in mass and composition) for their atmospheric components. How, under such conditions, can we hope to predict and model the atmospheres of terrestrial-type exoplanets? We shall see that understanding of the Earth's geochemical cycles have allowed us to define the concept of a habitable planet. Theoretical studies of the properties of habitable planets offer a point of departure for exploring the diverse types of terrestrial planets that may be expected, and also provides a fascinating link between exoplanets and astrobiology.

### 7.3.2.1 Habitability

A habitable planet is taken to be a terrestrial planet whose surface is covered, at least partially, by a stable layer of liquid water, over geological periods of time. The habitable zone around a star corresponds to the region around a star where such planets may exist. The presence of liquid water is a condition judged to be necessary for the existence of life, but which is probably not sufficient. This statement may, however, be called into question if it were to be shown that life can develop in the absence of liquid water, by making use of other fluids. This would mean a completely different biology from the one that we know, so for operational reasons, for the moment we will limit our considerations to life as we know it on Earth. Certain locations in the Solar System, other than Earth, undoubtedly conceal non-surface liquid water, such as the subsoils of Mars and the icy satellites (Europa and perhaps Callisto), or clouds on Venus and nothing *a priori* forbids forms of life from existing there. However, we are interested here in searching for signs of life on planets around other stars. The first space observatories that may allow us to detect habitable worlds around other stars, such as the Darwin and TPF projects, will have a chance of detecting signs of biological activity on a planet only if the biosphere is sufficiently widespread and active to alter the observable factors (the overall composition of the atmosphere or the surface). The presence of liquid water at the surface allows the existence, in particular, of photosynthetic activity (liquid water and starlight being available simultaneously). The possibility of using stellar photons and water as an electron donor allows the biosphere to transform the planetary environment in a way that is quantitatively and qualitatively distinct from other metabolisms (Rosing 2005). A planet harbouring a restricted biosphere that is relatively inactive or confined beneath the planetary surface without modifying the surface characteristics, will never reveal the presence of life to observation from a distance. In such a case, only exploration *in situ* would detect biological activity. Future exobiological missions for exploring Mars and Europa, will be directed towards this end. For this reason, an exoplanet that lacks the possibility of having liquid water at its surface would be considered 'uninhabitable', even though liquid water and life forms might actually exist in its interior.

### a. The circumstellar habitable zone

The habitable zone around a planet is defined as the region in which water can exist permanently in the liquid state at the surface of the planet. This implies a surface temperature $T_s$ above 273 K and sufficient atmospheric pressure. $T_s$ depends on the incident stellar energy and its spectral distribution, the radiative properties of the atmosphere (absorption, scattering, and emission) and the reflective properties (the albedo) of the surface and of the atmosphere (primarily through the existence of clouds). Kasting (1988), and then Kasting et al. (1993) have applied realistic atmospheric models and estimated the limits of this zone for different types of star. These authors have decided to tackle this complex problem by considering the Sun-Earth model: by modifying the orbital distance they have been able to determine the minimum and maximum distances at which the Earth's oceans are, respectively, completely evaporated or completely frozen.

### b. The inner limit of the habitable zone

Currently, most of Earth's surface water is contained in the oceans. Decreasing the Earth's orbital distance consequently vaporizes a greater fraction of this reservoir of water. The increase in the level of water vapour in the atmosphere produces a more effective greenhouse effect and thus creates additional heating of the surface, provoking the evaporation of yet more water. Happily, more water vapour in the atmosphere also implies more effective cooling of the surface by convection. This second effect is essential, because, without it, the greenhouse effect on our planet would run away, and would lead to all the oceans boiling away. Taking these two mechanisms (greenhouse effect + convection) into account in what is known as a radiative-convective model allows us to determine the temperature and water-vapour profile of the atmosphere.

From surface temperatures of about 370 K, corresponding to a distance of 0.95 AU (from the current Sun), water vapour becomes the dominant component in the atmosphere. If the planet is brought even closer to the Sun, the amount of water vaporized becomes very considerable and the energy emitted by the planet in the infrared reaches a plateau, because, thanks to the high opacity of an atmosphere of $H_2O$, the zone in which the thermal emission arises is displaced towards a radiatively isothermal region at a high altitude and becomes unaffected by the surface temperature. The thermal energy radiated away at $T_s = 550$ K or at $T_s = 1500$ K is almost the same. It is only when $T_s > 1500$ K that the thermal emission climbs once more, together with $T_s$, because the surface then radiates in the near infrared, and then cools directly to space through atmospheric windows that are transparent at such wavelengths. In terms of planetary evolution, the threshold corresponds to a dramatic transition: if a planet reaches $T_s = 550$ K, any increase, however minor, in the stellar luminosity (or decrease in the orbital distance) causes heating of the surface to 1000 K, and the complete evaporation of the reservoir of water. This is probably the fate that Venus has suffered early in its existence. This phenomenon, known as the runaway greenhouse effect is illustrated in Fig. 7.23.

The orbital distance of 0.84 AU calculated by Kasting should be considered as indicative only, because this estimate has been obtained for a cloud-free atmosphere, and also because there are uncertainties about the spectroscopic properties of water vapour at high pressure. The presence of clouds seems realistic for an atmosphere dominated by water vapour, and would limit the intensity of heating of the surface. Indeed, clouds effectively reflect solar radiation: Venus, for example, reflects 70 per cent of the energy received because of its global cloud blanket, whereas without clouds, the albedo would not exceed 30 per cent. An albedo of 70 per cent would allow a liquid ocean to be retained out to orbital distances of about 0.5 AU.

Moreover, for orbital distances less than 0.95 AU, water vapour becomes the main component of the atmosphere at all altitudes, except in the very highest layer, where $H_2O$ is dissociated into H and O by UV radiation. In Earth's present-day atmosphere, water vapour is an extremely minor component above the tropopause. The tropopause is the boundary between the troposphere and the stratosphere, where the temperature reaches a minimum (*see* Sect. 4.4.2.2). It forms a 'cold trap' that water vapour cannot cross, where it condenses as it rises. This cold trap keeps the loss of hydrogen to space to negligible values, this being limited by the molecular diffusion of H towards the exobase (the altitude at which collisions become negligible). In a hot atmosphere dominated by $H_2O$, photodissociation of $H_2O$ and the escape of H are no longer limited other than by the EUV irradiation of the upper atmosphere. Under these conditions, an amount of water equivalent to the terrestrial ocean may be lost in less than $10^9$ years. It is estimated that this is the fate that Venus has undergone. Early in its history, the solar luminosity was sufficiently weak for a very hot, liquid ocean to exist, as shown in Fig. 7.23. However, the erosion of this reservoir of water by UV radiation must have been very effective, and if Venus had a quantity of water less than that in Earth's oceans, the latter must have been lost even before the increase in solar radiation reached the threshold at which total vaporization occurred. The quantity of water present on a planet therefore needs to

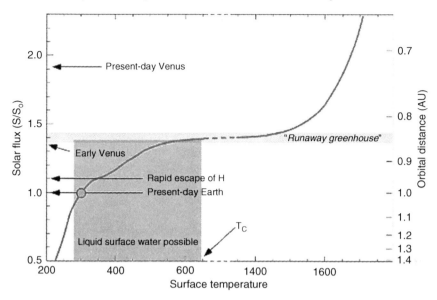

**Fig. 7.23** The runaway greenhouse effect. This graph shows the relationship between the solar flux received at the top of the atmosphere (relative to the flux currently received by the Earth: $S_0$) and the surface temperature $T_s$. When $T_s$ exceeds about 550 K, the infrared flux emitted by the planet reaches a threshold. This threshold corresponds to an orbital distance of 0.84 AU from the present-day Sun, and also corresponds to the inner boundary of the habitable zone. Beyond this threshold, any increase, even small, in the irradiation leads to a runaway greenhouse effect, which evaporates the whole reserves of water and takes the surface to temperatures above 1500 K. The flux received by present-day Venus is indicated, as is the flux $4 \times 10^9$ years ago (early Venus), the flux that corresponds to rapid escape of H, and the temperature/flux zone where liquid water can exist at the surface. The connection between the flux and distance from the Sun is given on the right-hand vertical axis. Note the break in the temperature axis between 600 and 1400 K (After Kasting, 1988)

be taken into account, for at least two reasons. The first is the escape of this reservoir to space, which may be represented as a relationship between orbital distance and the lifetime of the reservoir of water. The second relates to the fact that water's critical point corresponds to a temperature of 647 K and a pressure of 220 bar. In the case of the Earth, vaporizing the whole of the oceans would produce a pressure of 270 bar, above the critical point $T_c$: a temperature above $T_c$ is therefore necessary to vaporize the whole ocean. A less massive reservoir could, however, be entirely vaporized at lower temperatures. For example, an ocean 100 metres deep would be vaporized at a temperature of 450 K, and therefore at a distance of about 0.9 AU from the present-day Sun (assuming a cloud-free case).

### c. The outer limit of the habitable zone

On Earth, the principal greenhouse gas is water vapour. However, because water vapour is in thermodynamic equilibrium with the reservoir of liquid water at the surface, its abundance in the atmosphere, as well as $T_s$, is determined by distance from the Sun and by the abundance of the other major greenhouse gas, $CO_2$.

In the case of a planet such as the Earth, with continents that emerge from the ocean and active volcanism, the level of $CO_2$ is also dependent on the distance from the Sun, this being because of the carbonate-silicate cycle and is dependence on $T_s$. In fact, if the level of $CO_2$ is insufficient to keep $T_s > 273$ K, water freezes and $CO_2$ can no longer precipitate in the form of carbonate, although it is still being emitted by volcanism. It abundance therefore increases until the frozen surface turns back into liquid. This is how our planet has escaped from phases of worldwide glaciation. If the level of $CO_2$ continues to rise, $T_s$ will increase significantly above 0°C, increasing the quantity of water vapour in the atmosphere. Precipitation will become more and more significant, leading to strong erosion of the rocks and thus increasing the rate at which carbonates form. These two negative-feedback processes tend to stabilize the level of $CO_2$, at least over timescales of several million years, such that $T_s$ remains higher than 273 K. This mechanism has been described in detail by Walker et al. (1981), and probably regulated the Earth's climate throughout the course of its history, and stabilized $T_s$ despite the increase in the Sun's luminosity since it was formed (*see also* Sect. 4.4.3.3).

So, if we were to move the Earth to an orbit farther from the Sun, the carbonate-silicate cycle would lead to a higher level of $CO_2$. At shown in Fig. 7.24, a temperature of 273 K at 1.2 AU from the present-day Sun, for example, corresponds to a $CO_2$ partial pressure of about 50 mb (about 170 times the current level of $CO_2$).

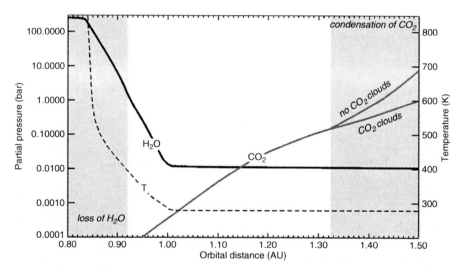

**Fig. 7.24** Variation in the surface temperature and the partial pressure of $H_2O$ and $CO_2$ with orbital distance for the Earth, within the habitable zone. The concept of a habitable planet and zone arose from the theoretical experiment of modifying the orbital distance $a$ of the Earth, It is assumed here that the carbonate-silicate cycle stabilizes the level of $CO_2$. The partial pressure of $CO_2$ thus increases with orbital distance. At 1.3 AU the formation of clouds of water ice takes place in the atmosphere, and the exact relationship between $PCO_2$ and $T_S$ remains to be determined (Forget et Pierrehumbert, 1997). Towards 0.93 AU, water vapour becomes the principal atmospheric component even in the upper atmosphere, which results in the photodissociation of large quantities of water and the loss of hydrogen to space

Determining the outer limit of the habitable zone therefore consists of determining the distance above which $T_s < 273$ K, whatever the level of atmospheric $CO_2$.

Beyond an orbital distance of 1.3 AU (from the present-day Sun), clouds of $CO_2$ ice form in the atmosphere and have a complex influence. The increase in the albedo that they produce tends to decrease $T_s$, but by scattering the thermal emission emitted by the surface, they produce an additional greenhouse effect, which heats up the surface. According to Forget and Pierrehumbert (1997), the heating dominates and $CO_2$ clouds allow habitability to be maintained beyond 1.3 AU, and perhaps as far out as 2.2 AU. It may be noted that Mars (1.5 AU), although not 'habitable' now, lies inside this limit. If the surface of Mars is not currently habitable, it is mainly for two reasons. On the one hand, the carbonate-silicate cycle does not operate because the internal energy flux is too weak to maintain active volcanism. On the other hand, the weak gravity of Mars, and the absence of a magnetic field have allowed a significant portion of its atmospheric components to escape. The low mass of Mars is thus responsible for its current desert state, and we may assume that a planet with a greater mass (at least half that of the Earth) would have remained habitable at Mars' distance. Numerous geological pieces of evidence appear to show that more than 3800 million years ago, liquid water flowed or lay as open water (or both) on the surface of Mars for long periods, implying a warmer climate (*see* Sect. 4.4.3.3). According to our definition, Mars was then habitable, or quasi-habitable if the surface temperature was less than 273 K and water was liquid only seasonally or daily. Because of the lower luminosity of the Sun at that time, this situation corresponds to a distance of 1.75 AU from the present-day Sun, which seems to demonstrate that habitability may be maintained beyond 1.3 AU. We may therefore consider the value of 1.75 AU (or 0.32 times the solar energy currently received by Earth) as the outer limit of the habitable zone.

### d. The continuously habitable zone

The luminosity of a star increases over the course of its existence, progressively pushing the boundaries of the habitable zone outwards. It is for this reason that Venus probably lay in the habitable zone in the past, but is not included in it at present. So we may also define the Continuously Habitable Zone, over a duration $t$, as the region that remains habitable for a period of time $> t$. The choice of $t$ is not obvious, particularly if one wants to compare the continuously habitable zones of different types of star having a wide range of lifetimes. Stars twice as massive as the Sun only live for $10^9$ years, while the lifetime of stars half the mass of the Sun is (theoretically) over $8 \times 10^{10}$ years. Generally, the boundaries of the continuously habitable zone are calculated for durations of $10^9$ to $5 \times 10^9$ years. Figure 7.25 shows the limits of the continuously habitable zone, obtained for $t = 10^9$ years for different types of star, determined by using the stellar evolution models by Baraffe et al., 1998.

### e. Habitable zones around other stars

Knowing the limits of a habitable zone, calculated for the present-day luminosity of the Sun, can we use this to calculate the boundaries of a habitable zone for any star? This is possible for stars of spectral type (and thus temperature) close to that of the Sun. Here it suffices to use the relationship $D_* = D_\odot (L_*/L_\odot)^{1/2}$, where $D_*$ is the distance at which the energy flux from a star of luminosity $L_*$ is the same as the solar flux received at a distance $D_\odot$ from the present-day Sun (of luminosity $L_\odot$). By replacing $D_\odot$ by the boundaries of the habitable zone for the present-day Sun, we obtain a good approximation of the boundaries for another star, or for the Sun at a different age.

For stars that have an effective temperature (in other words, colour) that is very different from the present Sun, this approximation is not valid. In fact, the albedo of a planet depends noticeably on the spectral distribution of the energy. The Earth's albedo is about 0.3, but this is not an intrinsic value: the albedo would be different if, with the same atmospheric composition, the Earth were subject to radiation from a hotter (type F) or a colder (type K) star. The hotter the star, the greater the contribution from short wavelengths (UV and visible). Conversely, the cooler the star, the more the maximum emissivity in its spectrum moves towards the near infrared. Because Rayleigh scattering of light varies as $\lambda^{-4}$, the scattering of incident radiation back into space is more effective, and the albedo is greater, for hot stars. By contrast, the atmospheric components on which habitability depends ($H_2O$ and $CO_2$) absorb moderately in the visible, but strongly in the near infrared. So a greater fraction of the incident radiation is absorbed by the planet if the star is cooler. This effect has been studied by Kasting et al. (1993), and is included in calculations of the continuously habitable zone in Fig. 7.25.

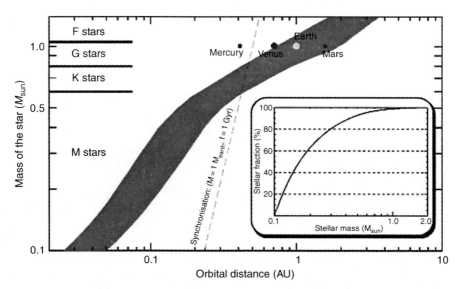

**Fig. 7.25** Boundaries of the Continuously Habitable Zone as a function of the mass of the central star. The oblique line indicates the orbital distance at which a planet with the Earth's mass would be trapped in synchronous rotation, and always turn the same side towards the star, in less than $10^9$ years. (This relationship is valid for a circular orbit only.) The insert shows the relative populations of stars in the Galaxy as a function of their mass. [Diagrams after F. Selsis (limits of the habitable zone) and J.-M. Grießmeier (synchronization time)]

The habitable zone defined previously does not take account of anything other than the luminosity of the star. To be habitable, a planet must orbit within this zone, but this does not necessarily mean that a planet within the zone is inevitably habitable, nor that planets will actually be found there. We have already mentioned the planetary mass as a decisive factor, but other parameters must be taken into account.

The Sun is a G-type star and its lifetime is approximately $10^{10}$ years. Stars more massive than the Sun have shorter lifetimes: stars that are more than $2\,M_\odot$ explode as supernovae in less than $10^9$ years. Such stars are, however, very much in the minority. Most stars have a low, or even very low, mass. The habitability of possible planets orbiting these stars is therefore a very important point that we ought to discuss. Within the range 0.1–2 $M_\odot$, 95 per cent of stars have a mass that is below $0.6\,M_\odot$. These a M-type stars. Because of their low luminosity, the habitable zone is extremely close to the star, which results in the rotation of the planets being braked by tidal effects. So, if habitable planets exist around M-type stars, they are in synchronous rotation, that is, they always present the same face to the star (or approximately so, if their orbits are eccentric, as in the case of Mercury).

The climate of such planets in synchronous rotation have been studied by Joshi et al. (1997) and Joshi (2003). The results obtained show that the very significant greenhouse effect and the atmospheric circulation generated in a very dense ($> 1.5\,\mathrm{bar}$) $CO_2$ atmosphere tend to reduce the temperature contrast between the day and night sides. Without this redistribution of the incident stellar energy, the atmosphere and the water would all completely condense on the dark side of the planet. However, in this case, it is more difficult to explain how the carbonate-silicate cycle could maintain the level of $CO_2$ at its high, beneficial level. In the most recent studies, Joshi has shown that if the planet possessed an ocean, its thermal inertia and circulation would enable habitability to be maintained with distinctly lower levels of $CO_2$ (a few tens of millibars). It remains to be seen, however, whether from an evolutionary point of view such a state could be attained and maintained.

Another problem calls into question the habitability around M stars. The extremely strong activity of these stars, which appears in the form of powerful X-ray and EUV emission, a significant mass loss (in a stellar wind), and violent coronal mass ejections, would result in considerable erosion of any atmosphere in a potentially habitable zone. If we were to place the present-day Earth in the habitable zone of an M-type star, the X-ray radiation would raise the upper atmosphere to such temperatures that it would escape violently. Atmospheric oxygen and nitrogen would be endangered. $CO_2$, however, provides good protection against heating by X-rays and we have just seen that a habitable planet around an M-type star should anyway have an atmosphere that is much richer in $CO_2$ than the Earth. The problem of X-rays is therefore solved by climatic aspects: either the atmosphere is condensed on the night side, in which case any escape is minor, or the partial pressure of $CO_2$ is very high, and limits the temperature of the upper atmosphere and thus thermal escape. Nevertheless, the most formidable danger to the habitability of planets orbiting M stars is coronal mass ejections. The violent eruptions of material from the stellar corona occur frequently in low-mass stars and the atmospheric loss that they would cause on planets in the habitable zone might be considerable (Khodachenko et al., 2007). This erosion is made all the more effective by the slow rotation of the planets, which probably has the result in a weak magnetic field when compared with that of the Earth, thus weakening the protection against outbursts of the stellar wind.

### f. A model for a habitable planet or Exo-Earth

The parameter space defining the atmosphere of a terrestrial planet is drastically reduced if we restrict ourselves to an Earth-like planet. The amount of atmospheric water vapour and $CO_2$ are no longer free parameters, but depend on the orbital distance and the luminosity of the star, as shown in Fig. 7.25. We can then model the atmospheric structure of a habitable planet as a function of orbital distance, and produce the synthetic spectra shown in Fig. 7.26. One unknown remains the partial pressure of $N_2$. It is generally accepted (although disputed by some) that practically all of the Earth's nitrogen was degassed in the form of $N_2$ very early in the Earth's history (in the first 50 million years). By fixing nitrogen, life introduced a transfer of N from the atmosphere to the crust, but it seems that this flow involves only a tiny fraction of atmospheric nitrogen. The atmosphere of Venus contains about 4 times

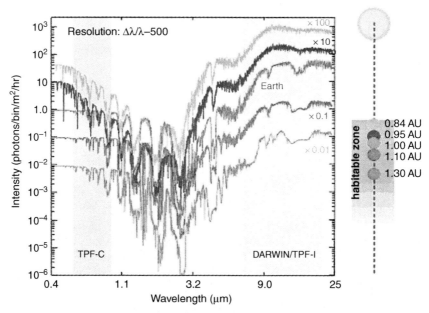

**Fig. 7.26** The effect of orbital distance on the synthetic spectra of an exo-Earth. These synthetic spectra have been calculated for an Earth located at different points in the Sun's habitable zone. The atmospheric profiles (pressure, temperature, and composition) correspond to the conditions shown in Fig. 7.24. Beyond 1.3 AU, the $CO_2$ condenses in the atmosphere, forming clouds whose radiative properties are little known. So the outermost region of the habitable zone has not been considered in the particular study. In configurations at less than 1 AU, atmospheric oxygen has not been included, because the surface conditions are very different from those on our planet. The spectral windows of the future space observatories TPF-C (visible region – visible) and Darwin/TFP-I (IR region – thermal emission) are shown (After Paillet, 2006)

the amount of nitrogen found on Earth, and the $CO_2:N_2$ ratios in the atmospheres of Venus and Mars are practically identical. If we assume that this ratio remains true for Earth, there would be a significant amount of terrestrial nitrogen in the mantle – which is a very controversial suggestion. We shall not discuss this question of $N_2$ any further, but it is important to note that the partial pressure of $N_2$ is likely to vary from one habitable planet to another. In the majority of simulations of exo-Earths that have been carried out, an atmosphere of 0.8 or 1 bar of $N_2$ has been adopted, by analogy with Earth. But it should be borne in mind for future modelling that lesser or greater pressures are also realistic and that, in particular, higher pressures may modify the spectral properties of the atmosphere (by increasing the albedo, and also by collisional broadening of the absorption lines).

The ideal model of a 'habitable planet' certainly does not reflect the great diversity of atmospheres of terrestrial planets, nor even those of habitable planets. We have just seen this in the case of $N_2$, which remains a free parameter. But we should also consider the more-or-less massive presence of volcanic gases (such as sulphur compounds, for example), the accumulation of atmospheric $H_2$ in the case of a massive planet where escape is ineffective, or the production of abiotic methane by the reaction between $CO_2$ and $H_2O$ in cases where hydrothermal activity is very significant. These remain paths that need to be explored in future models.

In the atmosphere of a habitable planet, not modified by life, and consisting of $N_2$, $CO_2$, and $H_2O$, photochemistry does not significantly alter the atmosphere's structure and spectral properties, except in some extreme cases studied by Selsis et al. (2002), where the abundance of $O_2$ or $O_3$ (or both) reaches values that cannot be neglected. Two cases should be noted: the first is that of a planet in the inner region of the habitable zone, where the abundance of stratospheric water vapour becomes sufficient large to generate a gravitational loss of H as a result, leading to the accumulation of $O_2$. In this situation, $O_3$ does not become a significant component, because it is destroyed by the hydrogenated radicals produced by the photolysis of $H_2O$. The second case is that of a planet in the outer region of the habitable zone, where the partial pressure of $CO_2$ is around 1 bar. If the planet is no longer volcanically active and is no longer releasing hydrogen or if H escapes effectively (or both) – these two conditions are found in the case of a planet that is significantly less massive than the Earth – then the $CO_2$ photochemistry may generate an $O_2$ partial pressure of about 10 millibars or more and an ozone layer that is sufficiently dense as to alter the structure of the upper atmosphere. However, the presence of a dense $CO_2$ atmosphere and the absence of effective volcanism are two contradictory hypotheses. To conclude this discussion, let us simply say that the possibilities for the accumulation of abiotic $O_2$ and $O_3$ are restricted to fairly exotic cases.

Where the composition of the atmosphere is modified by the presence of a biosphere and deviates from our model of a 'habitable planet', which is dominated by abiotic geochemistry, photochemistry may then play a predominant role. In Earth's present-day atmosphere, it is the layer of ozone, produced from photolysis of $O_2$, that creates stratospheric warming by absorbing UV and visible radiation. In the early atmosphere on Earth, at the time when oxygen was just a minor component, methane produced by the biosphere was probably the principal greenhouse gas and

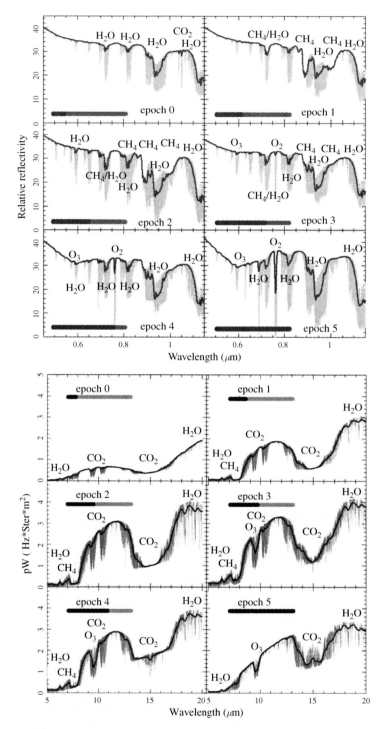

**Fig. 7.27** (continued)

its atmospheric abundance profile was determined by photochemistry. The $CO_2$ was not regulated by the carbonate-silicate cycle, and Earth must then have been hotter. Estimating the structure and the detailed composition of the atmosphere on a primitive Earth at different stages in its evolution is an extremely difficult task in the absence of any firm constraints. We can, however, model the Earth's spectral evolution based on a standard atmospheric evolution, although this is likely to be modified by future work on the primitive terrestrial atmosphere. It is this spectral evolution that is shown in Fig. 7.27. The spectra there are shown at a moderate resolution (high resolution spectra are shown in a grey tint) and in the spectroscopic windows that will be available to the future space observatories TPF-C (visible and near infrared: from 0.6 to 1.2 μm) and Darwin/TPF-I (thermal infrared: from 0.5 to 20 μm).

### 7.3.3 Hot Neptunes, Super-Earths, and Ocean Planets

Before the discovery of exoplanets and their diversity, when the Solar System was our sole reference for understanding planetary formation, there was a strong theoretical distinction between giant planets and terrestrial planets. A terrestrial planet was assumed to have formed between the star and the ice line, in a region where ices could not contribute towards the accretion of solid cores, and where only small planets of a few Earth masses, maximum, could form. Beyond the ice line, more massive cores of rocky material and ices could grow until they reached the critical mass above which a giant planet formed through rapid accretion of gas from the disk. In this scenario, the ice giants, Uranus and Neptune, remained aborted giant planets that formed from a disk that was already depleted in gas, because of the longer accretion times at their orbital distances (*see* Sect. 4.3.2.4).

This scenario has now been completely shattered by the diversity of objects that have been revealed by observation, and also by the now obvious role of planetary migration. Henceforth, we will have to take account of the fact that migration is not only necessary to understand the existence of giant exoplanets, but that it is also needed to understand the formation of Jupiter and Saturn (Alibert et al., 2005), and the architecture of the whole Solar System beyond Jupiter (Tsiganis et al., 2005 and Sect. 4.3.2.6). Planetary migration within protoplanetary disks, which evolve and

---

**Fig. 7.27** (continued) Possible evolution of terrestrial spectral from 3900 million years ago to the present day. Based on a 'standard' model of the Earth's atmosphere, describing the evolution of the surface temperature and the atmospheric components ($H_2O$, $CO_2$, $O_2$, $O_3$, $CH_4$), Kaltenegger et al. simulated the evolution of the apparent spectrum of our planet over the course of its history. The latter was broken down into 6 epochs before the present (BP): $3.9 \times 10^9$ BP (0); $3.5 \times 10^9$ BP (1); $2.4 \times 10^9$ BP (2); $2 \times 10^9$ BP (3); $2 \times 10^9$ BP (4); $0.8 \times 10^9$ BP (5). The 6 diagrams on the top show the visible spectrum (reflected light), while the 6 diagrams on the bottom show the infrared spectrum (thermal emission) (After Kaltenegger et al., 2007)

disperse rapidly (in a few million years for the gas and dust), may result in an astonishingly diverse range of planets, in terms of mass, chemical composition, and orbital distance. In addition, it may give rise to a population of planets with masses between those of giants and terrestrial planets. Such planets must be abundant, because radial-velocity detections of planets below 15 $M_{Earth}$ are accumulating despite the observational bias that does not favour low masses (Fig. 7.28). A striking example is the HD 69830 system, which contains three planets, whose minimum masses all lie between 10 and 18$M_{Earth}$, and where one of them, the largest, orbits within the habitable zone. Among the thousands of models of planetary formation (including migration and the evolution of the disk) that were tested by Alibert et al. (2006) by varying the initial conditions, the one shown in Fig. 7.29 reproduces a system similar to the one that is observed. Here, migration is the factor that allows massive planets to be found within the inner region of the system. It is the progressive dissipation of the disk of gas and the initial density of solids that prevent these planets from evolving into giant planets. It may be noted that at the end of its formation, the

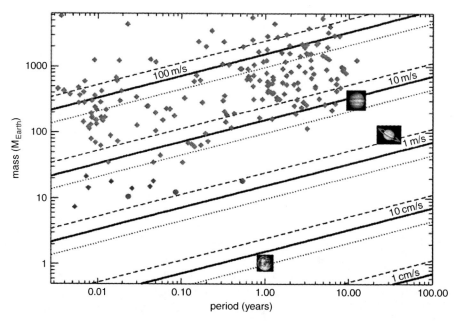

**Fig. 7.28** The mass and orbital period of known exoplanets. This diagram shows the approximately 209 currently known, detected either by radial-velocity or transit observations. For planets detected solely by radial velocity, the mass is the minimum mass (the inclination not being known). The straight lines indicate the amplitude of the variations in radial velocity produced by a star of 1 $M_\odot$ (*continuous line*), 0.5 $M_\odot$ (*dashed lines*), and 2$M_\odot$ (*dotted line*). This gives an idea of the instrumental accuracy required for detection. Currently the HARPS instrument has a sensitivity better than 1 m/s, revealing a new population of planets (known as hot Neptunes or Super-Earths, depending on the authors), of less than 20 $M_{Earth}$ (in *violet*). The fact that these detections have been made despite the strong observational bias favouring high masses demonstrates that these planets are very common. The violet circles mark the HD 69830 system (Fig. 7.29)

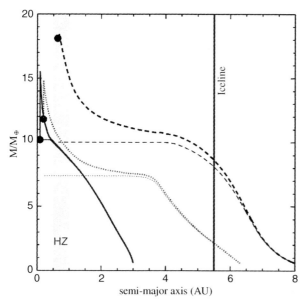

**Fig. 7.29** The scenario for the formation of the triple system around HD 69830. The star HD 69830 has three planets, lying at distances of 0.0785, 0.186 and 0.63 AU, which were detected by radial-velocity measurements (Lovis et al., 2006). The minimum masses of these planets are 10.2, 11.8, and 18.1 $M_{Earth}$, respectively. The curves shown in a thin line show the evolution of the mass of the solid core, while the thicker lines indicate the overall mass (After Alibert et al., 2006). For the 2 inner planets, the decrease in mass is caused by evaporation of the atmosphere through EUV irradiation

third planet lay within the habitable zone, but, in this model the envelope of gas ($H_2$ and He) would amount to about $8\,M_{Earth}$. The layer of water that lies below this is therefore in the form of a super-critical fluid at a high temperature and pressure.

These 'intermediate' planets, which populate the short and medium periods may have various origins. The hottest may be giant planets greatly eroded by atmospheric loss caused by EUV radiation from the star (Baraffe et al., 2005). For the planets below the critical mass for the accretion of gas ($\sim 10 M_{Earth}$), they may be 'Super-Earths', formed between the ice line and the star, but having gained significant mass through the migrations of the embryo planet inwards in the system. It may also be that these planets consist of 50 per cent ice, having formed beyond the ice line and subsequently migrated inwards. The latter may have ended their migration in the habitable zone, and thus given rise to ocean planets (Léger et al., 2004; *see* Sect. 7.2.2). Knowing just the mass and periods of these objects is not sufficient to define their chemical composition, nor any possible atmosphere. Apart from their initial, variable amounts of different elements in their composition, degassing, photochemistry, capture of nebular gas, condensation, and escape all combine to give a vast, diverse range of possible atmospheric compositions. Theoretical studies of these 'new planets', of their atmospheres, and their spectral signatures are only just beginning.

# Bibliography

## Internal structure of the planets

Beaulieu, J.-P., Bennet, D.P., Fouqué, P., et al., 'Discovery of a cool planet of 5.5 Earth masses through gravitational microlensing', *Nature*, **439**, 437–440 (2006)

Guillot, T., 'Physics of sub-stellar objects interiors, atmospheres, evolution', in *Extrasolar Planets*, (eds) Cassen, P., Guillot, T. and Quirrenbach, A., Saas-Fee Advanced Course 31, 243–361, Springer-Verlag, Heidelberg (2006)

Guillot, T., Chabrier, G., Gautier, D. and Morel, P., Effect of radiative transport on the evolution of Jupiter and Saturn, *ApJ*, **450**, 463–472 (1995)

Marley, M.S., 'Interiors of the giant planets', in *Encyclopedia of the Solar System*, (eds) Weissman, P., McFadden L.-A. and Johnson, T.V., 339–355, Academic Press, New York (1999)

Pollack, J.B. and Hollenbach, D., et al., 'Composition and radiative properties of grains in molecular clouds and accretion disks', *Astrophys. J.*, **421**, 615–639 (1994)

Sotin, C., Grasset, O. and Mocquet, A., Curve mass/radius for extrasolar earth-like planets and ocean planets, *Icarus*, 191, 337–351 (2007)

Valencia, D., O'Connell, R. and Sasselov, D., 'Internal structure of massive giant planets', *Icarus*, **181**, 545–554 (2006)

Zharkov, V.N. and Trubitsyn, V.P., *Physics of Planetary Interiors*, Astronomy and Astrophysics Series, (ed.) Hubbard, W.B., Pachart, Tucson (1978)

## Atmosphere

Alibert, Y., Mordasini, C. and Benz, W., et al., 'Models of giant planet formation with migration and disc evolution', *Astron. Astrophys.*, **434**, 343–353 (2005)

Alibert, Y., Baraffe, I. and Benz, W., et al., 'Formation and structure of the three Neptune-mass planets system around HD 69830', *Astron. Astrophys.*, **455**, L25–L28 (2006)

Baraffe, I., Chabrier, G., Allard, F. and Hauschildt, P.H., 'Evolutionary models for solar metallicity low-mass stars: mass-magnitude relationships and color-magnitude diagrams', *Astron. Astrophys.*, **337**, 403–412 (1998)

Baraffe, I., Selsis, F. and Chabrier, G. et al., 'The effect of evaporation on the evolution of close-in giant planets', *Astron. Astrophys.*, **419**, L13–L16 (2004)

Baraffe, I., Chabrier, G. and Barman, T., 'Hot-Jupiters and hot-Neptunes: a common origin?' *Astron. Astrophys.*, **436**, L47–L51 (2005)

Baraffe, I., Alibert, Y., Chabrier, G., et al., 'Birth and fate of hot-Neptune planets', *Astron. Astrophys.*, **450**, 1221–1229 (2006)

Barman, T.S., Hauschildt, P.H. and Allard, F., 'Phase-dependent properties of extrasolar planet atmospheres', *ApJ*, **632**, 1132–1139 (2005)

Burrows, A., Sudarsky, D. and Hubeny, I., Theory for the secondary eclipse fluxes, spectra, atmospheres, and light curves of transiting extrasolar giant planets, *ApJ*, **650**, 1140–1149 (2006)

Chabrier, G., Barman, T. and Baraffe, I., et al., 'The evolution of irradiated planets: application to transits', *ApJ*, **603**, L53–L56 (2004)

Chabrier, G., Baraffe, I. and Selsis, F., et al., 'Gaseous planets, protostars and young brown dwarfs: birth and fate', in *Protostars & Planets V*, (eds) Reipurth, B., Jewitt, D. and Keil, K., Arizona Univeristy Press, Tucson (2007)

Charbonneau, D., Brown, T.M. and Noyes, R.W., et al., 'Detection of an extrasolar planet atmosphere', *ApJ*, **568**, 377 (2002)

Charbonneau, D., et al., 'Detection of thermal emission from an extrasolar planet', *ApJ*, **626**, 523–529 (2005)

Cooper, C.S. and Showman, A.P., 'Dynamics and disequilibrium carbon chemistry in hot Jupiter atmospheres, with application to HD 209458b', *ApJ*, **649**, 1048–1063 (2006)

Deming, D., Seager, S. and Richardson, L.J., et al., 'Infrared radiation from an extrasolar planet', *Nature*, **434**, 740–743 (2005)

Deming, D., Harrington, J. and Seager, S., et al. 'Strong infrared emission from the extrasolar planet HD 189733b', *ApJ*, **644**, 560–564 (2006)

De Pater I., Lissauer J.J., 'Planetary Sciences', Cambridge University Press (2001)

Ehrenreich, D., Tinetti, G. and Lecavelier des Etangs, A., et al., 'The transmission spectrum of earth-size transiting planets', *Astron. Astrophys.*, **448**, 379–393 (2006)

Forget, F. and Pierrehumbert, R.T., 'Warming early Mars with carbon dioxide clouds that scatter infrared radiation', *Science*, **278**, 1273 (1997)

Fortney, J.J., Saumon, D. and Marley, M.S., 'Atmosphere, interior, and evolution of the metal-rich transiting planet HD 149026b', *ApJ*, **642**, 495–504 (2006)

Guillot, T., 'The interiors of giant planets. Models and outstanding questions', *Ann. Rev. Earth Plan. Sci.*, **33**, 493–530 (2005)

Guillot, T., Santos, N.C. and Pont, F., et al., 'A correlation between the heavy element content of transiting extrasolar planets and the metallicity of their parent stars', *Astron. Astrophys.*, 453, L21–L24 (2006)

Harrington, J., Hansen, B.M. and Luszcz, S.H., et al., 'The phase-dependent infrared brightness of the extrasolar planet u Andromedae b', *Science*, **314**, 623–626 (2006)

Hubeny, I., Burrows, A. and Sudarsky, D., 'A possible bifurcation in atmospheres of strongly irradiated stars and planets', *ApJ*, 594, 1011–1018 (2003)

Ikoma, M., Guillot, T. and Genda, H., 'On the origin of HD 149026b', *ApJ*, 650, 1150–1159 (2006)

Joshi, M.M., Haberle, R.M., Reynolds, R.T., 'Simulations of the atmospheres of synchronously rotating terrestrial planets orbiting m dwarfs: conditions for atmospheric collapse and the implications for habitability', *Icarus*, **129**, 450–465 (1997)

Joshi, M.M., 'Climate model studies of synchronously rotating planets', *Astrobiology*, **3–2**, 415 (2003)

Kaltenegger, L. and Fridlund, M., 'The Darwin mission: search for extra-solar planets', *Advances in Space Research*, **36**, 1114–1122 (2005)

Kaltenegger, L., Traub, W.A. and Jucks, K.W., 'Spectral evolution of an earth-like planet', *ApJ*, **658**, 598–616 (2007)

Kasting, J.F., 'Runaway and moist greenhouse atmospheres and the evolution of Earth and Venus', *Icarus*, **74**, 472 (1988)

Kasting, J.F., Whitmire, D.P. and Reynolds, R.T., 'Habitable zones around main sequence', *Icarus*, **101**, 108 (1993)

Khodachenko, M.L., Ribas, I. and Lammer H. et al., 'Coronal mass ejection (CME) activity of low mass M stars as an important factor for the habitability of Terrestrial Exoplanets. I. CME impact on expected magnetospheres of earth-like exoplanets in close-in habitable zones', *Astrobiology*, **7**, 167–184 (2007)

Léger, A., Selsis, F., Sotin, C., Guillot, T., Despois, D., Mawet, D., Ollivier, M., Labéque, A., Valette, C., Brachet, F., Chazelas, B., Lammer, H., 'A new family of planets? "Ocean-Planets"', *Icarus*, **169**, 499–504 (2004)

Levrard, B., Correia, A.C.M. and Chabrier, G., 'Tidal dissipation of hot Jupiters: a new appraisal', *Astron. Astrophys.*, **462**, L5–L8 (2007)

Liang, M.-C., Seager, S. and Parkinson, C.D., 'On the insignificance of photochemical hydrocarbon aerosols in the atmospheres of close-in extrasolar giant planets', *ApJ*, **605**, L61–L64 (2004)

Lovis, C., Mayor, M. and Pepe, F., et al., 'An extrasolar planetary system with three Neptune-mass planets', *Nature*, 441, 305–309 (2006)

Marley, M.S., Fortney, J. and Seager, S., et al., 'Atmospheres of extrasolar giant planets', in *Protostars & Planets V*, Univeristy of Arizona Press, Tucson (2007)

Martin, H., Claeys, P. and Gargaud, M., et al., 'From suns to life 6. Environmental context, *Earth, Moon and Planets*, **98**, 205–245 (2006)

Mischna, M.A., Kasting, J.F., Pavlov, A. and Freedman, R., 'Influence of carbon dioxide clouds on early martian climate', *Icarus*, **145**, 546–554 (2000)

Paillet, J., 'Caractérisation spectrale d'exoplanètes telluriques', Thesis submitted to l'Université Paris XI (in French) (2006)

Pavlov, A.A., Kasting, J.F., Brown, L.L., Rages, K.A. and Freedman, R., 'Greenhouse warming by CH4 in the atmosphere of early Earth', *J. Geophys. Res.*, 105, 11981 (2000)

Pollack, J.B., Hubickyj, O. and Bodenheimer, P., et al., 'Formation of the giant planets by concurrent accretion of solids and gas', *Icarus*, **124**, 62–85 (1996)

Rasool, S.I. and De Bergh, C., 'The runaway greenhouse and accumulation of $CO_2$ in the Venus atmosphere', *Nature*, **226**, 1037 (1970)

Raymond, S.N., Quinn, T. and Lunine, J.I., 'Making other Earths: dynamical simulations of terrestrial planet formation and water delivery', *Icarus*, **168-1**, 1–17 (2004)

Raymond, S.N., Quinn, T. and Lunine J.I., 'High-resolution simulations of the final assembly of Earth-like planets 2: water delivery and planetary habitability', *Astrobiology*, **7**, 66–84 (2007)

Reynolds, R.T., MacKay, C.P. and Kasting, J.F., 'Europa, tidally heated oceans, and habitable zones around giant planets', *Adv. Space Res.*, **7**, 125 (1987)

Richardson, L.J., Harrington, J. and Seager, S., et al., 'A spitzer infrared radius for the transiting extrasolar planet HD 209458b', *ApJ*, **649**, 1043–1047 (2006)

Rosing, M.T., 'Thermodynamics of life on the planetary scale', *Int. J. Astrobiol.*, **4**, 9–11 (2005)

Selsis, F, 'Review: physics of planets I: Darwin and the atmospheres of terrestrial planets, in *ESA SP-451: Darwin and Astronomy: the Infrared Space Interferometer*, 133 (2000)

Selsis, F., Despois, D., Parisot, J-P., 'Signature of life on exoplanets: Can Darwin produce false positive detections?', *A & A*, **388**, 985–1003 (2002)

Selsis, F., 'The atmosphere of terrestrial exoplanets: detection and characterization' in *ASP Conf. Ser. 321: Extrasolar Planets: Today and Tomorrow*, 170 (2004)

Selsis, F., 'Planetary evaporation', in *Formation planétaire et exoplanètes*, (eds.) Halbwachs, J.-L., Egret, D. and Hameury, J.-M, Strasbourg: Observatoire astronomique de Strasbourg and SF2A, 271–306 (http://astro.u-strasbg.fr/goutelas/g2005 – in French (2006)

Showman, A.P. and Guillot, T., 'Atmospheric circulation and tides of "51 Pegasus b-like planets"', *Astron. Astrophys.*, **385**, 166–180 (2002)

Sotin, C., Grasset, O., Mocquet, A., 'Mass radius curve for extrasolar Earth-like planets and ocean planets', *Icarus*, **191**, 337–351 (2007)

Tsinganos, K., Gomes, R. and Morbidelli, A., et al. 'Origin of the orbital architecture of the giant planets of the Solar System', *Nature*, **435**, 459–461 (2005)

Vidal-Madjar, A., Lecavelier des Etangs, A. and Désert, J.-M., et al., 'An extended upper atmosphere around the extrasolar planet HD209458b', *Nature*, **422**, 143–146 (2003)

Vidal-Madjar, A., Désert, J.-M. and Lecavelier des Etangs, A., et al., 'Detection of oxygen and carbon in the hydrodynamically escaping atmosphere of the extrasolar planet HD 209458b', *ApJ*, **604**, L69–L72 (2004)

Walker, J.C.G., Hays, P.B. and Kasting, J.F., 'A negative feedback mechanism for the long-term stabilization of Earth's surface temperature', *J. Geophys. Res.*, **86**, 9776 (1981)

Williams, D.M., Kasting, J.F. and Wade, R.A., 'Habitable moons around extrasolar giant planets', *Nature*, **385**, 234–236 (1997)

Williams, P.K.G., Charbonneau, D. and Cooper, C.S., et al., 'Resolving the surfaces of extrasolar planets with secondary eclipse light curves', *ApJ*, **649**, 1020–1027 (2006)

Zahnle, K., Arndt, N., Cockell, C., Halliday, A., Nisbet, E., Selsis, F., Sleep, N.H., 'Emergence of a habitable planet', *Spa. Sci. Rev.*, **129**, 35–78 (2007)

# Chapter 8
# Present and Future Instrumental Projects

The whole range of instrumental projects to observe and describe extrasolar planets fits within the framework of an overall scheme, whose different stages are clearly established:

- investigation and statistical analysis of giant planets: this objective may be met by use of indirect methods of observation from the ground (radial-velocity measurements and observation of transits);
- investigation and statistical analysis of terrestrial planets: this objective may be met by the use of indirect methods, primarily gravitational lensing and observation of transits from space and, in cases where the planets are very close to their parent stars, by high-precision radial-velocity measurements;
- the spectral analysis of giant planets and terrestrial planets to determine the atmospheric composition and possibly to study their habitability: this objective may mainly be tackled by use of direct methods of observation and by spectroscopic analysis.

The aim of this chapter is to describe most of the instrumental projects that are current or likely to be implemented in the next couple of decades. This chapter is thus complementary to Chap. 2, which describes the principles and instrumental concepts already implemented. Rather than just describing the principles and techniques of detection, here we lay stress on the experimental implementation of these techniques, the main difficulties, and the solutions advanced to deal with these and to obtain the necessary accuracy. The list of projects described is by no means exhaustive, nor minimal, but is in any case indicative of the different types of projects that are currently being considered:

- at the time of writing, some projects have not yet been completely defined, or have not yet emerged onto the instrumental landscape;
- certain instruments and concepts have not yet been financed, nor incorporated into the plans of the appropriate institutes or agencies, but already occupy a significant place in the overall view of the next few years, and in the minds of the astrophysical community.

Table 8.1 lists the different classes of projects with their objectives, as well as some projects representative of each class.

M. Ollivier et al., *Planetary Systems*. Astronomy and Astrophysics Library,
DOI 978-3-540-75748-1_8, © Springer-Verlag Berlin Heidelberg 2009

**Table 8.1** The various classes of instrumental projects and their scientific aims, and various representative projects for each class

| Type of project | Objectives | Specific projects |
|---|---|---|
| Doppler velocimetry | Detection of giant planets and (marginally) terrestrial ones | See Sect. 8.1.1 |
| Astrometry from the ground | Detection of long-period giant planets | PRIMA, VLT-I, LBT |
| Astrometry from space | Detection of long-period, giant and terrestrial planets | GAIA, SIM |
| Transits from the ground | Detection of short-period giant planets | See Sect. 8.1.3.1 |
| Transits from space | Detection and analysis of short-period, giant and terrestrial planets | CoRoT, Kepler, Plato, Spitzer, JWST |
| Gravitational micro-lensing | Detection of terrestrial planets | Planet, MPF, Euclid |
| Ground-based imagery | Observation of young objects or giant planets around cool stars | NAOS-CONICA, SPHERE, ELT, Carlina |
| Imagery from space | Spectral analysis of giant and terrestrial planets | Spitzer, JWST, TPF-C, Gemini, Keck |
| Ground-based interferometry | Investigation and spectroscopic analysis of giant and terrestrial planets | VLT-I, LBT-I |
| Space interferometry | Investigation and spectroscopic analysis of giant and terrestrial planets | Darwin, TPF-I ALADDIN, GENIE |
| Radio astronomy | Detection of auroral emission from giant planets | UTR-2, LOFAR, SKA |

Finally, and bearing in mind the way in which the technical teams react, certain projects may come to be completely redefined for various reasons (either technical or financial). The GAIA mission (an astrometric mission) was initially an interferometer project and, after several years, was altered to a more classical concept.

## 8.1 Indirect Methods of Detection

Indirect methods of detection include:

### 8.1.1 Velocimetry

Velocimetry, which has enabled the detection of most of the known exoplanets, should continue to provide a series of results in future, all the more so, given that the technique has not yet been pushed to its limits. In addition, long-term monitoring of stars is essential to detect long-period planets. There are about ten current instruments or instrumental projects to detect planets by the radial-velocity method (cf. Chap. 2). Although each instrument is based on specific technical choices and consequently achieves different results, there are certain similarities between all the instrumental concepts. Before discussing the principal projects more specifically, let us consider how, in practice, one measures radial velocities.

The radial velocity of an object is measured from the Doppler shift of its spectral lines. These lines are all shifted towards the blue (short wavelengths) when the object is approaching the observer and shifted towards the red (long wavelengths) when it is receding. The formal description of this effect in given in Chap. 2. In practice, the shifts are extremely small, and it is for this reason that it is necessary to have many spectral lines to detect the overall effect on the whole spectrum. This equally supposes that any stellar variability is not too significant, which is why this technique is difficult to apply to hot stars. In the case of the ELODIE instrument at the Observatoire de Haute Provence (OHP), the shift of the lines to be detected is about 0.002 pixel on the CCD.

Measurement of radial velocities therefore implies being able to measure accurately the spectrum of the source and to compare the wavelengths of emission lines and their evolution over time, with a reference spectrum that is believed to be stable and invariant over time. In practice, the instrument capable of this operation is a high-resolution spectrograph (typically with a spectral resolution of several tens of thousands), linked to a spectral reference (Fig. 8.1). The light from the star is captured at the focus of the telescope by an optical fibre, and then sent to the spectrograph. The latter also receives the flux from the reference source through another optical fibre, such that the spectrum of the star is recorded simultaneously with that from the reference source.

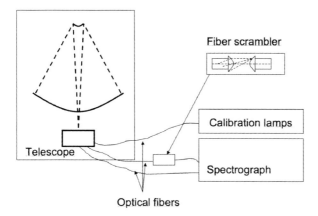

**Fig. 8.1** Schematic diagram of a spectrograph for measuring radial velocities (see the main text for a description of the principles involved)

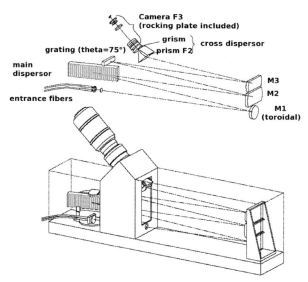

**Fig. 8.2** A schematic diagrams of the optics of the ELODIE spectrograph. The spectral dispersion is produced by the diffraction grating, while the orders are superimposed on the CCD by a grism (grating-prism) giving cross dispersion (After Baranne et al., 1996)

To obtain such high spectral resolutions, the spectrograph consists of a diffraction grating and a crossed-dispersion system which allows the various orders from the grating to be superimposed on a CCD. (Sometimes several tens of orders are involved.) A schematic diagram of the principles behind the ELODIE instrument is given in Fig. 8.2.

The reference source is generally one of the two following types:

- a cell containing iodine vapour,
- a thorium-argon spectral lamp.

The spectrum of these two references is shown in Fig. 8.3. These two reference sources are used independently, or sometimes, as with the AFOE spectrograph, simultaneously, to guarantee stability of the reference spectrum in the short term (with the spectral lamp) and in the long term (with the iodine cell).

The advantage of the spectral lamp over the iodine cell is that the emission spectrum covers a wider spectral range.

The level of accuracy required to measure the Doppler shift efficiently in the radial-velocity method requires perfect stabilization of the instrument during the measurement, and the possibility of being able to guarantee effective calibration from one measurement to the next.

In practice, spectrographs dedicated to the measurement of radial velocities are both thermally and mechanically controlled. (The spectrograph is placed in a temperature-controlled enclosure, generally in a vacuum to avoid distortions caused by variations in atmospheric pressure.) One of the keys to stability also lies in the fact that the conditions determining the illumination of the entrance pupil of the

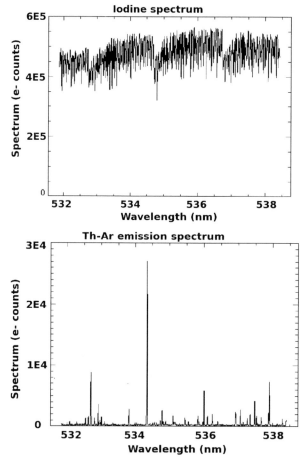

**Fig. 8.3** *Left:* reference spectrum of iodine vapour; the structure of the spectrum corresponds to the spectral lines and not to noise. *Right:* emission spectrum of a thorium-argon lamp

spectrograph should be as stable as possible, and in any case, independent of the telescope's guidance, and of atmospheric turbulence (and thus of the position of the image spot). This stability is obtained by an image scrambler on each of the optical fibres in the instrument (Fig. 8.1).

The highest precision available with current instruments (with the HARPS instrument on ESO's 3.6 m telescope at La Silla) is about $0.8\,\mathrm{m.s^{-1}}$. Stellar noise (movement of the stellar surface) is generally regarded as the ultimate limit of this method. The temporal frequencies at which the effect appears are, however, very different from those at which extrasolar planets are detected. So it is not unreasonable to think that by increasing the stability of spectrographs still further, it may be possible, in the near future, to obtain accuracies of about $10\,\mathrm{cm.s^{-1}}$, and thus detect large terrestrial-type exoplanets with this method (cf. Chap. 2). Finally, the size of the telescope itself is involved in evaluating the stability of measurements of the light flux that reaches the spectrograph.

Processing of the spectra is carried out by complex programs which take account of the various perturbing factors, and which subtract the Earth's motions from the radial-velocity measurements (cf. Chap. 6).

Table 8.2 lists the principal spectrographs currently in use or shortly to be commissioned. Each of the teams has drawn up a list of sample stars which it will undertake to monitor for radial velocities. The choice of the sample sets, more-or-less selectively chosen, meet certain criteria (proximity of parent stars to Earth, spectral type, magnitude, metallicity, etc.). It is estimated that several thousands of stars are followed by the different teams. Some stars are included in several samples. We therefore have data from several instruments, all with different limitations, which allows comparisons to be made.

The Keck Exoplanet Tracker (KET) is a new type of instrument, allowing relative radial velocity measurements. Instead of classical high resolution spectroscopy coupled with reference spectrum correlation, the KET is based on a fibre-fed, dispersed, fixed-delay interferometer. This instrument is a wide-angle Michelson interferometer followed by a medium-resolution spectrometer (R = 7000). The interferometer creates an interference pattern within each spectral channel, which allows the measurement of phase shift. With an accuracy of about 20 m/s on 8th magnitude stars, this low-cost instrument is designed for all-sky surveys.

## 8.1.2 Astrometry

Historically, astrometry was the first method used to try to detect exoplanets, but without success. A new generation of projects, both ground-based and (in particular) in space should enable the first detections by this method to be obtained.

### 8.1.2.1 Ground-Based Astrometric Instruments

PRIMA-VLTI

The PRIMA (Phase Reference Imaging and Microarcsecond Astrometry) instrument on the VLTI is the last VLTI first-generation instrument (along with VINCI, AMBER, and MIDI). It is due to be commissioned in 2008. The acronym actually describes an instrument with two very different functions:

- PRI (Phase Reference Imaging): this is, in effect, the recombiner plus two telescopes, transforming the VLTI into an imaging instrument;
- MA (Microarcsecond Astrometry): the astrometric instrument proper.

PRIMA has been greatly delayed, because its commissioning depends on the availability of a fast fringe sensor unit.

**Table 8.2** Principal spectrographs in service or being developed for velocimetry

| Instrument / commission date | Telescope | Spectral range | Spectral resolution | Reference spectrum | Radial-velocity accuracy | Reference / project web site |
|---|---|---|---|---|---|---|
| AFOE / 1992 | 1.5 m Whipple Obsy. (Arizona) | 392–664 nm | 56 000 | Thorium-Argon Lamp | | 1 |
| UCLES / 1998 | AAT 3.9 m | | | Iodine cell | | 2 |
| ELODIE / 1993 | 1.93 m OHP (France) | 390.6–681.1 nm | 42 000 | Thorium-Argon Lamp | 15 m.s$^{-1}$ at m$_V$=9 in 30 min | 3 |
| CORALIE | 1.2 m Euler (La Silla) | 390.6–681.1 nm | 42 000 | Thorium-Argon Lamp | 15 m.s$^{-1}$ | 4 |
| HARPS / 2002 | 3.6 m La Silla (ESO) | 380–690 nm | 90 000 | Thorium-Argon Lamp | < 1 m.s$^{-1}$ (long term) | 5 |
| SOPHIE / 2006 | 1.93 m at Obs. Haute Provence | 387.2–694.3 nm | 70 000 | Thorium- Argon Lamp | ~ 2 m.s$^{-1}$ | 6 |
| UVES | 8.2 m | 300–1100 nm | 80 000 | Iodine cell | | 7 |
| HRS / HET | Hobby Eberly telescope | 420–1100 nm | 120 000 | | < 10 m.s$^{-1}$ | 8 |
| California and Carnegie Planet Search | Keck – Hawaii | | | | | 9 |

References
1. http://cfa-www.harvard.edu/afoe/espd.html
2. http://www.aao.gov.au/local/www/cgt/planet/aat.html
3. Baranne et al., 1996. A&ASS, 119, 373–390
4. http://obswww.unige.ch/~udry/planet/coralie.html
5. http://obswww.unige.ch/Instruments/harps/Welcome.html
6. http://www.obs-hp.fr/www/guide/sophie/sophie-eng.html
7. http://www.eso.org/instruments/uves/
8. http://www.as.utexas.edu/mcdonald/het/instruments.html
9. http://exoplanets.org/

The concept has been revitalized by the addition of Swiss teams to the instrumental consortium, and should allow astrometric measurements to be carried out with an ultimate accuracy of about 70 µas. Such accuracy does not, however, allow the observation of anything other than giant planets far from their parent stars. This method is therefore an excellent complement to the radial-velocity method, which requires extremely long-duration tracking to detect planets in distant orbits.

## LBT

By virtue of the large equivalent aperture (a mirror 20 m in diameter), the LBT (Large Binocular Telescope) in interferometer-imaging mode (Fizeau mode given by the LINC NIRVANA instrument, see Sect. 8.2.2) allows measurement of the positions of targets with an accuracy of about 100 marcsec. Such accuracy is enough to allow the observation of giant planets. The LBT saw first light in 2005, and the equipment for the interferometer mode should shortly be available.

### 8.1.2.2 Space-Borne Astrometric Instruments

GAIA

The GAIA (Global Astrometric Interferometer[1] for Astrophysics, Fig. 8.4) is an astrometric mission planned by ESA for 2011, which has numerous astrophysical objectives, primarily concerned with understanding the structure and dynamics of the Galaxy (the Milky Way). In particular, GAIA is expected to measure the positions and motions of several tens of millions of stars to a high degree of accuracy.

Possessing from the very start both a catalogue of about 1000 million stars, with 340 000 down to V magnitude 10, and 26 million to V magnitude 15, and with an astrometric accuracy of 4 µas at V magnitude 10, and 11 µas at V magnitude 15, GAIA is an instrument that is perfectly suitable for searching for giant planets. Its sensitivity and its accuracy should enable it to detect a large number of objects down to the mass of Uranus. In contrast to the radial-velocity method, which is most sensitive to massive objects close to the parent star, the astrometric method is all the more sensitive, the more massive the object and the farther from the star (cf. Chap. 2). However, the one major constraint is that the observations must be carried out during a significant portion of the planet's orbit for the astrometric variations to be detected. In practice, and considering the duration of the mission (5 years minimum, of which 4 will be devoted to observation, and with the possibility of the mission being extended for 6 more years), the mission should detect objects with periods less than 30 years (one-third of the period being observed during the

---

[1] As initially conceived, the GAIA mission was a Fizeau interferometer and was included in ESA's Horizon 2000+ programme, under the category of 'space interferometer'. A revision of the mission's objectives meant that it was no longer an interferometer, because the desired aims could be achieved with a classical form of telescope. Nevertheless, the name of the mission was retained.

**Fig. 8.4** Artist's impression of the GAIA mission (image credit: courtesy ESA)

satellite's lifetime). Relative to its predecessor, the Hipparcos mission, GAIA should improve the accuracy of measurements by a factor of more than 100.

From the instrumental point of view (Fig. 8.5), GAIA consists of two identical, astrometric telescopes pointing in two directions, separated by 106°. Each telescope

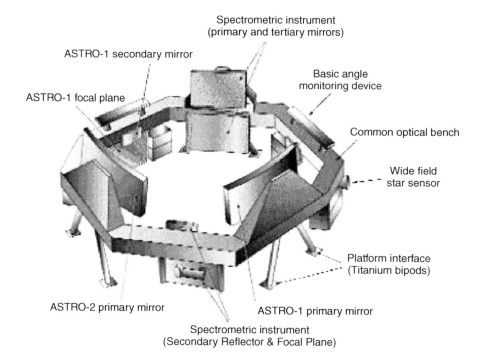

**Fig. 8.5** Schematic diagram of the instrumentation on board GAIA (ESA)

has a focal length of 50 m, a field of 0.32 square degrees, and observes in the spectral range of 300–1000 nm. GAIA is also equipped with a spectrometer having a field of view of 4 square degrees and a focal length of 4.17 m. It provides radial-velocity measurements with an accuracy of about 1–10 km.s$^{-1}$ of objects of V magnitude 16–17, and also spectrophotometry.

The instrument functions in spin mode (rotation around a satellite axis with a rotation rate of 120 arcsec/s), the axis of which undergoes precession (the precession period is 76 days), thus allowing gradual coverage of the whole sky during the mission.

GAIA will be placed in a Lissajous orbit around the L2 Lagrangian point by an Ariane 5 launcher.


SIM

In its initial version SIM (Space Interferometry Mission) was the American equivalent of GAIA. Unlike GAIA, SIM was an astrometric interferometer which would function in targeted mode, capable of observing a far more limited number of objects (several thousand at most) but with a much better accuracy than GAIA (intended accuracy: 1 μas). Such an accuracy would have allowed the detection of giant planets but was not sufficient to detect Earth-like planets at distances comparable to Earth's from the parent star. The astrometric performance proposed for SIM would be obtained by the use of an astrometric interferometer, in conjunction with two guide interferometers. The principal difficulty with the mission was the necessity of measuring and permanently controlling the geometry of the astrometric interferometer to an accuracy of a few tens of picometers over the baseline.

Because of funding difficulties, the initial SIM program was postponed *sine die* in 2006.

The SIM-LITE (Fig. 8.6) concept proposed in 2007, is a revision of initial SM concept that is devoted, almost exclusively to the detection of extrasolar planets.

The SIM-LITE concept is a response to one of the recommendations of the American roadmap towards direct detection and description of terrestrial-type extrasolar planets. The concept behind the instrument was revised to reduce the number of targets to a few tens of nearby stars but also to increase the accuracy of the astrometric mode to one tenth of microarcsecond. Such an accuracy clearly permits the detection of Earthlike planets at orbital distances comparable to that of the Earth – the astrometric signal equals 3 μas for an Earth at 1 AU at 10 pc – and the identification of the targets for future missions determining characteristics directly. At the same time, the reduction of initial costs was possible by limiting the instrument's capability to a narrow field, and thus reducing the interferometric loops.

The technology required to reach the sub-microarcsecond scales (picometer metrology mentioned above) is already available. At present, the SIM team is attempting to find international partners to warrant funding of the project.

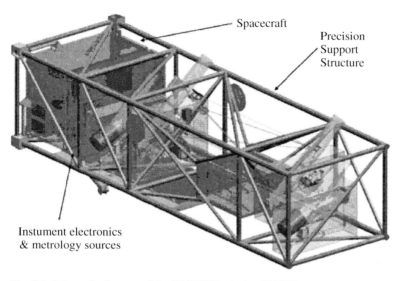

Spacecraft

Precision
Support
Structure

Instument electronics
& metrology sources

**Fig. 8.6** Schematic diagram of the SIM-LITE mission (NASA)

Other Observatories

As mentioned in Chap. 2, radial-velocity measurements lead only to a rough estima-
tion of the mass of the candidate because of the lack of information on the inclination
of the system (one measures M.sin(i)). To estimate the value of sin(i) and discrim-
inate between a planetary candidate with a small inclination and a brown dwarf or
a low mass star with a high inclination, astrometric data, when available, are often
used.

The Hipparcos mission data have a mean accuracy of several milliarcseconds.
This is not sufficient to detect a planetary candidate but it does allow the detection
of low-mass stars and heavy brown dwarfs.

The HST Fine Guidance Sensor has also been used for astrometry and the determination of characteristics of planetary systems (specifically, the inclination). Combined with Hipparcos data and with radial velocity measurements, it is a powerful
tool to differentiate between stars and planets.

## 8.1.3 The Study of Planetary Transits

### 8.1.3.1 Ground-Based Studies

Searching for transits of giant exoplanets from the ground is, in principle, a relatively easy exercise, because it does not have very stringent instrumental requirements. The amplitude of a transit of a giant planet is at the per cent level, so it is

necessary to be able to obtain photometric curves to an accuracy better than one per cent (see Sect. 2.2). To achieve this end, a telescope of a few tens of centimetres in diameter and a good CCD are all that are necessary. This is why some twenty-odd teams are conducting such programmes. The principal features of the projects that are current or in development are gathered together in Table 8.3. (Projects that have now ceased are not included in the table.)

It is interesting to note the significant differences between the methods employed to detect transits, and the few results obtain to date (5 planets identified by OGLE, 1 by STARE (TrES-1) and 3 planets detected by radial velocities, confirmed by transits). This is despite estimates that were clearly optimistic. The discrepancy between predictions and results may be explained in several ways:

- the number of light-curves that may be effectively used in any stellar field is clearly more limited than previously estimated;
- the limitations on continuous coverage (because of daylight, weather or clouds), enormously reduced the discovery effectiveness;
- uncorrected, remaining bias and systematic effects on the light-curves (correlated noise) which prevent the identification of transits.

This is why later transit-search programmes tend to favour the quality of the observing site and of the photometry, as well as the coverage over time, as being able to optimize the return from the observational programme. Let us mention the A-STEP project, for example, which aims to install a 40-cm telescope in Antarctica to benefit from the good photometric conditions and continuous coverage over a period amounting to several tens of days (during the southern polar night).

### 8.1.3.2  Observations From Space

Space allows us to avoid atmospheric effects, especially, as far as photometry is concerned, the detrimental effect of scintillation. Access to space is therefore indispensable for particularly stable photometry.

### CoRoT

The CoRoT space mission (Convection, Rotation, and Transits, Fig. 8.7) is a small mission under the leadership of the Centre national d'Etudes Spatiales (CNES: the French space agency), the major contribution being French, but achieved in cooperation with Austria, Belgium, Brazil, Germany, and Spain, as well as Europe, through the European Space Agency (ESA). The objectives of this mission cover two main themes:

- asteroseismology (the study of the interior of stars by observation of their pulsations, Baglin et al., 2006)
- the search for exoplanets from space by the transit method.

**Table 8.3** Principal features of ground-based programmes searching for giant planets by the transit method

| Project | Location | Telescope | Field | No. targets | Limiting magnitude | Results | Status |
|---|---|---|---|---|---|---|---|
| APT | Siding Spring (Australia) | 80 cm | $2° \times 3°$ | | 13 | 0 | Current |
| ASAS-3 | Kitt Peak (Arizona) | 7.1 cm | $8.8° \times 8.8°$ | 8000 | 14 | 0 | Current |
| ASP | | 0.9 m then 1.3 m | $2 \times 59' \times 59'$ | ~5600 | 18.5 | 0 | Current |
| A STEP | Antarctica | 0.4–0.45 m | 1–2 degrees square | A few thousand | | | Under development |
| BEST | Germany | 20 cm | $3.1° \times 3.1°$ | 30 000 | 13 | 0 | Current |
| GITPO | Israel | 1 m | $27–30°$ | | | | Under development |
| HATN (ex HAT-1) | Mt Hopkins (USA) | 6.4 cm | $9° \times 9°$ | 20 000 | 13 | 7 planets detected | Current |
| MONET North | Mt Locke (Texas) | 1.2 m | $11.3' \times 11.3'$ | | | | Current |
| MONET South | Sutherland (S. Africa) | 1.2 m | $11.3' \times 11.3'$ | | | | Current |
| OGLE III | Las Campañas (Chile) | 1.3 m | $35' \times 35'$ | | | 7 planets detected | Current |
| PASS North | | $15 \times 5$ cm | The whole northern sky > 30° | | | | Current |
| PASS South | | $15 \times 5$ cm | The whole southern sky > 30° | | | | Current |
| STARE | Tenerife (Canaries) | 10 cm | 6.1 degrees square | 25 000 | | 1 transit detected | Current |
| STELLA | Tenerife (Canaries) | 1.2 m | $22' \times 22'$ | | | | Under development |

**Table 8.3** (continued)

| Project | Location | Telescope | Field | No. targets | Limiting magnitude | Results | Status |
|---|---|---|---|---|---|---|---|
| SuperWASP North | La Palma (Canaries) | 8 × 20 mm | 8 × 7.8° ×7.8° | 400 000 | 13 | 15 planets detected | Current |
| SuperWASP South | South Africa | 8 × 20 mm | 8 × 7.8° ×7.8° | 400 000 | 13 | | Under development |
| STEPPS | Kitt Peak (Arizona) | 2.4 m et 1.3 m | 25' × 25' and 43' × 43' | | | 0 | Current |
| TAPT | Arizona | 80 cm | | | | 0 | Current |
| TrES | California, Arizona + STARE | 10 cm | | | | 4 planets found | Current |
| WHAT | Tel Aviv (Israel) | 20 cm | 8.2° × 8.2° | | | 0 | Current |
| Vulcain | | 4.5 cm | 7° × 7° | 6000 | 13 | 0 | Current |
| XO | Haleakala (Hawaii, USA) | 2 Canon EF200 lenses at f/1.8 | 7.2 degrees square | | | 3 transits detected | Current |

© CNES - Mai 2004/Illus. D. Ducros

**Fig. 8.7** Artist's impression of the CoRoT space mission (CNES)

As far as instruments are concerned, CoRoT consists of an afocal off-axis telescope of about f/4, with two parabolic mirrors having a collecting surface equivalent to that of a telescope 27 cm in diameter. This images a star field 7 square degrees in area, half of which (for the exoplanet survey) is imaged by 2 CCDs, with $2048 \times 2048$ pixels, cooled passively to $-40°C$, and working in photon-counting mode in the visible and near infrared (passband 400–900 nm).

CoRoT was launched on 27 December 2006 by a Soyuz rocket, and placed in a polar orbit at an altitude 896 km. Such an orbit allows the same field to be observed continuously for a maximum of 5 months, after which a rotation by 180° is necessary to avoid being blinded by the Sun. Another field is then observed for the next 5 months. Between long-duration observations, small, shorter programmes are carried out. Taking the chosen orbit and the constraints imposed by reversing the satellite because of the relative positions of the Earth and Sun into account, CoRoT will observe in two principal directions ('CoRoT's eyes') directed towards the galactic centre and the anticentre (Fig. 8.8).

CoRoT should be able to observe 12 000 targets simultaneously, with magnitudes between 11 and 16 per field for 5 months (150 days), and at least 5 fields during the mission's overall duration (about 3 years), making a total of more than 60 000 targets. The photometry of each star is measured every 8.5 min (512 s) with a photometric precision of a few times $10^{-4}$ in 1 h. Such accuracy should enable the detection of terrestrial-type objects a few Earth radii in size. Given the observational time for each field, and the necessity of detecting several transits to determine the period of the object, only 'hot' planets (close to their stars) are likely to be detected in sufficient quantity to be of use statistically. The first exoplanet was detected on 3 May 2007.

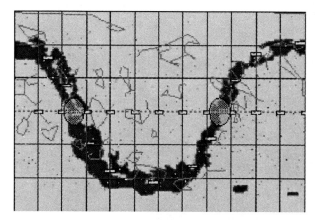

**Fig. 8.8** 'CoRoT's eyes': the regions of the sky towards the galactic centre (Right Ascension = 104.5°) and the anticentre (Right Ascension = 284.5°) which CoRoT will observe continuously for 5 months

The specific feature of the CoRoT instrument is that it provides photometry of 80 per cent of objects (the brightest) in three bands (Fig. 8.9). To do this, a system of prismatic dispersers is located between the objective and the focal plane. This gives each object an image in the form of a small spectrum. To avoid transferring the whole contents of a CCD each time it is read (every 32 s to avoid saturation), each CCD is partially read (over about 5–10 per cent of the surface), thanks to the use of photometric masks for each object (Fig. 8.9). These masks have a shape appropriate to each object. They are calculated by optimization of criteria based on a maximum signal-to-noise ratio and also taking the following into account:

- the magnitude of the object
- its temperature (its spectral type and thus its colour)
- its position on the CCD (the image spot for any one object varies across the field)
- the environment of each star (i.e., any possible contamination)

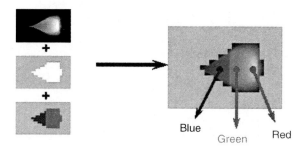

**Fig. 8.9** The principles involved in CoRoT's colour photometry. For each star, the image is a small spectrum (*top left*, where blue is to the left and red at the right), which is observed through a photometric mask (*centre left*), the three zones of which define three colours (*bottom left*)

The colour photometry of each object is obtained by selecting three distinct zones in the readout mask for the mini-spectra. So this is not spectral photometry in the classical sense of the term (with filters whose passband is known), but rather relative colour information about each object.

Colour information is of interest on two counts:

• It allows us to distinguish a stellar event from a planetary transit in the object's photometric curve. To a first approximation, a transit is achromatic (the relative variation in the luminosity is independent of the wavelength of observation), whereas a stellar fluctuation caused by a spot, a facula, or any other event is mainly chromatic. A sunspot, for example – where we can crudely model the effects such as a local cooling of the surface – is extremely chromatic: the relative variations in intensity vary by a factor of two between the blue and the red.

• It enables us to explore the image spot spatially. Bearing the resolution of the instrument in mind (the image spot covers about 300 arcsec$^2$), we should be able to distinguish a transit across a target star in an eclipsing binary if the components are spatially separate, by seeing if the effects are equally present in the three photometric channels.

After correction for the various sources of instrumental bias (sky background, electronic gain, periodic perturbations, fluctuations in the satellite's guidance, etc.) it should be possible to obtain a photometric curve where the residual noise is less than 1.4 times the photon noise of the target over a period of 8.5 min. To obtain such a performance, it is necessary to eliminate light scattered by the Earth. To do this, CoRoT's telescope has a shield that contains several optical baffles that block scattered light. The baffling reduction transmission of scattered light to less than $10^{-13}$.

The photometric curve, if planetary transits are present, shows regular extinctions over the period of the transit (Fig. 8.10). The data processing computers are able to identify transits and detect objects down to a size of about two Earth radii. For each candidate transit, it is possible to determine the period (and thus distance from the star), the duration and the intensity of the transit (and thus the size of the object). Several methods are used by the computers for detecting and analyzing the transits: variable-pitch Dirac comb, least-squares fit, morphological analysis, wavelet analy-

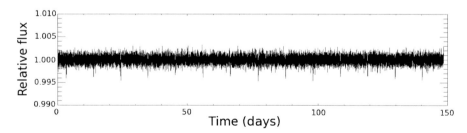

**Fig. 8.10** Photometric curve corrected for various errors for a star of solar type and of magnitude 14, representing the transit of an object 0.5 Jupiter diameters, with a period of 10.54 days

sis, etc. The results from the different data-processing algorithms for investigating transits primarily depend on:

- the depth of the transits
- their number
- the stability of the stellar photometry

The various algorithms are currently being evaluated in the CoRoT community.

Finally, we should also mention that a comprehensive, ground-based programme, following the transit candidates, will be necessary to confirm the planetary nature of the events. The follow-up programme, using various methods (high-angular-resolution photometry, and mainly radial-velocity measurements) will allow us to eliminate from the list of candidate transits 90 per cent of the alerts consisting of:

- transits of eclipsing binaries, of greater magnitude and in the background to the principal target, which would mimic a transit of the main target
- transits of grazing binaries where the principal target is an eclipsing binary.

CoRoT should enable us to identify several tens of hot Jupiters, or other objects with sizes comparable to Uranus and Neptune. Equally, CoRoT should enable us to carry out the first statistical analyses of objects whose size is equal to a few times that of the Earth, provided several tens of such objects are detected. An accurate estimate of the number of detections of this type of object is difficult to obtain, because their existence is not yet unequivocally confirmed.

**Fig. 8.11** Artist's impression of the Kepler mission (NASA)

KEPLER

The Kepler space mission (Fig. 8.11) being prepared by NASA is similar in principle to CoRoT. In this context it may be considered a second-generation mission relative to CoRoT, which should be seen as a pioneer in this technique. Kepler should be launched by a Delta II rocket in early 2009.

The principal differences between CoRoT and Kepler are:

- The duration of observations of single target: CoRoT observes the same field for 150 days, Kepler for 4 years. The aim of Kepler is to detect true terrestrial analogues, i.e., planets 1 AU from solar-type stars, orbiting in one Earth year. To detect and confirm the object and its period at least three transits are required: two to identify the period, and the third to confirm the nature of the event and the periodicity. The possibility of detecting long-period terrestrial-type planets (maximum period about two years), will allow – as CoRoT will do for short-period objects – a statistical analysis of the distribution in size, distance, and of the orbital parameters of the planets. CoRoT and Kepler are therefore, from a scientific point of view, the terrestrial-planet counterpart of the radial-velocity observational programmes for giant planets.

- The size of the telescope: Kepler has a telescope 95 cm in diameter (against 27 cm for COROT). The photometric curve's optimum signal-to-noise ratio (photon noise) is attained in a few minutes by Kepler, whereas CoRoT requires an hour's observation.

- The field of the telescope: Kepler observes a complete field of 105 square degrees in the constellations of Cygnus and Lyra, as against 3.5 square degrees for CoRoT. This large field allows Kepler to observe simultaneously about 100 000 targets on the Main Sequence between magnitudes 9 and 15. The same field is observed for the four years of the nominal mission. The major advantage that Kepler's large field offers is that it allows a choice of targets and avoids contamination by neighbouring objects (this is the same for CoRoT, which has to point close to the galactic plane to increase the number of objects in its significantly smaller field).

On the technical side, Kepler consists of a telescope (Fig. 8.12), whose focal plane it covered by 42 CCDs, each containing $2200 \times 1024$ pixels. The observational passband extends from 430 to 890 nm. Photometric performance, including overall noise, stellar variability, and photon noise, is better than $2 \times 10^{-5}$ for a star of magnitude 12.

The orbit chosen for Kepler is a heliocentric orbit drifting behind the Earth. This orbit allows the same field to be observed throughout the year, but still avoiding the relative positions of the Sun and Moon, which are the principal sources of parasitic light. Just like CoRoT, Kepler has a baffle that allows very little scattered light to pass.

Table 8.4 summarizes the results expected from Kepler as a function of the size of the objects, assuming that:

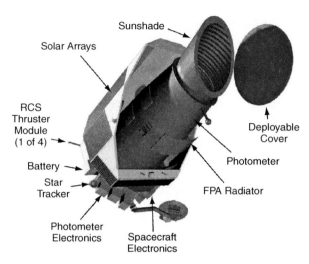

**Fig. 8.12** Functional diagram of Kepler (NASA)

**Table 8.4** Number of objects expected to be discovered by Kepler as a function of size

| Average size of planets | Number of objects detectable |
| --- | --- |
| 1 Earth radius | $\sim 50$ |
| 1.3 Earth radii | $\sim 185$ |
| 2.2 Earth radii | $\sim 640$ |

- large numbers of the objects concerned exist around the stars being observed (the probability of occurrence being several tens of per cent)
- the majority of objects are of the size specified (medium size)
- the orbital period of the objects is close to one year
- observation takes place over 4 years, with 4 transits detected
- theories about stellar variability and detection criteria are conservative.

The problems facing identification of transits in photometric curves obtained by Kepler are identical to those encountered by CoRoT. Kepler needs the same algorithms to correct for instrumental bias, to search for transits, and similarly needs a comprehensive, ground-based photometric and radial-velocity programme to distinguish true planetary transits among the candidates from events that may have the same photometric signature (binaries with total eclipses, for example).

CoRoT and Kepler, combined with all the observations already made of radial velocities should enable us, in no more than 10 years' time, to have an extremely accurate picture of the statistical distribution of exoplanets, their sizes and their periods, around nearby stars. These missions should also provide statistical information concerning the multiplicity of such systems.

## PLATO

PLATO is a post-CoRoT and post-Kepler mission among the call for proposals for ESA's 2015–2025 (Cosmic Vision) programme. The main idea is that the formation and evolution of stars, their magnetic fields and their planets are intrinsically linked and should be studied simultaneously. PLATO therefore proposes:

- to select a large sample of stars of all masses and ages (the sample including more than 100 000 objects);
- to search for planets and define their characteristics by the transit method;
- to study the rotation and internal structure of the same stars by asteroseismology;
- detect, describe and map the magnetic fields of the same stars by UV monitoring and by tomography techniques.

Unlike CoRoT or Kepler, the idea for PLATO is to try to observe over the widest possible field, so as to be able to select not just the brightest objects. PLATO's field may therefore be more than $30° \times 30°$ (i.e., nearly 1000 square degrees!).

The basic features of PLATO are currently being defined, but observing such a vast field may be envisaged as consisting of a mosaic of small telescopes (10 cm in diameter), together covering the whole field, with a photometric performance comparable with that of a telescope 1 m in diameter.

## Other Space-Borne Observatories

Several space-borne observatories were used to study transiting planets.

The MOST mission (Microvariability and Oscillations of STars) is the first Canadian space observatory. It is a micro-satellite with a 15-cm telescope allowing high-accuracy photometry designed to detect stellar oscillations and the reflected light from transiting planets. The observation is performed thanks to an array of Fabry lenses which project a large, stable image of the telescope pupil. The photometry is obtained by measuring the stellar flux within the telescope's pupil. Because of stray light problems, the accuracy of MOST did not permit a definitive conclusion about the reflected light from HD 209458 b.

Even though the Hubble Space Telescope (HST) was not particularly designed to the search for planets by the measurement of transit dimming, it was the first space facility to give the lightcurve of a transiting system (HD 209458 b, Brown et al. 2001; see references for Chap. 2). HST instruments have been used for many observations, particularly the Wide Field Planetary Camera, the STIS and the NIC-MOS instruments. HST was used to perform high-accuracy photometry of a globular cluster (47 Tucanae) but did not detect any transits there. HST was also used in the detection of sodium in the atmosphere of HD 209458 b (Charbonneau et al. 2002) and in obtaining an accurate determination of the chromatic diameter of this object, enhancing the role of the atmosphere.

The Spitzer observatory (see Sect. 8.2.1.2) and particularly the IRS spectrograph, is currently used to observe secondary transits of transiting stars. Comparison with models of hot planets allowed the announcement of the detection of several gaseous species, such as $H_2O$ and $CH_4$, in the atmospheres of HD 209458 b and / or HD 189733 b. It is currently one of the productive tools in determining the characteristics of the atmospheres of hot planets, because the observation of a secondary transit allows differentiation between the stellar and the planetary flux (this is thus a direct detection of the atmosphere), allowing direct detection of molecular species. These kinds of observation should be performed at a higher spectral resolution using the JWST (see Sect. 9.2.1.2).

## 8.1.4 Searching for Microlensing Events

### 8.1.4.1 Observation from the Ground

Searching for planets by means of gravitational microlensing events (see Sect. 2.2) requires monitoring gravitational amplification events photometrically as accurately as possible, with sampling every half-hour. Given the time required to obtain accurate photometry of objects, the programme to monitor events requires the provision of photometric alerts so that efforts are concentrated only on events that are in progress. Initially, gravitational amplification events were used to investigate dark matter in the universe, which was believed to be in the form of cold objects, of low mass and dark. Several teams (EROS in Europe, DUO, MACHO and OGLE in the USA) regularly provided photometric alerts. Currently two teams are still providing details of microlensing events (the OGLE team, which is equally interested in studying planets through transits, and which has discovered several candidates, and MOA). Accurate photometric monitoring of these events reveals photometric artefacts caused by the presence of planets.

The PLANET Collaboration

Photometric monitoring of microlensing events requires the greatest observational continuity possible. In the PLANET (Probing Lensing Anomaly NETwork) collaboration, several telescopes around the world (Table 8.5), together with open-access photometry software are being used to detect 'classical' gravitational amplification anomalies (photometric curves with Gaussian distributions). The location of instruments around the world is such that, in principle, an event may be monitored practically continuously (subject to the weather). Given the duration of artefacts created by passing across (or near to) a planetary caustic (typically a few hours at most), this continuous monitoring is necessary.

So far, the PLANET collaboration (Beaulieu et al., 2006) has detected one object whose mass is about 5.5 Earth masses. This is one of the least massive exoplanets yet detected. Several other events are currently being analyzed (cf. the collaboration's web page: http://planet.iap.fr).

**Table 8.5** List of telescopes used by the PLANET collaboration to monitor microlensing events (listed by site)

| Location | Longitude | Latitude | Diameter of telescope |
|---|---|---|---|
| Perth (Australia) | 116°8' | −32° | 0.6 m |
| Hobart (Tasmania) | 147°32' | −43° | 1 m |
| Sutherland (S. Africa) | 20°49' | −32° | 1 m |
| Bloemfontein (S. Africa) | 26°43' | −29° | 1.5 m |
| La Silla (Chile) | 289°16' | −29°18' | 1.54 and 2.2 m |

### The MICROlensing Follow-up Network (MICROFUN)

The MICROFUN collaboration is another network dedicated to the follow-up of microlensing events. Like PLANET, MICROFUN uses microlensing events announced by the OGLE and MOA consortia, and provides accurate and, if possible, continuous light-curves of on-going events. The instruments used by MICROFUN are listed in Table 8.6.

**Table 8.6** List of telescopes and sites used by the MICROFUN collaboration to monitor microlensing events (from www.astronomy.ohio-state.edu/~microfun/)

| Location | Longitude | Latitude | Telescope |
|---|---|---|---|
| Pakuranga (New Zealand) | 174°53'37″ E | 36°53'37″ S | 0.35 m f/10 Meade |
| Auckland (New Zealand) | 174°46'37″ E | 36°54'22″ S | 0.4 m f/10 Meade |
| Kumeu (New Zealand) | 174°31'29″ E | 36°48'23″ S | 0.35 m f/6.3 Celestron |
| Blenheim (New Zealand) | 173°50'21″ E | 41°29'30″ S | 0.4 m f/15 |
| Perth (Australia) | | | 0.25 m |
| Canberra (Australia) | 149°06'36″ E | 35°09'45″ S | 0.35 m f/10 Meade |
| Nelson (New Zealand) | 117°53'29″ E | 38°37'26″ E | |
| Patutahi (New Zealand) | 117°53'29″ E | 38°37'26″ E | 0.4 m f/5.2 Meade |
| Mitzpe Ramon (Israel) | 34°45'44″ E | 30°35'50″ N | 1 m f/10 |
| Pretoria (South Africa) | 28°26'18″ E | 25°24'32″ S | 0.35 m f/8 Meade |
| San Pedro de Atacama (Chile) | 68°10'49″ W | 22°57'10″ S | 0.5 m f/3 |
| Cerro Tololo (Chile) | 70°48'17″ W | 30°10'06″ S | 1.3 m f/13 |
| Hereford (Arizona, USA) | 110°14'15″ W | 31°27'08″ N | 0.35 m f/10 Meade |
| Mount Lemmon (Arizona, USA) | 110°47'20″ W | 32°26'35″ N | 1 m f/16 or f/45 |
| Kitt Peak (Arizona, USA) | 111°36'56″ W | 31°57'05″ N | 2.4 m f/7.4 or 1.3 m f/7.6 |
| Mount Palomar (California, USA) | 116°51'36″ W | 33°21'26″ N | 60-inch |
| Faaa (Tahiti, French Polynesia) | 149°35'15″ W | 17°33'04″ S | 11-inch f/10 Celestron |

The MICROFUN collaboration has announced 3 planetary candidates: OGLE-2005-BLG-071, OGLE-2005-BLG-169 and OGLE-2006-BLG-109 (www.astronomy.ohio-state.edu/~microfun/).

### 8.1.4.2  Observation from Space

The Microlensing Planet Finder Mission (MPF)

The MPF mission concept (Fig. 8.13) is an American project dedicated to the search for extrasolar planets by the microlensing method. The efficiency of the microlensing method, and thus the number of planetary candidates, depends strongly on both source and lens stellar-field density. The microlensing effect occurs specifically when source and lens are aligned. Detection from the ground is thus limited by the angular resolution achievable, taking into account the effects of atmospheric turbulence (about 1 arcsec). Planets are effectively detectable when they create a high-magnification event (when the planet is close to the Einstein ring i.e. 2–3 AU from the star).

Compared to ground-based observations, space surveys allow a higher angular resolution (to the diffraction limit of the telescope), and the lack of atmospheric turbulence allows a better determination of photometric variations and thus a higher sensitivity (Fig. 8.14), particularly towards closer objects (with distances of less than 1 AU).

Technically speaking, the MPF (Fig. 8.13) concept is a 1.1-m TMA telescope (Three Mirror Anastigmat). This optical configuration enables observation of a field of 0.65 square degrees. The focal plane is filled by a 145 Mpixel focal-plane array made of a HgCdTe detector, allowing observation in three spectral

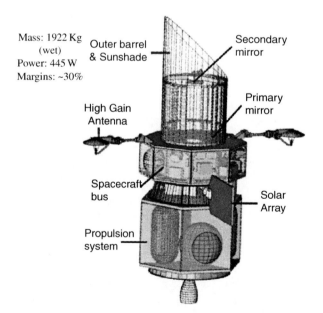

**Fig. 8.13** Schematic diagram of the Microlensing Planet Finder satellite (see the text for description) (after Bennett et al., 2007)

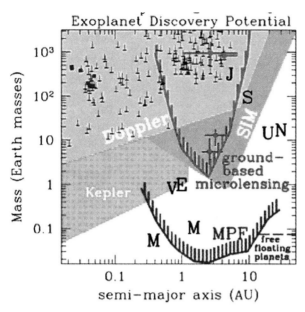

**Fig. 8.14** Sensitivity of space-based microlensing event detection compared with several other methods (ground-based microlensing, radial velocimetry, space-based astrometry and transit detection) (after Bennett et al., 2007)

bands from 600 to 1700 nm at an angular resolution of 0.24 arcsec per pixel. This design should lead to a photometric accuracy of better than 1 per cent at magnitude J=20.5.

Such performance will permit the detection of both planetary systems (star + planets) and free-floating objects. The sensitivity to each type of objects is given in Fig. 8.15.

The EUCLID Mission (Formerly Named DUNE)

The EUCLID mission is the result of the merger of the DUNE and SPACE mission concepts proposed to ESA in answer to its call for 'Cosmic Vision' proposals, and pre-selected for A-phase studies. The main goal of EUCLID is concerned with cosmology. The aim of EUCLID is to search for dark energy by gravitational shear. In its previous version (DUNE), EUCLID's method of observation mode was a complete survey of the sky including, thanks to the proposed observational strategy, a 3 month-survey of the galactic plane, which would have allowed a search for extrasolar planets by microlensing. The sensitivity of EUCLID in the detection of extrasolar planets is comparable with that of MPF. At present, a complete revision of the initial instrumental concept (DUNE) is under way, so there is little point in describing it in detail. Among visible and near-infrared photometric wide-field imaging and

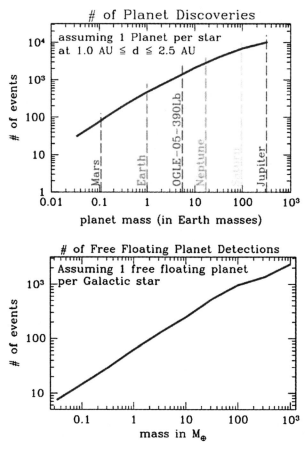

**Fig. 8.15** Sensitivity of MPF to several type of planetary objects, (*left*) in the case of planetary systems with a parent star, (*right*) free-floating objects (after Bennett et al., 2007)

spectroscopic instruments, the visible camera appears to be the instrument that is most suitable for this task.

## 8.2 Direct Methods of Detection

### 8.2.1 Imaging

Imaging exoplanets requires the use of highly capable telescopes with high angular resolution. Numerous programmes are under development.

### 8.2.1.1 Ground-Based Observatories

NAOS-CONICA

NAOS (Nasmyth Adaptive Optics System, Fig. 8.16) is the first set of adaptive optics mounted on one (Yepun) of the four 8.2-m telescopes of the European VLT on the summit of Mount Paranal in Chile. The instrument is coupled with the CONICA infrared camera (1–5 μm). The system, which has a deformable mirror with 185 actuators, a special tip/tilt wavefront correction system, and two wavefront analyzers (one in the visible region, and one in the near infrared), has been available to the scientific community since the end of 2002, and enables images to be obtained over a field that may amount to as much as 2 arcmin, with a Strehl ratio of up to 30–60 per cent, depending on seeing conditions. The instrument is used for all applications that require high angular resolution. The camera has a filter wheel which not only allows specific photometric bands to be chosen but also permits more specific modes of operation. In particular, a 4-quadrant coronagraph (cf. Chap. 2) is available and provides extinctions up to 500. It is this instrument that obtained what appears to be one of the first, if not the first, resolved image (Fig. 8.17) of a planet around its star (a brown dwarf, in fact), which in this case was 2MASS 1207, a young object, with a typical age of a few million years (and thus still hot) of about 5 Jupiter masses lying about 55 AU from the brown dwarf and where the contrast was favourable for detection (Chauvin et al., 2005).

SPHERE

The SPHERE (Spectro-Polarimetric High-contrast Exoplanet Research, Beuzit et al., 2006) project, currently in phase B, is a second-generation instrument for the VLT,

**Fig. 8.16** The NAOS instrument and the CONICA camera at the Nasmyth focus of Yepun, one of the four 8.2-m telescopes of the VLT on Mount Paranal in Chile (ESO image)

following-on from the NACO (NAOS-CONICA) equipment just described, and which it will replace. It should improve on the performance of NACO by at least an order of magnitude. SPHERE is designed to reach very high Strehl ratios (above 95 per cent), and to suppress the residual speckles to the level of $10^{-7}$ to $10^{-8}$ with the aim of detecting and carrying out spectral analysis of giant planets. SPHERE will have three scientific modules:

- an infrared module providing differential imagery (IRDIS),
- a module with an integral field spectrograph with a very low resolution (about 20) in the 0.95–1.35-μm spectral band (IFS),
- a module giving differential polarimetry in the visible (ZIMPOL).

The infrared modules with allow the detection of objects by their own emission and are thus particularly suitable for young, hot planetary systems. These modes of operation will occupy 80 per cent of SPHERE's observational time. The visible channel is more suitable for evolved systems, which are cooler, and where the companions are detected by their reflected radiation (light from the star reflected and polarized by the planet's atmosphere). In all cases, SPHERE will be imaging planetary systems around nearby stars, where the star–planet distances are between 1 and 100 AU, and where the contrast is about 12–14 magnitudes. The limiting magnitude for the detection of objects is about 24 in the H band. The aim is to be able to examine about 300–400 target stars. To carry out such a wide-ranging programme, SPHERE will be equipped with several coronagraphs, in particular (cf. Chap. 2) a

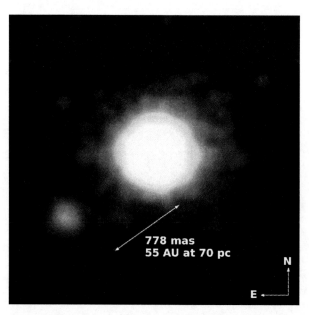

**Fig. 8.17** Image of the apparent planetary system 2M 1207 taken by NAOS-CONICA (after Chauvin et al., 2005)

4-quadrant half-wave phasemask coronagraph, a classic Lyot coronagraph, and an apodized Lyot coronagraph.

To obtain this performance, SPHERE will be coupled to a powerful adaptive optics system called SAXO (SPHERE Adaptive optics for eXoplanet Observation). This adaptive-optics system contains a deformable mirror 180 mm in diameter, controlled by $41 \times 41$ actuators with a maximum stroke of $\pm 3.5$ (m. The deformable mirror is controlled by a Shack-Hartman wavefront sensor with $40 \times 40$ lenslets, and operating in the visible and near infrared (0.45–0.95 µm). A 2-axis tip-tilt mirror with a resolution of 0.5 mas provides precise guidance control. In the coronagraphic mode, an additional tip-tilt mirror, whose sensor is located close to the focus of the coronagraph is used to enable thermo-mechanical drift to be eliminated.

The SAXO-SPHERE instrument should see first light around 2010.

## Gemini Planet Imager (GPI)

The Gemini Planet Imager (formerly called ExAOC: Extreme Adaptive Optics Coronagraph) is the Gemini Observatory's analogue of the SPHERE project. It aims to produce near-infrared images with a high dynamical range.

Technically speaking, GPI is an extreme AO system with 2000 actuators (MEMS technology), coupled with an apodised-pupil Lyot coronagraph. The wavefront sensing is performed at the nanometric level by a device that includes an infrared interferometer. The detection is performed by an integral field spectrograph working in the infrared spectral range. The system is designed to give a Strehl ratio better than 0.9 at the observing wavelength of 1.65 µm.

## The Keck Precision Adaptive Optics System (KPAO)

The KPAO project aims to provide the Keck Observatory with a new-generation adaptive-optics system. Like SPHERE and GPI, KPAO is to be a 1000-actuator-class AO system. To greatly reduce the residual wavefront errors and increase the image quality, the KPAO project would employ multiple laser guide stars (LGS) to reduce focal anisoplanetism by a factor of 10 compared with present LGS systems. The goal of KPAO is also to reach a Strehl ratio better than 0.9 in the K band.

## CARLINA

The CARLINA project is the first experimental trial aiming to demonstrate the feasibility of hypertelescopes, such as that proposed by A. Labeyrie. It is being carried out by the Observatoire de Haute Provence.

Several spherical mirrors, 250 mm in diameter and with a radius of curvature of 70 m are located on a sphere. The bundles of rays are combined at the focus

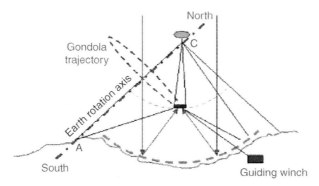

**Fig. 8.18** Schematic diagram of the CARLINA instrument. The gondola beneath the captive balloon carries the recombiner. The position of the whole assembly is controlled by cables

of the sphere where suitable instrumentation is held in place by a captive balloon (Fig. 8.18).

This arrangement means that the overall focal length is 35 m.

The first trials took place in May 2004, and were able to obtain fringes of Vega, using 2 mirrors, stopped down to 50-mm pupils, spaced 400 mm apart at the ground. The balloon was held in position at an altitude of 120 m.

The next stage, in February 2006, consisted of installing the spherical aberration corrector (the Mertz corrector) in the suspended gondola. The first images of Tania Australis were obtained with three mirrors on the ground on a sphere 10.5 m in diameter, together with the corrector, in a non-interferometer mode (superimposing the images at the focal plane, but without attempting to adjust the segments to correct for phase differences).

The next stage will consist of using an optical-adjustment method to superimpose the images, and equally, to ensure the segments are in phase (precise control of the position of the mirrors to ensure that they are precisely on the sphere of the virtual mirror that they are synthesizing). This stage is currently under way.

Carlina is a very long-term experiment, which anticipates the very large arrays of space telescopes, such as that proposed in the Exo Earth Imager concept (Labeyrie et al., 2003).

### 8.2.1.2 Space Observatories

SPITZER

The SPITZER space mission is an infrared telescope launched on 25 August 2003 by NASA, which is comparable in size to its European predecessor ISO. SPITZER consists of an 85-cm diameter, f/12 telescope, cooled both passively (its orbit maintaining it behind the Earth) and actively (360 l of liquid helium are carried on board to cool the instrument and the focal plane to about 5 K) (Fig. 8.19).

**Fig. 8.19** Artist's impression of the SPITZER observatory (image credit: courtesy NASA, JPL Caltech)

The telescope is fitted with three instruments:

- The IRAC camera (InfraRed Array Camera) with a detector matrix of $256 \times 256$ pixels, capable of recording the image of a field of $5.12 \times 5.12$ arcminutes in four spectral bands simultaneously: at 3.6, 4.5, 5.8, and 8 μm
- the IRS spectrograph (InfraRed Spectrograph) consisting of four modules, two giving low-resolution spectra (over the ranges 5.3–14 μm and 14–40 μm) and two giving high-resolution spectra (over the ranges 10–19.9 μm and 19–37 μm)
- the imaging spectrophotometer MIPS (Multiband Imaging Photometer for Spitzer) providing imagery at 24, 70, and 160 μm, as well as spectrometry over a range of 50–100 μm.

Initially, the lifetime of the mission was to be at least 2.5 years, with operation nominally expected to last 5 years.

Because of its size and limited angular resolution (about 5 as at 10 μm), SPITZER was not designed to image exo-systems. It may, however, be used to study the environment of certain stars and detect the presence of discs of dust from the excess infrared radiation.

SPITZER has also been used to study planetary systems that transit (HD 209458b and TrES-1 are examples). Observation of the star before and during a secondary transit (the planet passing behind the star) has enabled us, by photometric comparison, to determine the flux from the planet in a few spectral regions (Charbonneau et al., 2005).

## JWST

The James Webb Space Telescope (Fig. 8.20), initially known as the NGST (New Generation Space Telescope) is the planned successor to the Hubble Space Telescope, which has been in Earth orbit since 25 April 1990. JWST is being developed as a collaboration between NASA, ESA, and the Canadian space agency.

Unlike the HST, which was optimized for the visible and ultraviolet, the JWST is optimized for the visible, near and medium infrared regions. It consists of a telescope, 6.5 m in diameter, with a segmented mirror, passively cooled, and protected from sunlight by a thermal sunshade. Its launch is expected to be in 2013.

JWST will be fitted with 4 instruments:

- MIRI (Mid InfraRed Instrument): This instrument, the product of a European consortium and JPL, is an imaging spectrometer that covers the range 5–27 μm. It consists of two modules, one for wide-band imagery, and the other for medium-resolution spectroscopy. The imaging instrument is also fitted with a coronagraph, which will allow direct imaging of giant planets and sub-stellar objects. The operating temperature of the instrument is 7 K. This temperature cannot be reached by passive cooling of the telescope. The instrument is therefore fitted with additional cooling systems such as a Joule-Thomson expansion system.
- NIRcam (Near InfraRed Camera). This is a wide-field imager with a high angular resolution that exploits the size of the JWST in the wavelength range 0.6–5 μm (visible and near infrared). The instrument may also be used to check the quality of the wavefront for the whole telescope.

**Fig. 8.20** Artist's impression of the JWST (image credit: courtesy NASA, ESA)

- NIRspec (Near InfraRed Spectrograph). This is a multi-object spectrograph (100 objects simultaneously) covering a field 9 arcminutes square on the sky. NIRspec will obtain medium-resolution spectra in the 1–5 μm region, and low-resolution spectra in the 0.6–5 μm region.
- FGS (Fine Guidance Sensor). This is a wide-spectral-band camera (1–5 μm) with a wide field giving accurate guiding over 95 per cent of the sky, using a reference star.

The JWST's scientific programme is clearly geared to the study of 'origins' – of the Universe and of planetary systems. It is a multipurpose observatory which will cater, simultaneously, for extremely varied programmes in cosmology, galactic physics, the physics of the interstellar medium, and in planetology. The MRI instrument, in particular, which has coronagraphic capability, is the favourite for the study of giant planets, where it should be able to obtain some images and spectra in the middle infrared (5–30 μm). In addition, it should be able to image protoplanetary disks with great accuracy, thanks to the telescope's excellent angular resolution (which is diffraction-limited).

### TPF-C

The TPF-C mission (Terrestrial Planet Finder – Coronagraph) is a NASA project (Fig. 8.21). The aim is to build a space telescope operating in the visible and the near infrared that allows imaging of planetary systems consisting of terrestrial-type planets orbiting nearby stars (typically for some thirty targets), and that is also capable of carrying out spectral analysis of each of the components in a planetary system, over the range 0.6–2 μm, with a spectral resolution of at least 70. The idea is to study the composition of the atmospheres of terrestrial exoplanets to detect – as is the case with TPF-I/DARWIN – biological tracers. TPF-C fits into the same scheme as TPF-I/DARWIN and is complementary to it. The spectral information gained in the visible and near infrared on the one hand, and in the thermal infrared on the other, will allow us to set constraints on atmospheric models, and give an indication of the biological or abiotic origin of any atmospheric gasses that may be detected (cf. Chaps. 7 and 9).

To provide effective spectroscopy of terrestrial-type planets in the visible and near infrared, the instrument needs to have an extremely wide dynamical range (typically $10^{10}$, or 25 magnitudes!), which amounts to observing objects of magnitude 30–32 alongside objects of magnitude 5–7.

Such an instrumental concept cannot function without permanent control of the wavefronts and without a high-performance coronagraph. TPF-C, which is currently being defined, pushes space technology to its ultimate limits in terms of:

- control of the surface of the mirrors to better than a few tens of picometres. To achieve this requires an adaptive system to correct for deformation of the wavefront,
- superfine polishing (to control scattering) – almost to the atomic level,

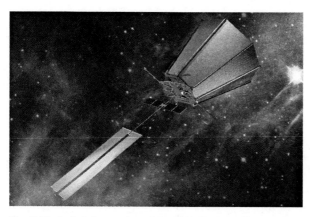

**Fig. 8.21** Artist's impression of the TPF-C mission (image credit: courtesy NASA)

- control and correction of mechanical and thermal distortions,
- guiding accuracy.

Initially foreseen for 2014 in its coronagraphic version and 2019 in the interferometer version, TPF-C, like all ambitious projects of this sort, faces extremely high development costs, incompatible with the current state of finances of the various space agencies. This project, like its interferometer counterpart, will be forced to adapt.

Preliminary versions of TPF-C, with scientific objectives limited to spectroscopy of giant planets and Super-Earths (i.e., bodies of several Earth-radii) are also being drawn up. They are Ellipse (in the USA) and SEE-COAST (Super-Earth Explorer, Coronagraphic Off-Axis Space Telescope (in Europe).

TFP-O

TPF-O (Terrestrial Planet Finder – Occulter, Fig. 8.22) concept came from the realization than an internal coronagraph (as in the TPF-C design) requires both the wavefront correction and control to be of very high quality to avoid leakage of the stellar light. The consequence is a complex design with adaptive optics to correct wavefront residuals. By contrast, the use of an external occulter permits the elimination of the major part of the light before it enters the telescope.

The main difficulty of this 'pure' separate-occulter version is the requirement for the manufacture of a perfect apodizing and occulting device, so that the stellar residual is reduced to a level that allows off-axis observation. The most intensively studied design is a mix of an occulting device and a coronagraphic device.

Formation flying of the two spacecraft is also a key issue in obtaining optimal pointing of the instrument.

**Fig. 8.22** An artist's impression of TPF-Occulter project. The occulter and the telescope are separate spacecraft (NASA image)

## 8.2.2 Interferometry

If one wishes to carry out direct observation in the infrared of objects very close to their stars (hot Jupiters for example) or more distant planets, the angular resolution available with a monolithic telescope (which is diffraction limited) is not sufficient to resolve the system. So it is necessary to resort to interferometry (cf. Chap. 2)

### 8.2.2.1 Ground-Based Observatories

VLT-I, KECK-I, CHARRA

Apart form the astrometric methods and instruments mentioned earlier, classical interferometer arrays can selectively provide observations of extrasolar planets. Here, in the absence of dark-fringe recombination interferometry, measurement of the modulations of visibility caused by the presence of a companion is the method used (cf. Chap. 2). The principal limitation of this method is the dynamical range that results from the star/planet contrast. So this technique is limited to giant planets.

Several instruments (AMBER and MIDI on the VLT-I) are capable of measuring visibility with sufficient accuracy, which is about 0.1 per cent. However, such results are available only at the price of several tens of nights of observation, which limits this method to an extremely limited number of candidates. In practice, it is solely if one wants to carry out spectroscopy of the object that such observations are necessary. It is then essential to have a method of translating the chromatic information on visibility into chromatic information about the companion. In the majority of cases, however, information obtained by this method may also be used by other, less restrictive methods.

A nulling instrument (KIN: Keck Interferometer Nuller) has been installed on the Keck-Interferometer, combining the two 10-m Keck telescopes. The KIN allows interferometric nulling of the on-axis object better than 100 in the N spectral region. This instrument is thus particularly suited to the observation of stellar environments and disks.

LBT

The LBT (Large Binocular Telescope) is an unusual project, consisting of two 8.4-m telescopes on the same mount. The two primary mirrors are separated by 14.4 m (Fig. 8.23). Such an arrangement, where all spatial frequencies are measured with a pupil equivalent to 20 m in diameter, has several recombination modes, namely:

- a direct Fizeau imaging mode (the LINC-NIRVANA instrument). Here, the PSF is not an Airy disk but patch of interference fringes caused by interferometer recombination. In this mode, and thanks to the fringe structure of the PSF, it is possible to measure the relative positions of the stars in the field, with an accuracy of about 100 μas.
- A dark-fringe interferometer mode. With a universal recombiner, two instruments may be used: NUL (Nulling Interferometer for the LBT), a dark-fringe interferometer with a typical rejection of $10^4$, and the NOMIC camera (Nulling Optimized Mid-Infrared Camera), capable of operating in the N band (at 10 μm), among others.

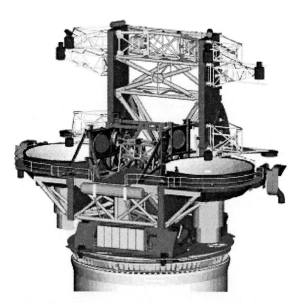

**Fig. 8.23** The LBT and the UBC universal recombiner for interferometry

The unique feature of the LBT is that is specifically optimized for dark-fringe interferometry in the N (10 μm) band. Unlike the VLT-I and other classical interferometers, the optical path between the point at which it enters the telescope and the instrumentation is reduced, and the number of mirrors is low. Given the distance between the two primary mirrors (14.4 m, centre to centre), and the spatial resolution that results, this instrument is specifically suitable for observation of stellar environments (at least for nearby stars), and of debris disks out to distances of several AU from their central stars. This instrument is equally well-adapted to the observation and spectroscopy of giant planets.

GENIE

The GENIE project (Ground-based European Nulling Interferometer Experiment) is in line with preparations for the DARWIN mission (described later). GENIE, initially being considered for use on the VLT-I had several aims:

- to validate, by means of a 'life-size' project, the concept of dark-fringe interferometry;
- to train the European community in the techniques for nulling interferometry;
- to achieve part of the preparatory scientific work for DARWIN (in particular, describing the zodiacal environment of DARWIN's targets).

Preliminary studies carried out both by industrial consortia and by ESA have come to the following conclusions:

- the choice of spectral band L (3.8 μm): this is a compromise between band N (10 μm), where atmospheric turbulence has less effect than at shorter wavelengths, but where the thermal background predominates, and band K (2.2 μm), where the sky background is less significant, but turbulence is more critical.
- Inadequacies between the current structure of the VLT-I and dark-fringe interferometry. Before reaching the recombination laboratory, the interferometer beam has to undergo more than 10 reflections by mirrors, which limits the propagation efficiency, and increases the relative proportion of emission from the instrument.

As a result, the final results are mediocre, and the project of installing GENIE on the VLT-I has been abandoned. The concept might, however, be considered at a better site, in particular, in Antarctica.

ALADDIN and Antarctic Projects

The basic principles that have led to the ALADDIN concept are identical to those for GENIE. Scientifically, it aims at describing the zodiacal environment around DARWIN's targets, to a sensitivity about 20 times better than our Solar System's

**Fig. 8.24** Artist's impression of the ALADDIN instrument. The two dark-fringe recombination telescopes are mobile along a rail, the orientation of which may also be changed (Alcatel Alenia Space)

integrated zodiacal emission. Because the site and infrastructure of the VLT-I would not enable us to achieve the desired results, it is a question of finding a site, and a design that are optimum for the purpose.

The Antarctic, and more particularly the location of Dome C, offers some major assets:

• The humidity if low, allowing excellent observations in the infrared.
• Because of the low temperature, the first turbulent layer appears to lie within 30 metres of the surface. Equipment installed above this height would therefore avoid the effect of this layer.
• The coherence time set by the turbulence is significantly longer than at Paranal, for example. This particular characteristic means that the passband of the adaptive systems may be reduced, gaining sensitivity or more accurate correction.
• There is an infrastructure at Dome C (the Franco-Italian Concordia base), which would allow new equipment to be located at this site.

The ALADDIN instrument (Antarctic L-band Astrophysics Discovery Demonstrator for Interferometric Nulling), makes use of the scientific specification for GENIE, without having to include the VLT-I's specifications and technical constraints. The resulting concept has been put forward by an industrial and scientific consortium. It involves two 1-m-class telescopes, mounted on a base that allows the spacing to be altered (from a few metres to 40 m), and operating as a Bracewell interferometer. Rotation of the base would enable the transmission map of the instrument to be altered (Fig. 8.24).

Again, spectral band L is a good compromise between the sky background and atmospheric turbulence.

It is estimated that ALADDIN could observe 30 per cent of DARWIN's potential targets. A direct-recombination mode also means that it is equally suited to the more classical type of observations in stellar physics (measurement of stellar diameters, study of environments close to stars, etc.).

The site is currently being tested to confirm the points mentioned above (Agabi et al., 2003). The nature of the environment (polar night, impossibility of resupply during the southern winter, logistical difficulties, etc.) means that any project at Dome C has additional difficulties which mean that it somewhat resembles space-borne projects. The site is equally coveted for other projects, most of which are also of interest for the discovery and study of extrasolar planets: searching for planets by the transit method (A-STEP), asteroseismology projects (SIAMOIS), and extreme adaptive-optics projects.

In the more distant future, the possibility of installing an Extremely Large telescope or a kilometric interferometer array (the KEOPS project) has also been suggested.

The results from the site-testing campaigns currently under way should enable rational decisions about the use of the site to be taken.

### 8.2.2.2 Space Observatories

DARWIN and TPF-I

The DARWIN space mission (Fig. 8.25) and the interferometer version of its American counterpart TPF-I (Terrestrial Planet Finder) are based on the Bracewell interferometer principle (Fridlund et al., 2000, Beichman et al., 1999) described earlier in Chap. 2.

The object of this mission is to detect exoplanets directly (both giant and terrestrial-type) around nearby stars, with the aim of carrying out spectroscopic analysis of the composition of any likely atmospheres. The primary idea is to carry out a comparative study of different systems, and equally to try to detect signs of biological activity.

DARWIN is an interferometer, that is, the beams of light from several physically separated telescopes are recombined to give information with a spatial resolution that is equivalent to that given by a monolithic telescope of comparable size to the distance between the telescopes in the array.

Several designs have been advanced or are in the process of being evaluated in minimal versions with just three telescopes, or in more complex versions with five or six telescopes. The number of telescopes is a compromise between:

- the complexity of recombination;
- the performance of the central extinction: DARWIN's special feature is that it works in a high-angular-resolution, conjugate-recombination mode with a great dynamic range, known as 'dark-fringe interferometry', the aim of which is to extinguish the central source of light to examine a nearby faint object;
- the capacity to extract a planetary signal from its surroundings (the zodiacal light in the exosystem, for example). We may recall that in the Solar System, Earth's

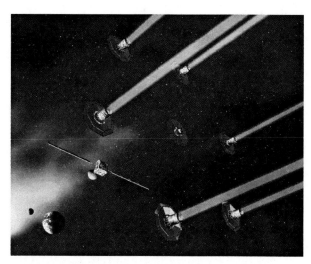

**Fig. 8.25** Artist's impression of the DARWIN mission, where six telescopes are arranged in a circle. The central satellite is the recombination laboratory. An eighth satellite, not shown here, handles transmission of information to Earth and monitors the spacing of the overall array (image credit: courtesy ESA)

own emission is one-300th part of the integrated emission from the zodiacal light over the whole celestial sphere.

In any scheme, each telescope, as well as the recombination laboratory and the communications equipment, is carried by an independent, free-flying satellite, automatically manoeuvred by small thrusters, and whose position is controlled by a measurement system capable of ensuring a constant (but alterable) distance between the telescopes, to an accuracy of a few millimetres. The array's phase is controlled by delay lines, with a path length of a few centimetres. Configurations with more than two telescopes allow sub-interferometers to be isolated within the array, and these may be recombined by applying a rapid phase modulation between each sub-interferometer. This technique of internal modulation enables the planetary signal to be modulated relative to that from the star, to achieve what is the equivalent of synchronous detection, without turning the whole array, as was initially proposed by Bracewell (Mennesson et al., 2005).

DARWIN/TPF has been proposed to ESA's new Cosmic Vision programme for the period 2015–2025. It would possibly be launched in 2025 and placed in orbit around the L2 Lagrangian point. This point is particularly suitable for orbiting several satellites in formation, because it has weak gravity gradients. It is equally suitable in terms of the thermal environment. The principal instrument on DARWIN/TPF will be a dark-fringe interferometer recombiner, the scientific results of which will be maps of the planetary systems, derived from analysis of the chromatic, modulated signals from sub-interferometers, as well as spectra of the different components of the planetary systems (Fig. 8.26).

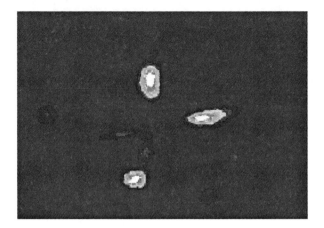

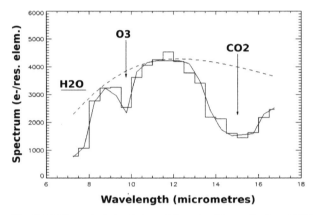

**Fig. 8.26** Numerical simulation of the scientific results from DARWIN's dark-fringe interferometer. *Left*, a map of the Solar System such as might have been observed by DARWIN on 1 January 2001, 10 pc above the north pole of the Sun (centre in the image). Three objects are visible and these are Venus, Earth, and Mars. *Right*, the spectrum of the Earth as it would be observed by DARWIN. The histogram represents the spectrum, reconstructed after simulation of the observation. The continuous line shows the Earth's actual spectrum, sampled at the instrument's resolution, and a dashed line shows the spectrum of a blackbody at 300 K. This diagram clearly shows that DARWIN/TPF will be able to detect the presence of certain components of biological interest in the atmospheres of terrestrial-type planets, even at a low spectral resolution (20 in this image). $H_2O$, $CO_2$, and $O_3$, in particular, are clearly visible relative to the blackbody at 300 K (After Mennesson and Mariotti, 1997)

The results expected from DARWIN/TPF-I utterly depend on the arrangement adopted for recombination, and in particular the number and size of the telescopes. Figure 8.27 shows the performance in terms of the signal-to-noise ratio for a linear arrangement of four telescopes, 3 m in diameter. (A study made for TPF-I.)

On the technical side, dark-fringe interferometry proves to be difficult because it requires extreme accuracy in terms of the coherence of the light-paths. In particular, to eliminate the extreme contrast between the star and the planet (typically $10^6$ in the thermal infrared region), it is necessary for the interferometric extinction of the star be at least $10^5$, with a relative stability of 1 in 10 000. Table 8.7 summarizes the principal characteristics of the DARWIN recombiner. By way of comparison, the characteristics of PEGASE, a possible predecessor to DARWIN (see the next section) are also given in this table.

Given the complexity of designing and constructing a dark-fringe interferometer recombiner, numerous studies are being undertaken all over the world to construct laboratory demonstration models (Ollivier et al., 2001; Brachet et al., 2003; Serabyn, 2003a; Flatscher et al., 2003; Serabyn, 2003b; Gondoin et al., 2003). The demonstrator of the Institut d'Astrophysique Spatiale at Orsay is shown in Fig. 8.28. We may finally mention that industrial research is being carried out to define the parameters and performance of the mission.

Precursors to DARWIN and TPF-I

Two missions have been proposed as precursors and are in the study phase, evaluating the instrumental concepts that will lead to DARWIN/TPF-I:

- NASA's FKSI mission (Fig. 8.29). This is a Bracewell interferometer with two telescopes mounted on a rigid base, with a fixed baseline.
- The PEGASE mission (Fig. 8.30), proposed to CNES in response to a call for proposals for a flight in formation, PEGASE is a Bracewell interferometer consisting of three satellites, two carrying siderostats, 30–40 cm in diameter, and one carrying the recombination laboratory and the detection chain.

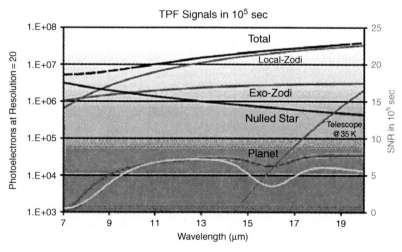

**Fig. 8.27** An estimate of the various factors contributing to the performance of the TPF-I instrument in a 4-telescope configuration (After Beichman et al., 1999)

**Table 8.7** Principal technical characteristics of the instruments on the DARWIN/TPF missions and a possible precursor, PEGASE

| Characteristic | DARWIN/TPF-I | PEGASE |
|---|---|---|
| Spectral band | $7-18\,\mu m$ | $2.5-5\,\mu m$ |
| Cophasing accuracy (difference in residual path-lengths) | 3 nm rms | 2.5 nm rms |
| Maximum interferometric extinction | $10^5$ | $10^4$ |
| Dephasing accuracy over the whole spectral band | A few $10^{-3}$ rad | $5 \times 10^{-3}$ rad |
| Interferometer baselines | 50–500 m | 20–500 m |
| Maximum angular resolution | 3 mas (at $7\,\mu m$) | 1 mas (at $2.5\,\mu m$) |
| Total instrumental transmission (including detection) | Depends on the configuration used | 7 per cent |
| Satellite guiding accuracy | A few arcsec | A few arcsec |
| Fine guiding | 8 mas | 20 mas |
| Fine propulsion and guiding | FEEP | Cold gas |
| Size of telescopes | 1.5–3 m | 0.3–0.4 m |
| Number of spacecraft | 4–8 | 3 |
| Temperature of the instrument | 40 K | $100\,K \pm 1\,K$ |
| Temperature of the detector | 10 K | $55\,K \pm 1\,K$ |
| Orbit | L2 | L2 |

**Fig. 8.28** Laboratory bench demonstrator of dark-fringe interferometry at the Institut d'Astrophysique Spatiale at Orsay

PEGASE has two main objectives:

- Scientific: to study low-mass companions, either stellar or planetary, and in particular those that are very close to their star (i.e., the Pegasids);
- Technological: to validate certain instrumental techniques foreseen for DARWIN, notably space interferometry and flying in formation.

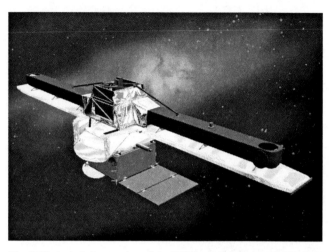

**Fig. 8.29** Artist's impression of the FKSI mission (image credit: courtesy NASA, GSFC)

**Fig. 8.30** Artist's impression of the PEGASE mission (image credit: courtesy Thales Alenia Space, CNES)

## 8.2.3 Direct Detection of Radio Waves

### 8.2.3.1 Low Frequency Array (LOFAR)

The LOFAR project (LOw Frequency ARray) is a Dutch project undertaken by the ASTRON Institute, which brings together in the same consortium 18 European research organisations, the majority Dutch and Germany. The aim of this project is not just astrophysical, because institutes carrying out research into geophysics and agronomy are equally involved. The array actually consists of radio aerials, but also includes 3-axis geophones and microbarometers. From the astrophysical point of view, the aim is to construct a giant aperture-synthesis radiotelescope, the size and equivalent area of which are 100 times that of the largest array currently in existence in this frequency region (Kharkov in the Ukraine).

The principle on which the array is based consists of a core, and more distant stations, thus creating several baselines (Fig. 8.31). The first version consists of 32 stations installed at Drenth (in the Netherlands), and 45 peripheral stations providing maximum baselines of about 100 km. Table 8.8 describes the parameters of this so-called 'nominal' configuration, and Table 8.9 gives the performance as a function of frequency. It should be noted that in terms of equivalent surface, LOFAR has four times the extent of the UTR array at Kharkov.

**Table 8.8** Characteristics of the nominal configuration of the LOFAR array (Source: project site)

| | | |
|---|---|---|
| Frequency band | Low frequencies | 30–80 MHz[a] |
| | High frequencies | 110–240 MHz |
| Number of antennae | Low frequencies | 7700 |
| | High frequencies | 7700 |
| Configuration | Core | 3200 antennae (LF and HF) at 32 stations. Baselines from 100 m to 2 km |
| | Extended array | 4500 antennae (LF and HF) at 45 stations. Baselines up to 100 km[b] |

[a]This passband is given for the nominal initial configuration. Studies are under way to extend the band to the 10–20 MHz region.
[b]Studies are under way to extend the baselines to 400 km by adding distant antennae to the array.

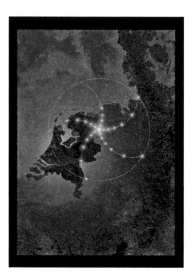

**Fig. 8.31** Schematic diagram showing the arrangement of the LOFAR array of antennae (image credit: courtesy ASTRON)

**Table 8.9** Performance of the nominal configuration of LOFAR as a function of the observational frequency (Source: project website)

| Frequency (MHz) | Point-source sensitivity | | Effective collective surface | | Angular resolution (size of the PSF) | |
|---|---|---|---|---|---|---|
| | Core (mJy) | Complete array (mJy) | Core | Complete array | Core | Complete array |
| 30 | 4.8 | 2.0 | $7.9 \times 10^4\,\mathrm{m}^2$ | $1.9 \times 10^5\,\mathrm{m}^2$ | 21' | 25" |
| 75 | 3.3 | 1.3 | $1.2 \times 10^4\,\mathrm{m}^2$ | $3.0 \times 10^4\,\mathrm{m}^2$ | 8.3' | 10" |
| 120 | 0.17 | 0.07 | $7.9 \times 10^4\,\mathrm{m}^2$ | $1.9 \times 10^5\,\mathrm{m}^2$ | 5.2' | 6.0" |
| 200 | 0.15 | 0.06 | $2.9 \times 10^4\,\mathrm{m}^2$ | $6.9 \times 10^4\,\mathrm{m}^2$ | 3.1' | 3.5" |

**Fig. 8.32** Some of the antennae forming the LOFAR array (Source: project site)

As with optical interferometry, the idea behind such arrays is to combine a large number of small-sized and poorly directional antennae. Combining the various antennae, through electronic delay lines allows the signals to be combined to:

- gain sensitivity (by increasing the collecting surface area),
- gain resolution (by reducing the size of the beams of individual antennae),
- pointing the instrument. The antennae are fixed (Fig. 8.32) and the array's lobe is positioned on the sky by adjustment of the different delays between antennae (the phased array principle).

One of the main technological challenges is being able to transfer, in real time, information from 77 stations (32 in the core and 45 in the rest of the array) to enable interferometric recombination, in contrast to classic long-baseline radio interferometry, where the signals from various antennae are recorded and only correlated subsequently.

The LOFAR array, when it becomes operational, will be able to study the auroral emission from giant exoplanets, following the example of the UTR-2 array at Kharkov, but with an increased sensitivity relative to existing programmes.

## 8.2.3.2 The Square Kilometric Array (SKA)

The SKA project aims to construct an interferometer array with a surface area that will reach one million square metres, i.e., about 200 times that of a classic radio telescope like the Lovell Telescope at Jodrell Bank. The sensitivity of such an array should also allow it to capture signals as faint as those of television broadcasts at the distances of the nearest stars to the Sun (a few parsecs). Research into auroral radio emissions from extrasolar planets is, of course, only one of the scientific aims of the astrophysical instrument.

SKA, in its high-frequency mode, should be able to detect the gaps created by the migration of giant planets (or even terrestrial ones) in protoplanetary disks.

The project is about to enter a study phase, which will last about four years, with the aim of defining the concept and the associated equipment. Currently, the construction timeline of this array has not been defined. The final choice of instrumentation should be made by the end of the decade, and that of the core site (Western Australia or the Karoo in South Africa) after a site assessment, and in any case, again not before 2010.

# Bibliography

Agabi, A. et al., 'First whole atmosphere nighttime seeing measurements at Dome C, Antarctica', PASP, **840**, 344–348 (2003)

Baglin, A. et al., in *The CoRoT Mission – Pre-Launch Status – Stellar Seismology and Planet Finding*, ESA SP-1306 (2006)

Baranne, A., Queloz, D., Mayor, M., Adrianzyk, G., Knispel, G., Kohler, D., Lacroix, D., Meunier, J-P., Rimbaud, G., Vin, A., 'ELODIE: A spectrograph for accurate radial velocity measurements', *A & A supp.*, **119**, 373–390 (1996)

Barillot, M. and Laramas, C., 'Alcatel space involvement in Darwin', *ESA-SP*, **539**, 345–347 (2003)

Beaulieu, J-P., et al., 'Discovery of a cool planet of 5.5 Earth masses through gravitational microlensing', *Nature*, **439**, 437–440 (2006)

Beichman, C.A., Woolf, N.J. and Lindensmith, C.A., The Terrestrial Planet Finder: A NASA Origins Program to Search for Habitable Planets, JPL Publication, Pasadena, **99–3** (1999)

Bennett, D.P., Anderson, J., Beaulieu, J-P., Bond, I., Cheng, E., Cook, K., Friedman, S., Gaudi, B.S., Gould, A., Jenkins, J., Kimble, R., Lin, D., Rich, M., Sahu, K., Tenerelli, D., Udalski, A., Yock, P., 'An extrasolar planet census with a space-based microlensing survey', White Paper Submitted to the NASA/NSF ExoPlanet Task Force, eprint arXiv:0704.0454 (2007)

Beuzit, J.-L. et al., 'A planet finder instrument for the VLT' in *Direct Imaging of Exoplanets: Science & Techniques*. Proceedings of the IAU Colloquium #200, Edited by Aime, C. and Vakili, F. Cambridge University Press, Cambridge, 317–322 (2006)

Brachet, F., Labèque A. and Sekulic, P., 'Polychromatic laboratory test bench for Darwin/TPF: first results', *ESA-SP*, **539**, 385–387 (2003)

Charbonneau, D., Brown, T.M., Noyes, R.W., Gilliland, R.L., 'Detection of an extrasolar planet atmosphere', *Astrophys. J.*, **568**, 377–384 (2002)

Charbonneau, D. et al., 'Detection of thermal emission from an extrasolar planet', *ApJ*, **626**, 523–529 (2005)

Chauvin, G., Lagrange, A.-M., Dumas, C., Zuckerman, B., Mouillet, D., Song, I., Beuzit, J.-L., Lowrance, P., 'Giant planet companion to 2MASSW J1207334-393254', *A & A*, **438**, L25–L28 (2005)

Flatscher, R., Sodnik, Z., Ergenzinger, K., Johann, U. and Vink, R., 'Darwin nulling interferometer breadboard I: system engineering and measurements', *ESA-SP*, **539**, 283–291 (2003)

Fridlund, M. et al., *DARWIN, the Infrared Space Interferometer*, ESA-SCI(2000)12 (2000)

Gondoin, P., et al. 'Darwin ground-based European nulling interferometry experiment (GENIE)', *Proc. SPIE*, **4838**, 700–711 (2003)

Hinz, P.M., Angel, J.R.P., McCarthy, D.W. Jr., Hoffman, W.F. and Peng, C.Y. The large binocular telescope interferometer, *Proc. SPIE*, **4838**, 108–112 (2003)

Labeyrie, A., 'Detecting exo-Earths with hypertelescopes in space: the exo-Earth discoverer concept' in EAS Publications Series, Volume 8, *Astronomy with High Contrast Imaging*, Proceedings of the Conference held 13–16 May, 2002 in Nice, France. Edited by Aime, C. and Soummer, R., 327–342 (2003)

Mennesson, B. and Mariotti, J.-M., 'Array configurations for a space infrared nulling interferometer dedicated to the search for earthlike extrasolar planets', *Icarus*, **128**, 202–212 (1997)

Mennesson, B., Léger, A. and Ollivier, M., 'Direct detection and characterization of extrasolar planets: the Mariotti space interferometer', *Icarus*, **178**, 570–588 (2005)

Ollivier, M., Mariotti, J.-M., Léger, A., Sékulic, P., Brunaud, J. and Michel, G., 'Interferometric coronography for the DARWIN space mission – laboratory demonstration experiment', *Astron. Astrophys.*, **370**, 1128 (2001)

Serabyn, E., 'Nulling interferometry progress', *Proc. SPIE*, **4838**, 594–608 (2003a)

Serabyn, E., 'An overview of the Keck interferometer nuller', *ESA-SP*, **539**, 91–98 (2003b)

# Chapter 9
# The Search for Life in Planetary Systems

The discovery of numerous planets orbiting solar-type stars has given new currency to a question that is as old as the hills: are there other inhabited worlds? This question was posed by the Greek philosophers Democritus and Epicurus, and by the Roman poet Lucretius. In the 16th century, the idea of extraterrestrial life was taken up by Giordano Bruno, who paid with his life for his numerous iconoclastic views, but it continued to gain ground among many astronomers and philosophers such as Kepler, Kant, Huygens and Fontenelle. Thoughts initially turned to searching for traces of life in the Solar System itself, in particular on the planet Mars. In parallel with this, the first conceptions of a chemical evolution of life appeared around the 1920s, when the Russian chemist Oparin first advanced the theory that complex organic molecules could evolve, through a long series of chemical transformations, into micro-organisms. An essential step was taken in 1953, when the American chemists Miller and Urey first succeeded in the laboratory in synthesizing amino acids from plausible prebiotic conditions. From the 1970s, the discovery of more and more, and increasingly complex interstellar molecules proved that a prebiotic chemistry did indeed exist in the Universe. After the discovery of prebiotic molecules in Titan's atmosphere at the beginning of the 1980s, it seemed that a rich prebiotic chemistry was active in the most varied types of environment, from the interstellar medium to circumstellar envelopes and to planetary atmospheres.

And now, with the discovery of many exoplanets orbiting stars near to the Sun, and comparable with it, the debate over the possibility of extraterrestrial life takes on a new dimension.

## 9.1 What is Life?

### 9.1.1 How Should Life be Defined?

Before undertaking a discussion on the search for extraterrestrial life, it is just as well to define what we mean by life. Perforce, the criteria are chosen based on the sole example at our disposal, that of life on Earth. Biologists agree on using the following criteria:

M. Ollivier et al., *Planetary Systems*. Astronomy and Astrophysics Library,
DOI 978-3-540-75748-1_9, © Springer-Verlag Berlin Heidelberg 2009

- auto-reproduction (i.e., reproduction of identical organisms)
- evolution by mutations (corresponding to accidental changes in the reproduction mechanism)
- self-regulation with respect to its surroundings (which ensures the growth and survival of the individual life form).

The second criterion allows organisms to adapt to environmental conditions and to their evolution (such as a change in the ambient temperature or the quantity of water available). The species resulting from natural selection are those that have proved most capable of survival.

## 9.1.2  The Role of Carbon and of Liquid Water

One chemical element is the basis of life as we know it: carbon, which is the basis of all organic molecules. What are this atom's specific properties? First, it is one of the most abundant in the Universe; it has four valence electrons, which enable it to create multiple bonds with itself and with other atoms (particularly with hydrogen, oxygen, and nitrogen, but also with sulphur and phosphorus). More than one hundred gaseous-phase molecules have been discovered in the interstellar medium, containing up to fifteen atoms, most of which are carbon. Carbon is also present in the Universe in the form of more complex molecules, containing tens or even hundreds of atoms, such as the polycyclic aromatic hydrocarbons (PAH).

Could we imagine another form of life based on an atom other than carbon? According to Mendeleyev's periodic table (Table 9.1), the most favourable element would be silicon, which also has four valence electrons. But this element is only one-tenth as abundant in the Universe as carbon. To this day, no interstellar molecule containing more than one atom of silicon has been discovered, whereas interstellar molecules have been observed to have as many as 11 atoms of carbon. So it appears reasonable to assume that if extraterrestrial life exists, it would also be based on carbon chemistry.

Liquid water is another element essential for life on Earth, because it provides a medium in which molecules may dissolve and react with one another. The $H_2O$ molecule is a polar solvent, that is to say that is has a bipolar moment, with the hydrogen atoms carrying the positive charge and the oxygen atom the negative charge. Polar solvents such as water are able to dissolve polar organic molecules, known as hydrophils, whereas non-polar molecules, known as hydrophobes, do not react with water. The hydrophilic or hydrophobic properties of different molecules are used by living organisms to produce specific functions.

Could we imagine the development of a form of life in another solvent than liquid water? A priori, nothing forbids it. Ammonia, $NH_3$, is also a polar molecule which could play the same part; chemical reactions within a liquid solution would, however, be slower, because the temperature would be lower, but, in principle, they could take place. However, we should note that water possesses a certain number of trumps that may explain the role that it has played in the development of life

**Table 9.1** Mendeleyev's periodic table of the elements

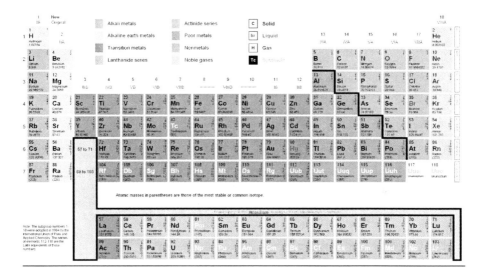

on Earth. Formed from hydrogen and oxygen, both of which are particularly abundant in the Universe (see Sect. 1.4.5), water is omnipresent in planetary, stellar, and galactic sources. In the interstellar medium, oxygen is primarily in the form of $H_2O$; a few percent are in the form of carbon monoxide, CO, methanol, $CH_3OH$, or other oxygenated molecules; molecular oxygen is notoriously absent. These abundances are repeated in the composition of cometary material. Water, then, because of its cosmic abundance, is particularly well placed to play the role of solvent. We should also, finally, mention that water possesses a unique property that may have proved decisive in preserving life, once the latter had appeared in the liquid medium. Water is, in fact, the sole molecule where ice formed by condensation from the liquid phase has a density less than that of the liquid itself. This is why icebergs float on the ocean. This feature undoubtedly played a key role in the history of life on Earth. During cooling episodes, the oceans became covered with a layer of ice, which allowed the oceans to survive, together with the living species within them. If the oceans had solidified from the bottom, as would be the case with hydrocarbons, for example, the solid phase would, bit by bit, have invaded the whole of the oceans, thus destroying any complex life forms that they may have sheltered. To summarize, although water is perhaps not a unique solvent capable of supporting the appearance and development of life, it obviously possesses specific features that render it particularly suitable for this role.

## 9.1.3 The Building-Block of Life: Macromolecules

Several types of macromolecules are involved in forming a living system: lipids, carbohydrates, proteins, and lastly, nucleic acids.

The lipids are a group of polymers, one end of which is hydrophilic, with the other by hydrophobic (Fig. 9.1). Overall, they are poorly soluble in water, and form aggregates that are weakly bound chemically, which allows them to store energy, while retaining a high degree of flexibility, which is particularly useful in the formation of membranes.

Carbohydrates are polar molecules, possessing hydroxyl groups (OH bonds), and are thus soluble in water. The sugars, in particular (Fig. 9.2), are carbohydrates, which, dissolved in water, adopt a circular form. They are able to link with one another to form polymers, such as the polysaccharides. These can store energy and provide structure to organisms.

|                              | Octadecanoic acid (stearic acid, $C_{18}$) | Octadecenoic acid (oleic acid, $C_{18}$) |                  |
|------------------------------|--------------------------------------------|------------------------------------------|------------------|
|                              | COOH                                       | COOH                                     | Hydrophilic end  |
|                              | \|                                         | \|                                       |                  |
|                              | $CH_2$                                     | $CH_2$                                   |                  |
|                              | \|                                         | \|                                       |                  |
|                              | ......                                     | ......                                   |                  |
|                              | \|                                         | \|                                       |                  |
|                              | $CH_2$                                     | $CH_2$                                   |                  |
|                              | \|                                         | \|                                       |                  |
|                              | ......                                     | CH                                       |                  |
|                              | ......                                     | \|\|                                     |                  |
|                              | ......                                     | CH                                       |                  |
|                              | \|                                         | \|                                       |                  |
|                              | $CH_2$                                     | $CH_2$                                   |                  |
|                              | ......                                     | ......                                   |                  |
|                              | \|                                         | \|                                       |                  |
|                              | $CH_2$                                     | $CH_2$                                   |                  |
|                              | \|                                         | \|                                       |                  |
|                              | $CH_3$                                     | $CH_3$                                   | Hydrophobic end  |

**Fig. 9.1** Examples of the molecular structure of two lipids, octadecanoic acid (stearic acid) and octadecanoic acid (oleic acid). Both have a hydrophilic head and a hydrophobic tail (After Gilmour and Sephton, 2003 © Cambridge University Press and The Open University)

Glucose

$$
\begin{array}{cc}
H & O \\
\backslash & // \\
 & C \\
 & | \\
H - C & - OH \\
 & | \\
OH - C & - H \\
 & | \\
H - C & - OH \\
 & | \\
H - C & - OH \\
 & | \\
 & CH_2OH
\end{array}
$$

Ribose

$$
\begin{array}{cc}
H & O \\
\backslash & // \\
 & C \\
 & | \\
H - C & - OH \\
 & | \\
H - C & - OH \\
 & | \\
H - C & - OH \\
 & | \\
 & CH_2OH
\end{array}
$$

**Fig. 9.2** Examples of two common sugars: glucose and ribose

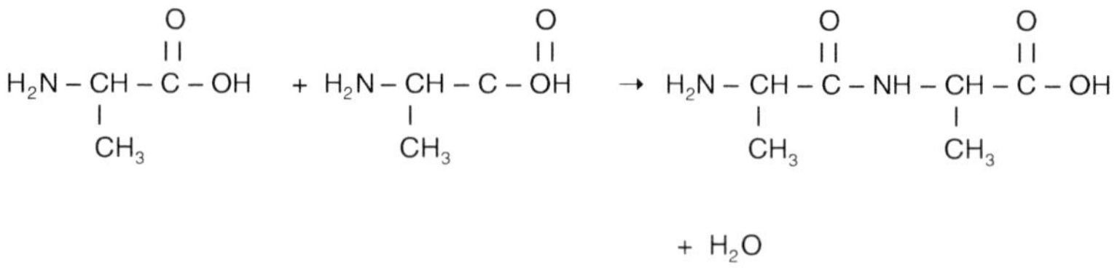

$$
\underset{\substack{| \\ CH_3}}{H_2N-CH-\overset{\displaystyle O}{\overset{\displaystyle \|}{C}}-OH} \;+\; \underset{\substack{| \\ CH_3}}{H_2N-CH-\overset{\displaystyle O}{\overset{\displaystyle \|}{C}}-OH} \;\rightarrow\; \underset{\substack{| \\ CH_3}}{H_2N-CH-\overset{\displaystyle O}{\overset{\displaystyle \|}{C}}-NH-}\underset{\substack{| \\ CH_3}}{CH-\overset{\displaystyle O}{\overset{\displaystyle \|}{C}}-OH}
$$

$$+ \; H_2O$$

**Fig. 9.3** An example of an amino acid. This may be polymerized by a reaction that links two monomers and undergoes the loss of a molecule of water. The process, repeated many times, leads to the formation of long chains that occur in the formation of proteins

The proteins, for their part, are the most complex macromolecules found in living organisms. They are associated with long polymerized chains known as amino acids (Fig. 9.3). Some twenty amino acids are found in living organisms, and it is the sequence of these amino acids that give each protein its function, which may take many different forms.

## 9.1.4 Nucleic Acids

Finally, nucleic acids are the largest macromolecules existing in living organisms. They consist of a series of interlinked nucleotides forming a long polymerized chain. The nucleotides contain a sugar molecule, one or more phosphates, and a nitrogen compound known as a base.

Two nucleic acids play a particularly important role in the formation of living things. Deoxyribonucleic acid (DNA) consists of four nucleotides, all containing the same sugar and phosphate groups, but with different bases (adenine, thymine, guanine, and cytosine, Fig. 9.4). In 1953, James Watson and Francis Crick discovered the double helix of DNA, which, by separating the two halves of its helix, is able to produce an identical copy of itself, the bases being associated in pairs (adenine—thymine and guanine—cytosine). The duplication of DNA molecules (Fig. 9.5) is the basis of the reproduction mechanism of living creatures; the sequence of bases in DNA defines the genetic code.

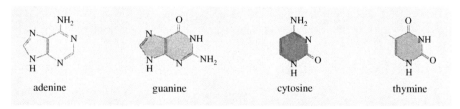

adenine       guanine       cytosine       thymine

**Fig. 9.4** The chemical composition of four bases found in DNA (After Gilmour and Sephton, 2003)

**Fig. 9.5** The double helix
of DNA. The bases can only
associate in pairs: A–T and
G–C. This property ensures
identical replication of the
double helix (After Gilmour
and Sephton, 2003)

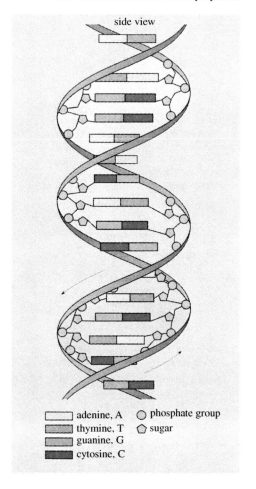

Another nucleic acid also plays an essential part: ribonucleic acid (RNA). RNA
is similar to DNA, with a few differences: the sugar it contains is ribose, and not
deoxyribose, and the base uracil replaces thymine.

RNA plays the role of transmitting the information contained within DNA. The
transmission takes place as follows: the DNA unwinds its two chains and these,
instead of reproducing themselves identically, use uracil instead of thymine to create
a chain of RNA. The latter then separates from the original DNA to transmit its own
version of the DNA sequence to another location.

### 9.1.5  The Role of the Cell

The chemical complexity required to pass from prebiotic chemistry to life requires
the concentration of a large number of compounds into a small space. The cell avoids

these molecules being scattered within the solvent. The cell membrane separates these molecules from the exterior medium. At the centre of a cell, DNA molecules store the genetic information; they are surrounded by a solution of proteins and amino acids, and are protected by the cellular membrane, itself consisting of lipids and proteins, carbohydrates and amino acids. The cell is the basic structure of all living organisms that we know about, whether we are dealing with monocellular organisms or complex systems such as humans, who each contain some $10^{12}$ individual cells. Viruses do not consist of cells, but parasitize them to achieve their own life-cycle.

## 9.2 Prebiotic Material in the Universe

### 9.2.1 Organic Material in the Universe

Observations carried out in recent decades at millimetre wavelengths by radio telescopes have shown that the interstellar medium is rich in complex organic molecules (Table 9.2). This wavelength region is particularly suitable for such studies because of the size of the molecules: relatively large, their fundamental frequencies are lower than those of smaller molecules.

Most of the polyatomic molecules have been detected in the interstellar medium, in particular in dense molecular clouds. The temperature there is sufficiently low (10—20 K) that any gaseous molecule that encounters a solid grain is immediately captured to form a layer of ice. The organic molecules incorporated in this layer are subject to ultraviolet radiation from nearby stars, which accelerates their evolution into even more complex refractory molecules. If the grain happens to become part of the denser and hotter environment of a star in the process of formation, these molecules, returning to the gaseous state, will enrich the ambient protoplanetary medium.

The circumstellar envelopes of evolved stars, in particular of stars rich in carbon, is also a medium where complex organic molecules are found in abundance, in particular the polycyclic aromatic hydrocarbons (PAH). When the envelope of one of these stars is ejected, these molecules enrich the surrounding interstellar medium. Currently more than 120 interstellar molecules are known (Table 9.2), the largest being $HC_{11}N$, consisting of 13 atoms, 11 of which are carbon. In addition to these molecules, which have been identified individually in their gaseous phase by their spectrum at millimetre wavelengths, since the 1980s, PAHs have also been found. These are cyclic organic molecules, identified from their infrared spectra (Fig. 9.6), in which several tens of atoms are bound to their benzene groups.

**Table 9.2** Molecules detected in the interstellar medium and in circumstellar envelopes [After Gilmour and Sephton, 2003]

**2 atoms**

| | | | | |
|---|---|---|---|---|
| $H_2$ | OH | $CH^+$ | NH | $SO^+$ |
| HF | HCl | $C_2$ | CN | CO |
| CSi | CP | CS | NO | NS |
| OS | NaCl | KCl | AlF | AlCl |
| PN | SiN | SiO | SiS | $CO^+$ |

**3 atoms**

| | | | | |
|---|---|---|---|---|
| $H_3^+$ | $CH_2$ | $NH_2$ | $H_2O$ | $H_2S$ |
| $C_2H$ | HCN | HNC | HCO | $HCO^+$ |
| $C_2O$ | $C_2S$ | $SiC_2$ | $CO_2$ | $N_2O$ |
| $SO_2$ | OCS | MgCN | MgNC | NaCN |

**4 atoms**

| | | | | |
|---|---|---|---|---|
| $NH_3$ | $H_3^+$ | $C_2H_2$ | $H_2CO$ | $H_2CN$ |
| $HCNH^+$ | $CH_2CN$ | $C_3H$ | $c-C_3H$ | HCCN |
| HNCO | $HOCO^+$ | HNCS | $C_3N$ | $C_3O$ |
| $C_3S$ | | | | |

**5 atoms**

| | | | | |
|---|---|---|---|---|
| $CH_4$ | $SiH_4$ | $H_2C_3$ | $c-C_3H_2$ | $CH_2NH$ |
| $H_2COH^+$ | $CH_2CN$ | $NH_2CN$ | $CH_2CO$ | HCOOH |
| $C_4H$ | $HC_3N$ | HCCNC | HNCCC | $C_5$ |

**6 atoms**

| | | | | |
|---|---|---|---|---|
| $C_2H_4$ | $CH_3OH$ | $CH_3SH$ | $CH_3CN$ | $CH_3NC$ |
| $NH_2CHO$ | $H_2C_4$ | $HC_3NH^+$ | $HC_2CHO$ | $C_5H$ |
| $C_5N$ | | | | |

**7 atoms**

| | | | | |
|---|---|---|---|---|
| $CH_3NH_2$ | $CH_3CCH$ | $CH_3CHO$ | $c-CH_2OCH_2$ | $CH_2CHCN$ |
| $C_6H$ | $HC_5N$ | $C_7(?)$ | | |

**8 atoms**

| | | | |
|---|---|---|---|
| $HCOOCH_3$ | $CH_3COOH$ | $CH_3C_3N$ | $H_2C_6$ |

**9 atoms**

| | | | | |
|---|---|---|---|---|
| $C_2H_5OH$ | $CH_3OCH_3$ | $C_2H_5CN$ | $CH_3C_4H$ | $C_8H$ |
| $HC_7N$ | | | | |

**10 atoms**

| | | |
|---|---|---|
| $CH_3COCH_3$ | $CH_3C_5N(?)$ | $NH_2CH_2COOH(?)$ |

**11 atoms**

$HC_9N$

**13 atoms**

$HC_{11}N$

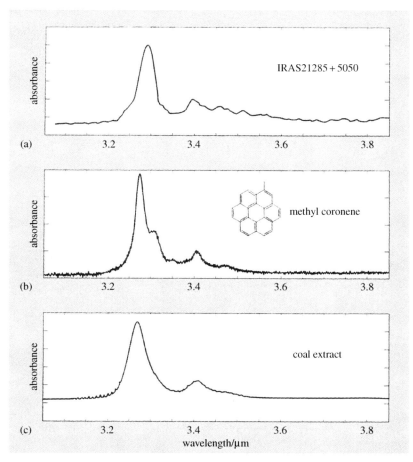

**Fig. 9.6** The infrared spectrum of a circumstellar envelope (IRAS 21282+5050) compared with a laboratory spectrum of a simple PAH molecule (methyl-coronene), and with a mixture of PAHs extracted from coal (After de Muizon et al., 1986)

## 9.2.2 The Synthesis of Organic Molecules: Miller and Urey's Experiment

How do we get from organic interstellar molecules to the amino acids necessary for the development of life? One element of the answer was provided by the decisive experiment carried out in 1953 by Stanley Miller and Harold Urey. Several decades earlier, Oparin had suggested that micro-organisms could appear, in a reducing medium, at the end of a long series of chemical reactions involving complex organic molecules. Miller and Urey undertook to simulate the atmosphere of the primordial Earth by means of a reducing medium, based on hydrogen, methane, and ammonia, in the presence of liquid water, and repeatedly subject to electrical discharges

**Fig. 9.7** The experiment by
Miller and Urey, used to
carry out abiotic synthesis
of amino acids. The lower
bulb, heated to boiling point,
simulates a liquid ocean.
The water vapour enters the
upper bulb which simulates
a reducing atmosphere based
on hydrogen, methane and
ammonia. The energy source
is provided by electrical
discharges

(Fig. 9.7). After a week, numerous complex organic molecules had formed, including amino acids.

Since Miller and Urey's historical experiment, numerous simulations have been
carried out in the laboratory, with different sources of energy and different initial
mixtures. All led to the formation of molecules that were biologically significant,
provided that the initial mixture was reducing. Here a difficulty appears: we have
seen (Chap. 3) that the Earth's primitive atmosphere, like that of Venus and Mars,
was definitely not reducing, but was dominated by carbon dioxide and nitrogen.
How can we explain the appearance of the first prebiotic molecules on Earth? Perhaps we need to envisage an external source, by way of meteoritic bombardment.

Assuming an external source of amino acids, did the primitive Earth have sources
of energy that could drive the chemical reactions that would evolve to life? Among
the possible sources (sunlight, electrical discharges from thunderstorms, shock-
waves from meteoritic impacts, etc.), solar ultraviolet radiation is by far the predominant source. Stellar evolution models suggest that solar radiation would have
been less than its current value by about 20–30 percent, but the energy contribution
of solar UV radiation to the primitive Earth was still more than one thousand times
that of the other sources of energy just mentioned.

## 9.2.3 Transport of Complex Organic Molecules to the Primordial Earth

### 9.2.3.1 The Meteoritic Bombardment

We have seen (Sect. 4.3.2.2) that the formation of planets by accretion favoured the presence of refractory, heavier elements, in relative proximity to the Sun, whereas elements such as C, N, and O were trapped in the form of ices at greater heliocentric distances, beyond the asteroid belt. This explains why carbon is more abundant at $r > 2$ AU (Fig. 9.8). Yet the presence of liquid water requires $r < 2$ AU. How can we reconcile these two factors that appear equally indispensable for the appearance of life? Meteoritic bombardment may offer a solution to the problem, as has been suggested by the astrobiologist Oro. The history of the Solar System's formation shows that an intense meteoritic bombardment occurred during the first 1000 million years, with a peak activity, the Late Heavy Bombardment (LHB) about 3800 million years ago. We see the traces of this bombardment most markedly on the surfaces of the Moon and Mercury, as well as on those of Phobos and Deimos, the two satellites of Mars, which are undoubtedly captured asteroids.

The fall of the Murchison meteorite (Fig. 9.9) in Australia in 1969, provided decisive support to the theory that prebiotic molecules arrived on the early Earth from outside. This primitive meteorite, a carbonaceous chondrite, proved to be rich in some twenty amino acids, in particular those that are found in proteins. The list of amino acids found in the Murchison meteorite, as well as their relative abundances, also shows a remarkable similarity to those formed in Miller and Urey's experiments (Table 9.3). This fundamental discovery reveals two important facts: (1) A complex prebiotic chemistry was at work from the beginning of the Solar System's history;

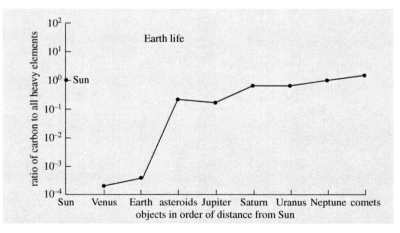

**Fig. 9.8** The abundance ratio of carbon to the combined total of heavy elements as a function of heliocentric distance (After Gilmour and Sephton, 2003)

**Fig. 9.9** The Murchison meteorite

**Table 9.3** The abundance of amino acids synthesized in Miller and Urey's experiment, compared with those found in the Murchison meteorite [After Gilmour and Sephton, 2003]

| Name of amino-acid | Abundance in the Miller-Urey experiment | Abundance in the Murchison meteorite | Found in terrestrial proteins (Y/N) |
|---|---|---|---|
| glycine | xxxx | xxxx | Y |
| alanine | xxxx | xxxx | Y |
| α-amino-N-butyric acid | xxx | xxxx | N |
| α-aminoisobutyric acid | xxxx | xx | N |
| valine | xxx | xx | Y |
| norvaline | xxx | xxx | N |
| isovaline | xx | xx | N |
| proline | xxx | x | Y |
| pipecolic acid | x | x | N |
| aspartic acid | xxx | xxx | Y |
| glutamic acid | xxx | xxx | Y |
| β-alanine | xx | xx | N |
| β-amino-N-butyric acid | xx | xx | N |
| β-aminoisobutyric acid | x | x | N |
| α-aminobutyric acid | x | xx | N |
| sacrosine | xx | xxx | N |
| N-ethylglycine | xx | xx | N |
| N-methyalanine | xx | xx | N |

(2) Prebiotic molecules necessary for the appearance of life could have been brought to the primitive Earth by meteoritic impacts, or even cometary impacts.

### 9.2.3.2 Chirality

There is another important property of organic molecules which enables us to distinguish organic molecules of biological origin from abiotic molecules: This is chirality. What is a chiral molecule? It is a molecule that has two isomers, i.e., two species with the same chemical formula but which have a different structural arrangement, depending on whether this is right-handed or left-handed. The isomer of any chiral molecule appears identical to its mirror-image (Fig. 9.10). With the exception of glycine, all the amino acids used in the construction of proteins are chiral molecules.

The special feature of living organisms on Earth is to use just one form of chirality, left-handed chirality. All proteins are built from amino acids with left-handed chirality. Only the left-handed form of amino acids can stabilize the $\alpha$-helix structure in proteins, because of the spatial position of side chains of amino acids in the polymer chains. Right-handed form of amino acids would lead to a configuration that prevents proteins from retain its helical structure, and would lead to a disruption of the molecule. With abiotic organic molecules, however, we observe equal number of both types of chirality. A mixture of amino acids of both types would lead to the formation of proteins that would not ensure their biological functions. So terrestrial life, consisting of left-handed chiral molecules, could develop subsequently only with that type of chirality.

What is the chirality of the amino acids found in the Murchison meteorite and other carbonaceous chondrites? There again, we find an excess of left-handed chirality, which tends to reinforce the theory that prebiotic molecules arrived on Earth from outside. It is purely chance that has favoured left-handed chirality in the prebiotic molecules that have arrived from space? One possible explanation is that it could be the effect of circularly polarized, ultraviolet stellar radiation which tended to preferentially destroy right-handed chiral molecules (whether these were abiotic or not), and thus create an excess of the left-handed form. This phenomenon may be explained by the fact that in circularly polarized light, the electric field direction rotates along the beam. It is thus a chiral phenomenon. Left- and right-handed molecules have different absorption intensities for left- and right-circularly

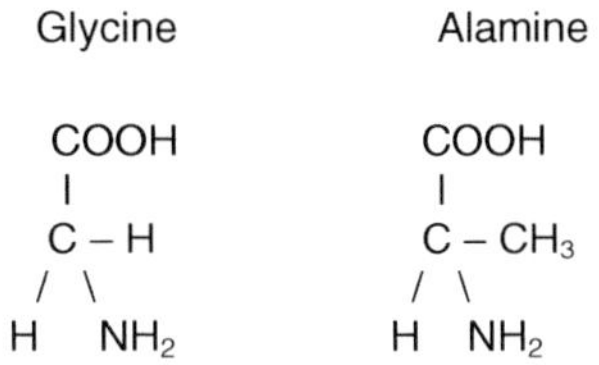

```
Glycine              Alamine

COOH                 COOH
  |                    |
C – H                C – CH₃
/ \                  / \
H   NH₂              H   NH₂
```

**Fig. 9.10** Examples of a non-chiral molecule (glycine, *left*) and a chiral molecule (alanine, *right*) (After Gilmour and Sephton, 2003)

polarized light. The resulting photolysis efficiency therefore depends both on the
light and on the types of molecules.


## 9.3 Stages on the Road to Complexity

### 9.3.1 Polymers and Macromolecules

We should perhaps first note that the mechanisms by which organic molecules as-
semble to become a living organism are far from being understood. In this section,
we follow the description of Gilmour and Sephton (2003) and a number of processes that are likely to have been involved in the development of ever more complex
molecules.

How can we generate macromolecules from simple organic molecules? The first
mechanism is polymerization, which enables monomers to combine, via a chemical
reaction that involves the loss of a water molecule. For example, the –OH groups
in two sugars may combine with the loss of $H_2O$ (Fig. 9.11). It is the same for the
–$NH_2$ and –COOH groups in two amino acids. The new molecules may themselves
combine to form increasingly complex chains. However, polymerization is the presence of liquid water is difficult, because the water tends to destroy polymers.

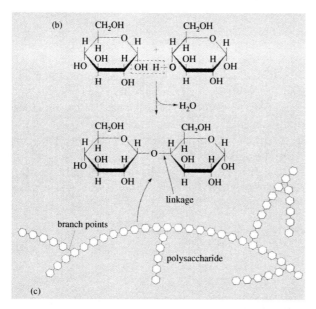

Fig. 9.11 An example of the polymerization of two sugars (glucose and ribose). The H and OH
bonds link to form a molecule of water, which is lost, giving rise to a bond between two monomers
(After Gilmour and Sephton, 2003)

## 9.3.2 The Formation of Membranes

The formation of membranes is an indispensable stage in increasing complexity, because they are able to isolate and protect macromolecules with specific functions from external influences. Their formation enables cells to be constructed.

How can a membrane form? Certain molecules have the property of being ambiphilic, that is, they have one end that is hydrophilic (polar) and the other hydrophobic (non-polar). Once immersed in liquid water the ambiphilic molecules form on the surface, the hydrophilic heads in contact with the liquid water and the hydrophobic tail pointing outwards. They thus form a monomolecular layer (Fig. 9.12). Several types of ambiphilic molecules are able to form more complex structures, such as multiple layers or spherical structures, which were without any doubt at the origin of cellular membranes (Fig. 9.13).

Laboratory experiments (Sydney Fox, 1958) have shown that polymerization of amino acids may lead to the formation of 'proteinoids', which once dissolved in hot water and then cooled, cluster together in microspheres with a double membrane. In 1985, David Deamer observed that the amino acids in the Murchison meteorite, once dissolved in water, also formed micro-bubbles with membranes. These structures therefore have the property of being able to withstand an episode of dehydration, and then be re-activated in an aqueous medium.

## 9.3.3 RNA and DNA

Life as we know it uses DNA to store genetic information and RNA to transmit this information within the cell. In addition, specific proteins, the enzymes, are necessary to catalyze reactions, which the nucleic acids cannot do. In what order did these three types of molecules appear?

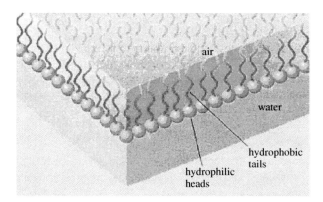

**Fig. 9.12** Formation of a monomolecular layer of a lipid (After Gilmour and Sephton, 2003)

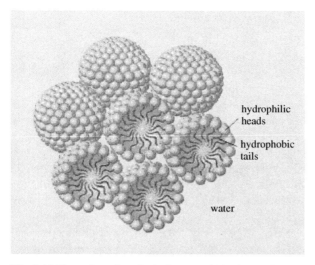

**Fig. 9.13** Formation of a spherical monocellular layer; the sphere is bathed in water, which surrounds it and isolates a pocket of air within its interior (After Gilmour and Sephton, 2003)

Following the work of Sydney Altman and Thomas Cech, biologists have come to the conclusion that in the first stage, only RNA was present. In fact, it appears that RNA may carry out some of the enzymatic functions required for replication. In addition, the RNA nucleotides are more easily synthesized than those of DNA. DNA and the proteins necessary for catalysis arose later, DNA having a far more effective replication mechanism than RNA.

## 9.4 The Appearance of Life on the Primitive Earth

### 9.4.1 Favourable Conditions

Lying 1 AU from the Sun, Earth has benefited, ever since its formation, with a temperature that enabled liquid water to exist. Subsequently, it also acquired the conditions that allowed it to preserve this favourable environment. One essential process is plate tectonics, which ensures that atmospheric carbon dioxide is recycled. Plate tectonics, driven by the Earth's internal energy (whose origin is the radioactive decay of unstable isotopes within the interior), recycles oceanic crust by carrying it down into the mantle at subduction zones, and regenerating it at mid-ocean ridges. The presence of a moderate amount of $CO_2$ in the terrestrial atmosphere ensures that there is a greenhouse effect sufficient to raise the planet's equilibrium temperature by some thirty degrees, which allows the water in the oceans to remain liquid. The mass of the Earth plays a vital role here, because it is what provides, via radioactivity, sufficient internal energy to drive plate tectonics and volcanism.

We may note that the Earth's geological record appears to show that the planet has, in the past, suffered from almost complete glaciation. The cooling process is one that is self-reinforcing: if the surface temperature decreases, the area of the polar caps increases, which increases the planet's overall albedo. So the planet absorbs less solar energy and its temperature tends to decrease even farther, which increases the glaciation. It seems that several episodes of global glaciation took place between 750 and 580 million year BP. It was probably a violent volcanic episode that re-established the greenhouse effect by injecting a large amount of carbon dioxide into the atmosphere.

## 9.4.2 The Environment of the Primitive Earth: The Hydrosphere and Atmosphere

### 9.4.2.1 A Primordial Reducing Atmosphere?

The question of the composition of the primordial atmosphere remains far from being settled. Sources of hydrogen and of reduced compounds must have been significant on the early Earth. There would have been emission of volcanic $H_2$; oxidation of iron by water, liberating $H_2$; and the formation of $CH_4$, $NH_3$, and more complex organic molecules in the presence of $H_2$ in hydrothermal systems. $CH_4$ and $NH_3$ are extremely fragile molecules once in the atmosphere, where they are oxidized and photodissociated by UV radiation. As for hydrogen, it escapes from the top of the atmosphere. It is the balance between the emission of $H_2$, $CH_4$, and $NH_3$, and the photochemical destruction of $CH_4$ and $NH_3$ and the loss of hydrogen from the top of the atmosphere that determines whether the atmosphere was reducing or not, and the effectiveness of organic synthesis of the sort that Urey and Miller carried out.

The estimate of the escape of H has been recently drastically revised downwards (Tian et al., 2005), which has revived the discussion about the composition of the primordial atmosphere, because those authors found that a rate of emission of $H_2$ comparable with current volcanic emissions, in an atmosphere of $N_2$ (0.8 bar) and $CO_2$ (0.2 bar) leads to an atmosphere containing more than 0.1 bar of $H_2$, instead of values below 1 mbar, found in work that was regarded as standard (Kasting, 1993). With even more significant emissions, for example at $3.8 \times 10^9$ years BP, Tian et al. found even greater abundances, and atmospheres where $H_2$ is the dominant component. The difference essentially arises from the fact that Kasting (1993) assumed that the loss of H to space was limited only by its diffusion through the atmosphere. So, if the planet's exosphere is too cold, it is the temperature that will limit any loss to space to far smaller amounts (unless non-thermal escape mechanisms are dominant) and maintain the abundance of atmospheric hydrogen at low values. This is a question that is still unanswered.

The time parameter is also very important. For a long time it has been believed that life could not have appeared and persisted until about $3.8 \times 10^9$ years ago, because the bombardment would have sterilized the Earth at earlier epochs. At a period $3.8 \times 10^9$ years ago the degassing of the Earth should be similar to present-day degassing, both in composition (because there are lavas of that age that have a 'modern' composition), and in intensity (to within a factor of 10), because the internal heat flux rapidly decreased during the first 500 million years. However, if we envisage a much earlier origin of life, around $4.3 \times 10^9$ years BP, then we can also imagine that the emission of gas from the primordial Earth was far more intense and rich in $H_2$, $CH_4$, and $NH_3$, producing an atmosphere of the Urey-Miller type. It transpires that the story of the bombardment of the early Earth has been recently rewritten (Gomes et al., 2005). The peak bombardment at 3.9–3.8 $\times 10^9$ years BP could have been far less intense than was thought, and be preceded by a much calmer phase between the end of the Earth's accretion phase and the peak, i.e., between about 4.4 and 3.9–3.8 $\times 10^9$ years ago. The peak bombardment was a traumatic episode for the Earth, but if some sufficiently complex microbial life was already present, and was capable of adapting, it could well have survived. In that case, there is nothing to prevent us from considering the theory of very precocious life-forms arising in a reducing atmosphere. However, if we consider a primordial atmosphere, we then need to take account of the effects of the intense X-ray and EUV radiation that was emitted by the young Sun. This radiation heated the upper atmosphere and photodissociated the molecules, accelerating the loss of H to space. Once again, a coherent model remains to be developed.

In the absence of any strong constraints on the time of the atmosphere's origin, and on its nature, it is important to consider the possibility that in-situ organic synthesis was ineffective, and that an external source of organic material did play a role.

### 9.4.2.2  The Origin of the Earth's Water

Meteorites contain a greater or lesser fraction of water as a function of the orbital distance of their parent bodies. Those that formed at distances greater than about 2.5 AU contain more than 10 percent of water in the form of hydrated minerals. Still farther out, beyond the ice line, the typical proportion of water reaches 50 percent, and this time essentially in the form of ice. In contrast, the asteroids that are found to be within 2.5 AU are almost completely depleted in water. The fact that the asteroid populations have not merged suggest that their distribution is a real sign of the properties of the protosolar nebula at distances where we observe them nowadays, and that there has been no major migration of these populations.

So, if the Earth itself formed from material that essentially formed within 2 AU, which is what the models of planetary formation suggested until recently, it should initially have been without water. Water must have been brought by a bombardment by bodies coming from more distant regions, either asteroids or micrometeorites rich in water, or comets. We know that cometary ices could have made only a limited contribution (less than 10 percent) because the water that they contain is richer in deuterium than the terrestrial oceans. The isotopic composition of carbonaceous chondrites (rocks formed at distances $> 2.5$ AU) and of micrometeorites (microscopic, rocky debris from comets or asteroids), is compatible with that of water on Earth.

At first it was thought that the water could have been brought by asteroids at the time of the late bombardment at $3.9 \times 10^9$ years BP. We now know, however, thanks to the analysis of terrestrial zircons dating from more than $4.3 \times 10^9$ years ago, that the Earth already possessed oceans at that time (Martin et al., 2006). So we have to explain an extremely early influx. A continuous addition by micrometeorites has been suggested (Maurette et al., 2000) but, in fact, the problem of the origin of the water is in the process of being resolved quite simply, thanks to new numerical models of planetary formation, which have been made possible by the power of modern computers. These N-body models are, in fact, now able to work with larger and larger values of N, and thus begin with far more realistic distributions of planetesimals. The most recent models show that the Earth did not form solely from material in a narrow ring around its orbit at 1 AU, but that planetesimals that originated much farther out were involved in the formation of the Earth. So an ad-hoc later influx of water is no longer necessary, and we now find ourselves with the opposite situation, because the models form 'Earths' that are too rich in water, sometimes by as much as a factor of 100. But this result is actually logical, because the models do not include the phenomena of the loss of volatile components through impacts between planetesimals, nor the degassing associated with their differentiation. In fact, the planetary embryos were sufficiently hot to degas an atmosphere, but were not sufficiently massive to retain it effectively. The composition of planetary atmospheres, notably in the rare gases and their isotopes, retains a trace of the violent events accompanying atmospheric escape (see Fig. 7.22, Chap. 7). For a long time it was thought that this escape must have taken place from the primordial atmospheres of the planets, once these had formed. But it now seems more logical to believe that these violent escape processes took place on the planetesimals and planetary embryos. So it is perfectly normal – and even essential – for the quantity of water and volatile components that were initially present in all the planetesimals that were involved in the formation of the Earth, to greatly exceed present-day reservoirs on the planets.

## 9.5 The Search for Habitable Locations in the Solar System

### 9.5.1 The Planet Mars

#### 9.5.1.1 The Story of the Search for Life

Apart from its distance from the Sun, Mars is, after Earth, the planet that is best placed to be searched for signs of life, and numerous past scientists and philosophers have raised the possibility. In 1877, the Italian astronomer Schiaparelli announced a discovery that caused a sensation: the presence on Mars of linear structures called 'canali' (Fig. 9.14). Although he himself was extremely cautious about the interpretation of his results, other astronomers, beginning with Percival Lowell, saw them as traces of an intelligent civilization. Despite the reservations of other astronomers such as Eugène Antoniadi (Fig. 9.15), the myth of intelligent life on Mars persisted until the 1960s. The first images returned by spaceprobes showed that the linear structures were no more than optical illusions.

In 1975, NASA launched Viking, an ambitious mission to Mars, the main objective of which was search for life on Mars. Two spaceprobes were placed in orbit around the planet, and continued to function for several years, while two descent modules landed on the surface of Mars, to carry out in-situ research (Fig. 9.16). A set of three biological experiments aimed to search for signs of biological activity by means of various tests, including analysis of the degassing of the soil in the presence of a nutrient medium, and search for potential photosynthetic activity of any possible carbon compounds. In terms of these investigations, scientists have concluded that biological activity is absent from the surface of Mars, at the two locations where the modules landed. The total absence of organic molecules on the surface of Mars has been attributed to the irradiation by ultraviolet solar radiation. Note that these experiments do not exclude the possible existence of past or present life at sites that have been protected from destructive solar radiation.

### 9.5.1.2 The History of Water on Mars

Following the Viking era, space missions to Mars multiplied. The most recent had the prime objective of searching for signs of past liquid water on the surface. We know, in fact (Chap. 4), that current temperature and pressure conditions do not permit liquid water to flow on Mars today. But several indications, acquired from the Viking images, bear witness to its presence in the past: branching networks of valleys in ancient terrains (Fig. 9.17); outwash valleys near the Vallis Marineris canyon system; craters with lobate ejects, possibly indicating the presence of subsurface liquid, etc. New, decisive evidence has been acquired during the course of recent years. In 2002, the Mars Odyssey probe, detected the presence of a significant quantity of ice beneath the poles. The result has been confirmed by the infrared spectrometer on the European Mars Express probe, which detected the spectral signature of water ice beneath the perennial $CO_2$ ice cap at the poles. In another remarkable result, the same instrument has detected, near Vallis Marineris, various types of sulphates, whose formation appears to have required the presence of liquid water. Similarly, the American robot Opportunity has discovered spherules consisting of sulfates at the Meridiani landing site.

Water therefore flowed on Mars in the distant past, but the central question is: in what abundance, and until what date? Did it last sufficiently long to have allowed the appearance and development of life? Scientists are reticent on this point. The

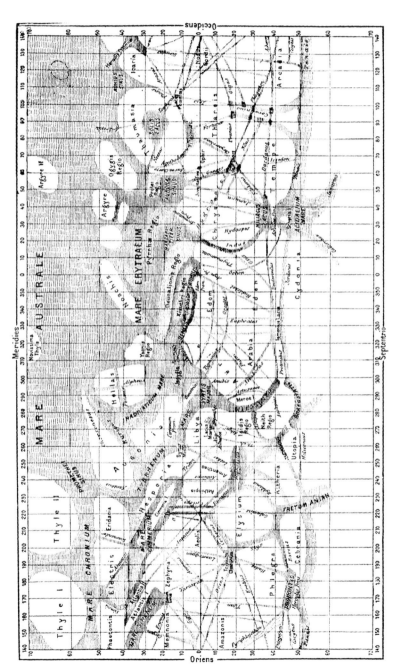

**Fig. 9.14** Map of the surface of Mars, drawn by Schiaparelli in 1879. The linear structures, called 'canali' were interpreted as signs of an intelligent civilization

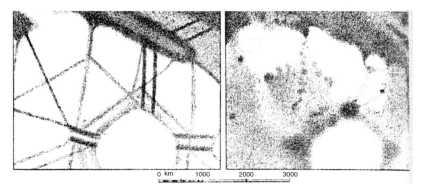

0 km      1000        2000        3000

**Fig. 9.15** The region of Elysium Planitia on Mars, as drawn by Schiaparelli between 1877 and 1890 (*left*) and by Antoniadi between 1909 and 1926 (*right*). Antoniadi's drawings showed that the quasi-linear structures seen by Schiaparelli were nothing like that when spatial resolution was increased (After Antoniadi, *La planète Mars*, 1930)

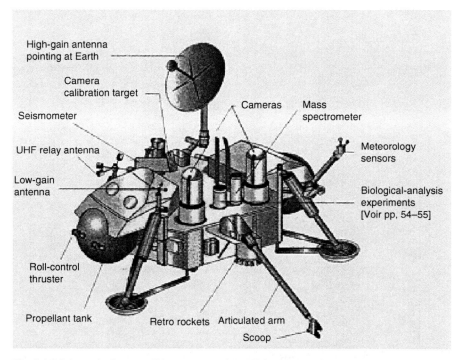

**Fig. 9.16** Schematic diagram of the structure of the Viking landers. Once on the surface, the probe used the energy from a plutonium-based radioactive source. The parabolic antenna allows communication with the orbiter or directly with Earth. The two descent probes carried identical equipment, including cameras, meteorological instruments, mass spectrometers and X-ray fluorescence spectrometers, a seismometer, and the biological experiments (NASA)

absence of hydrated minerals on the northern plains seems to question the theory that they were covered in liquid water and the theory of a possible northern ocean. Future Martian exploration missions will continue to give priority to the search for signs of liquid water, to measure the quantity of sub-surface water by the use of radar, as well as drilling into the surface.

## 9.5.2 The Satellites of the Outer Planets

Although too far from the Sun to support liquid water on the surface, the satellites of the outer planets may provide interesting niches for exobiology. In fact, the phase diagram for water shows that liquid water may exist at high pressure within the interiors of these objects, and models of the internal structure of the satellites predict this possibility.

(a)

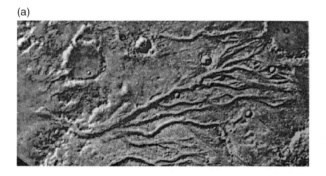

(b)

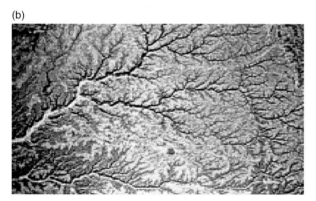

**Fig. 9.17** (**a**) A Martian network of branching valleys (near Chryse Planitia) compared with (**b**) a network of dry valleys in Yemen (After Forget et al., 2003)

### 9.5.2.1 Europa and the Galilean Satellites

The four Galilean satellites of Jupiter were first explored by the Voyager mission in 1979, and then by the Galileo orbiter between 1995 and 2003. With the exception of Io, which is very close to Jupiter and endowed with active volcanism, they are all covered in water ice. Europa (Fig. 9.18), the second Galilean satellite, is the most interesting from the point of view of exobiology. Like Io, but to a lesser extent, it is, in fact, subject to tidal forces produced by its proximity to Jupiter. It is also in resonance with Io and Ganymede. The result is a certain internal energy, sufficient, according to models of its internal structure, to keep water in liquid form, down to the silicate mantle.

The Voyager images, confirmed by those from Galileo, have shown that the surface of Europa consists of a complex network of linear, intersecting structures. The plates that they surround appear to have been displaced relative to one another on a viscous or even liquid medium. The best candidate for this medium is liquid water, probably saline (Fig. 9.19). Another sign appears to reinforce this theory: a magnetic field has been detected by Galileo; it could be generated by a dynamo effect within the medium.

How may the liquid ocean on Europa be explored? A first stage would consist of verifying its existence, by gravimetric measurements or radar sounding from an orbiter. It would then be necessary to evaluate the thickness of the layer of ice, which, according to the models, could be several tens of kilometres thick. The in-situ exploration of the interior of Europa therefore seems, from the outset, to be a long-term project.

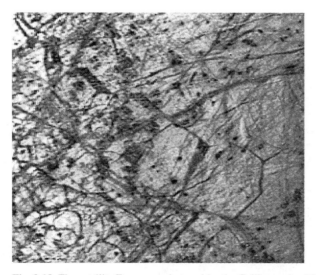

**Fig. 9.18** The satellite Europa as observed by the Galileo probe (NASA)

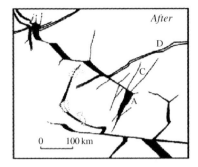 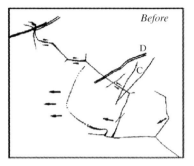

**Fig. 9.19** An example of the features on the surface of Europa: *left*, an image from the Galileo probe, *right*, interpretation of these features in terms of moving plates

Beyond Europa, Ganymede and Callisto are probably too far from Jupiter for tidal effects to generate sufficient energy to produce a liquid ocean deep enough to reach the silicate mantle.

### 9.5.2.2  Enceladus and the Icy Satellites of Saturn

In Saturn's environment, the satellites that are most likely to possess an internal, liquid ocean are those that undergo sufficiently strong tidal forces. In this respect, Enceladus is a special case. Confirming the first images taken by Voyager, the observations made by the Cassini probe have shown that the region around the south pole is much younger and hotter than the rest of the surface; plumes have also been detected, showing evidence for active cryovolcanism.

### 9.5.2.3  Titan

Titan, the largest of Saturn's satellites, is exceptional in the outer Solar System: it is the only satellite to possess a thick atmosphere. In addition, that atmosphere, like the Earth's, primarily consists of molecular nitrogen, and its surface pressure (1.5 bars) is close to that of the Earth's atmosphere.

It was the Voyager mission that revealed, in 1980, the nature of Titan, its atmospheric composition, and thermal structure. It also discovered the presence of a large number of hydrocarbon molecules and nitriles (Fig. 9.20). The hydrocarbons ($C_2H_2$, $C_2H_4$, $C_2H_6$, etc.) arose from the photochemistry of methane, as in the case of the gas giants; the presence of nitriles ($HCN$, $C_2N_2$) results from the dissociation of molecular nitrogen by energetic particles from Saturn's magnetosphere. Subsequently, $HC_3N$ and $CH_3CN$ were also detected. These molecules are precisely those that were predicted from laboratory experiments simulating the synthesis of complex organic molecules from simple mixtures, in the presence of energy sources.

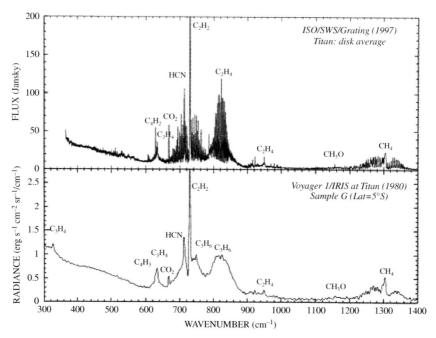

**Fig. 9.20** The infrared spectrum of Titan observed by the SWS instrument on the ISO space observatory (*top*), and by the IRIS instrument on Voyager 1 (*bottom*)

After this discovery, Titan appeared as a unique laboratory for prebiotic chemistry, capable of providing us with information about the first stages in the emergence of life, whether on Earth or in other environments. Note, however, that despite the similarities with Earth, the two objects reveal enormous differences. The first is Titan's low temperature (93 K at the surface): its first effect must be to slow down all chemical reactions by a considerable amount. The second is the atmospheric composition: that of Titan is reducing, which resembles the conditions used in Miller and Urey's simulations. Finally, the internal structure of the Earth and Titan are radically different. On Titan, there is permanent degassing of methane and perhaps active cryovolcanism based on $H_2O$ and $NH_3$.

On 14 January 2005, the Huygens probe successfully landed on the surface of Titan (Fig. 9.21). Its first images revealed a relatively flat surface, scattered with heavily eroded pebbles, which probably consist of water ice, and showing signs of fluvial outwash features, probably created by hydrocarbons (Fig. 9.22). The data from the Cassini orbiter and the Huygens probe are still currently being analyzed, and should give us a better understanding of the nature and origin this exceptional satellite.

**Fig. 9.21** The surface of Titan as observed by the DISR descent camera on the Huygens probe, 14 January 2005. The rounded pebbles probably consist of water ice (image credit: courtesy ESA; NASA, JPL University of Arizona)

**Fig. 9.22** Panorama of the surface of Titan, observed by the Huygens probe at an altitude of 8 km. Clearly visible are the traces of a dark fluvial network (ESA)

## 9.6 The Search for Life on Exoplanets

### 9.6.1 Exoplanets' Habitable Zones

The habitable zone in a planetary system (*see* Sect. 7.3.2.1) is generally defined as the region in which a terrestrial-type planet may have liquid water at its surface during part of its year (the period of revolution around its parent star). To calculate the location of this zone (a ring) as a function of the star's type, generally the following assumptions are made:

- the surface temperature of the planet is primarily caused by the heat flux from the star. As a result, all possible sources of internal energy (tidal forces, radioactivity, etc.) are ignored;
- the energy flux received by the planet corresponds to that emitted by the star, multiplied by a coefficient that lies between 0 and 1 and corresponding to the planet's albedo (reflection coefficient). The albedo has, in principle, a chromatic value, but it is considered to be constant over the whole range of radiation received from the star;
- thermodynamical equilibrium requires the incident flux to be equal to the flux emitted by the planet. From this we can deduce the effective temperature of the planet $T_{pl\ eff}$ from the following equation:

$$F(1-A)\ \pi\ \frac{r_{pl}}{D^2} = 4\ \pi\ r_{pl}^2\ \sigma\ T_{pl\,eff} \qquad (9.1)$$

where $F$ is the energy flux received from the star at distance $D$, $r_{pl}$ is the planet's radius, A is its mean albedo, and $\sigma$ is the Stefan-Boltzmann constant.

The temperature estimated by the method just described most certainly does not take account of the effects caused by the existence (or lack of) an atmosphere, in particular, greenhouse gases. Assuming an albedo of 0.5 for Venus, for example, theoretically we obtain an effective temperature of 277 K, which neither corresponds to the surface temperature of Venus (more than 700 K), nor to the atmospheric temperature. The notion of the habitable zone should not be considered as a very rigid one.

Using the definition just given, the habitable zone in the Solar System extends approximately from 0.9 to 1.3 AU (*see* Sect. 7.3.2.1).

As just mentioned, in calculating the habitable zone, no account is taken of sources of internal heat. However, Europa (the second Galilean satellite of Jupiter) and, more recently, Enceladus (a satellite of Saturn) have shown indications that lead us to believe that there is liquid water beneath their frozen surfaces (see Sect. 9.5.2). In the case of Europa, the presence of the liquid water is ascribed to the dissipation internally of energy from tidal effects. Although located well outside the habitable zone, Europa and probably Enceladus do correspond to the definition of the habitable zone that we have given. Several definitions of the habitable zone have been advanced that take account of possible internal sources of heat, but our initial definition that takes just stellar radiation into account is by far the most commonly-used.

The notion of a 'continuously habitable zone' may also be defined. This is the zone within which the surface temperature of a planet allows the presence of liquid water over a specific period (for example 1000 million years). The zone thus defined is narrower than the habitable zone proper, and assumes several factors regarding the orbits of planets. It therefore needs to be used with even greater caution than the simple notion of a habitable zone.

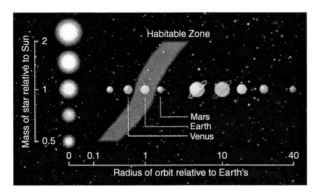

**Fig. 9.23** The extent of the habitable zone as a function of the mass of the parent star

Spectroscopic observation of a star enables us to determine its spectral type and its equivalent temperature. It is then possible to determine the extent of the habitable zone for each spectral type. Figure 9.23 shows the position of the habitable zone, as defined earlier, as a function of stellar type.

Even if it is not absolute, the notion of a habitable zone is extremely useful in determining the favourite or at least what are, on the face of it, the most favourable zones in which to search for habitable planets.

### 9.6.2 How May Life on an Exoplanet be Detected?

Given the distance that separates the Earth from even the closest stars (apart from the Sun, the star closest to the Earth, Proxima Centauri, is about 4.3 light-years away), it seem inconceivable that one could contemplate searching for life, in situ, at any time in the near future. Any such 'on site' exploration assumes:

- identification of a target where the presence of life has almost been established;
- mastery of sending high-speed interstellar missions – at one tenth of the speed of light, which would mean a real feat of interstellar navigation – it would take about 45 years to reach Proxima Centauri.

as a result, the lead time for such a mission would be several generations, with the risk that any probe would immediately become obsolete right after its launch or at least during its theoretical lifetime.

To detect life on an exoplanet, we will therefore consider only those methods that involve doing so from at distance, by observation of the object. That therefore supposes that life on a planet should be effectively detectable at a distance, and that it should have developed on a planetary scale. In particular, if life has developed in a very restricted ecological niche, such as purely locally, at the bottom of a lake, for example, or in a specific environment (with specific requirements for acidity, salinity or any other particular condition), it could well be undetectable. We will also assume that life as it developed on a planetary scale has modified the planet's environment, such as the composition of the atmosphere, or the electromagnetic environment. The latter depends on whether life has evolved into a form that is capable of communicating through electromagnetic radiation. (*see* Sect. 9.7.3).

Here, we shall deal solely with the question of detecting life through a search for spectral signatures (subsequently called 'biomarkers' or biological signatures) in the planet's atmosphere.

Such spectral signatures are obtained by spectroscopy from the light coming from the planet. This technique generally requires direct observation of the planet (separating the photons from the planet from those emitted by the star), and several projects have been described that would be capable of doing so (DARWIN, TPF, etc.). However, the differential spectroscopic analysis of the light from the star before and during a planetary transit – an indirect method of detection – has already allowed the detection of several species in the atmosphere of HD 20945B, thanks to the HST, in particular Na, H, C, and O (Charbonneau et al., 2002; Vidal-Madjar et al., 2004). The method may also be applied to secondary transits (behind the star), or through observations made in the thermal infrared in phase with the object's period of revolution. In this last case, the infrared excess is measured before the eclipse, relative to the star's flux during the eclipse. Deming et al. (2005) and Charbonneau et al. (2005) have used this technique with the Spitzer space telescope to carry out photometry in four spectral bands of HD 20945B and Tr-ES 1. This technique, if carried out by the JWST, should allow us to just about obtain low-resolution spectra of some giant planets that transit. Given the low signal-to-noise ratio of the method (because of the significant infrared flux from the star), it is limited to the observation of giant planets, and spectroscopy of terrestrial-type planets requires direct observation.

### 9.6.2.1  Spectral Signatures of Life

Any choice of biological signatures assumes a preliminary definition of what called life, and the theories on which the definition is based (see Sect. 9.1). This subject on its own could fill a complete book.

All the classic definitions of life cite a structured, material system, possessing a certain coded content of information. Chemical coding is one of the most obvious solutions, in particular, one based on the chemistry of carbon. The main reason is that on Earth, but also in the interstellar medium, carbon exists abundantly in several oxidized (e.g., $CO_2$) and reduced (e.g., $CH_4$), forms. This allows the formation of a considerable number of different combinations (molecules), which in turn simultaneously allow very diverse forms of coding, and of encoded structures. In addition, it is relatively easy to pass from one to the other by processes that require a limited amount of energy.

If we assume life where the coded information is based on carbon chemistry, then a large quantity of carbon in reduced form (i.e., organic molecules) is required.

However, the principal basic material is $CO_2$, that is, an oxidized form. The reaction required to turn oxidized carbon into an organic carbon compound is carried out on Earth by cyanobacteria and plants, and is photosynthesis, as mentioned earlier in this section. This reaction results in the release of oxygen.

Because the reaction is reversible (organic matter oxidizes), and because at the same time oxygen can oxidize other, highly reducing species (such as volcanic gases, metals, rocks, etc.) the massive presence of oxygen in an atmosphere may be considered as proof of the existence of biological activity. This point was put forward by Owen in 1980, and used in 1993 by Sagan and his colleagues when the Galileo probe flew past the Earth, to detect life on Earth from a distance by observing the simultaneous presence of oxygen (the oxidant) and methane (the reducing agent) – two species that could not exist in equilibrium – in the Earth's atmosphere. At the same time, if all biological production of oxygen were to cease, it would take only about 5 to 10 million years for the atmospheric oxygen to be consumed by the oxidation process and for it to disappear from the air.

Since 1980, several studies have been carried out to evaluate this criterion, by attempting to see if it is possible to generate, by abiotic means, large quantities of oxygen that could persist in the atmosphere of a planet. In particular, photolysis of $CO_2$ has been suggested (Rosenqvist and Chassefière, 1995), as well as the photolysis of water, through a runaway greenhouse effect. Selsis (2008) has shown, in particular by studying the sensitivity of chemical species to their environment (type of star, distance from the star, etc.) that the criterion should be modified and that to obtain a reliable criterion, rather than just the detection of $O_2$, the triple detection of $CO_2$, $H_2O$ and $O_2$, or even better, $O_3$, would be preferable as a marker for oxygen. Oxygen ($O_2$), in fact, has no spectral signature in the thermal infrared. Ozone ($O_3$), is a dissociation product of $O_2$ caused by ultraviolet photons from the star (via the Chapman cycle). Ozone is a good marker of the quantity of oxygen in an atmosphere. (There is a complete description of the discussion of the relevance of $O_3$ as a marker for $O_2$ in Léger et al., 1993, and in Selsis et al., 2008.)

Several other biological signatures have been suggested, but do not appear to be as effective as the triple detection of $CO_2$, $H_2O$, and $O_3$:

- As mentioned earlier, the simultaneous detection of species out of thermodynamic equilibrium may be proof of biological-type activity, for example $CH_4$ and $O_2$, or $CH_4$ and $H_2O$. It is, however, difficult to extract a simple, applicable, spectroscopic criterion, without an in-depth knowledge of the planet's atmosphere and without modelling the different interactions.
- The detection of technological gases (gases the synthesis of which is impossible or highly unlikely naturally), seems to be a good tracer of biological activity, but requires evolution of a technological civilization. Study of such tracers therefore resembles a SETI-type search (*see* Sect. 9.7.3).

The question of the composition of planetary atmospheres and the corresponding spectra has been treated in more detail in Sect. 7.3.

#### 9.6.2.2 Choice of Spectral Region

The choice of the spectral region for observation is a compromise between:

- the presence or absence of spectral signatures of the elements being sought in the region concerned,
- the ease with which these spectral signatures may be detected (width of lines, planetary flux, and stellar flux).

The contrast between the electromagnetic spectrum of Earth and the Sun has already been mentioned in Chap. 2. It may be shown that this contrast is minimal in the radio region, but in that case the planetary flux is extremely low.

The planetary flux received at the Earth is a maximum in the thermal infrared (about 10 photons per second per square metre of detector for a planet like the Earth at a distance of 10 parsecs, observing in the 6–20 µm spectral band). In such a case, the contrast with the star is about 7 million. The thermal infrared is also a region suitable for the detection of all the other atmospheric gases, because it is the wavelength region that corresponds to the vibration of the molecules. So it is possible to envisage the detection of $CH_4$, NO, $NO_2$, $SO_2$, and all the asymmetrical molecules that are present in large quantities in the atmosphere.

It is also possible to envisage being able to detect the three species $CO_2$, $H_2O$ and $O_2$ in the visible/near infrared region. Here, the number of chemical species that may be identified is reduced (being the band in which there are the harmonics of the fundamental frequencies of the various molecules), the planetary flux is weaker (by a factor of about 30 for the 0.5–2 µm band relative to the 6–20 µm band), and the contrast is higher by a factor of about 1000. However, certain projects with dedicated coronagraphs that are adapted to large-size telescopes (of a few metres) are capable of spectroscopic observation of terrestrial-type planets in the visible region.

## 9.7 The Search for Extraterrestrial Civilisations

### 9.7.1 The Drake and Sagan Equation

What is the probability that an exoplanet shelters a form of life? This question was posed long before the discovery of the first exoplanets. In the 1970s, the American astronomers Frank Drake and Carl Sagan tried to express the problem in figures. If N is the number of planets in the Galaxy that currently shelter some form of life, we may write:

$$N_{civil} = N^* \cdot F_{pl} \cdot F_{habit} \cdot F_{life} \cdot F_{civil} \cdot \frac{<T_{civil}>}{T^*} \tag{9.2}$$

In this equation, known as the Drake Equation, $N^*$ is the number of stars in our galaxy; $F_{pl}$ is the percentage of stars with planets; $F_{habit}$ is the fraction of these planets that are habitable; $F_{civil}$ is the fraction of inhabited planets where life has led to a technological civilization capable of communicating, $<T_{civil}>$ is the average lifetime of a technological civilization, and $T^*$ is the lifetime of stars.

The first two parameters are relatively well-known. If we take stars close to the Sun's spectral type, $N^*$ is about $10^9$. $F_{pl}$, according to the first results in the search for exoplanets, may be about 0.1 to 0.2; this parameter should be appreciably improved in the next decade. The lifetime of stars is known. It is a function of their spectral type. In contrast, the other parameters remain utterly unknown at the present time.

### 9.7.2 Communication by Radio Waves

The Earth, with its permanent source of emissions at radio wavelengths, betrays the presence of an advanced technological civilization. For several decades, radioastronomers have studied the possibility of communicating with radio waves with possible extraterrestrial civilizations. To carry their messages they have chosen a specific transition of neutral hydrogen, at a wavelength of 21 cm, which corresponds to a frequency of 1420 MHz. The choice of this frequency is explained by the fact that hydrogen is the most abundant element in the Universe; in addition, the atmosphere of the Earth, like that of all exoplanets that resemble it, is transparent in that spectral region. It is therefore possible to use very large radio telescopes from the surface of the Earth and to monitor the sky permanently (Fig. 9.24).

### 9.7.3 The State of SETI and CETI Searches

The SETI project (Search for Extraterrestrial Intelligence) arose in the United States in 1984 on the basis of this concept. Initially financed by NASA, it func-

**Fig. 9.24** The radio telescope at Arecibo in Puerto Rico

tions today on private funding. It represents a vast international collaboration, and finances numerous research projects, primarily oriented towards listening for possible 'intelligent' signals emitted in the cosmos. Although the search may have been unsuccessful so far, it nevertheless bears witness to the collective interest in this quest, which goes far beyond the scientific community. Note that the SETI project has an equivalent in the visible region, OSETI, which may benefit in future from technological advances linked to the development of high-powered pulsed laser systems.

Beyond simple monitoring, the research projects known as CETI (Communication with Extraterrestrial Intelligence) pose the question of the methods necessary to initiate contact with any possible extraterrestrial civilization. The first example is the coded message sent by Frank Drake, using the Arecibo radio telescope in 1974, in the direction of the globular cluster M13. Two frequencies were selected, close to 2380 MHz. The message was coded in 1679 bits to generate an image of 23 by 73 pixels (Fig. 9.25). The Arecibo message is of symbolic rather than real interest, but it represents a classic example of the encoding of information being sent to an unknown recipient.

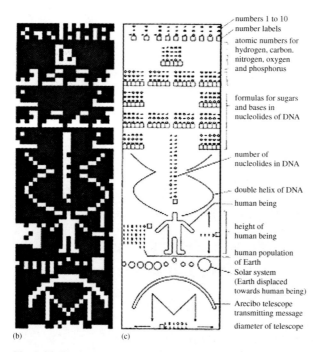

numbers 1 to 10
number labels
atomic numbers for hydrogen, carbon, nitrogen, oxygen and phosphorus

formulas for sugars and bases in nucleolides of DNA

number of nucleolides in DNA

double helix of DNA
human being

height of human being

human population of Earth
Solar system (Earth displaced towards human being)
Arecibo telescope transmitting message
diameter of telescope

(b)                (c)

**Fig. 9.25** The binary message sent by the Arecibo telescope in 1974

# Bibliography

Charbonneau, D., Brown, T.M., Noyes, R.W., Gilliland, R.L., 'Detection of an extrasolar planet atmosphere', *Astrophys. J.*, **568**, 377–384 (2002)

Deming, D., Seager, S., Richardson, L.J., Harrington, J., 'Infrared radiation from an extrasolar planet', *Nature*, **434**, 740–743 (2005)

De Muizon, J., Geballe, T.R., D'Hendecourt, L.B., Baas, F., 'New emission features in the infrared spectra of two IRAS sources', *Astrophys. J.*, **306**, L105–L108 (1986)

Forget, F., Costard, F., Lognonné, P., *La planéte Mars, histoire d'un autre monde, Bibliothéque Scientifique*, Belin, Paris, 144 p (2003)

Gilmour, I., Sephton, M.A., *An Introduction to Astrobiology*, Cambridge University Press, Cambridge (2003).

Gomes, R., Levison, H.F., Tsiganis, K. and Morbidelli, A., 'Origin of the cataclysmic late heavy bombardment period of the terrestrial planets', *Nature*, **435**, 466–469 (2005)

Kasting, J.F., 'Earth's early atmosphere', *Science*, **259**, 920–926 (1993)

Maurette, M., Duprat, J., Engrand, C., Gounelle, M., Kurat, G., Matrajt, G. and Toppani, A., 'Accretion of neon, organics, $CO_2$, nitrogen and water from large interplanetary dust particles on the early Earth', *Planetary and Space Science*, **48**, 1117–1137 (2000)

Martin, H., Claeys, P., Gargaud, M., Pinti, D. and Selsis, F., 'From suns to life 6. Environmental context', *Earth Moon and Planets*, **98**, 205–245 (2006)

O'Brien, D.P., Morbidelli, A. and Levison, H.F., 'Terrestrial planet formation with strong dynamical friction', *Icarus*, **184**, 39–58 (2006)

Raymond, S.N., Quinn, T. and Lunine, J.I., 'Making other earths: dynamical simulations of terrestrial planet formation and water delivery', *Icarus*, **168**, 1–17 (2004)

Rosenqvist, J., Chassefière, E., Inorganic chemistry of $O_2$ in a dense atmosphere, *Planet. Spa. Sci.*, **43**, 3 (1995)

Selsis, F., Paillet, J., Allard, F., 'Biomarkers of extrasolar planets and their observability, Extrasolar Planets: XVI Canary Islands Winter School of Astrophysics', Edited by Hans Deeg, Juan Antonio Belmonte and Antonio Aparicio. ISBN-13 978-0-521-86808-2 (HB). Cambridge University Press, Cambridge, UK, p.245 (2008)

Tian, F., Toon, O.B., Pavlov, A.A. and De Sterck, H., 'A hydrogen-rich early Earth atmosphere', *Science*, **308**, 1014–1017 (2005)

# Appendix A

## A.1 Star or Planet?

To simplify to the extreme, we may say that the physics of these objects is primarily governed by their masses, which are involved on two levels:

- the effect of gravity, which tends to compress the object, liberating gravitational energy,
- nuclear processes, which begin as and when the temperature rises in the object's core.

Mass is therefore an excellent parameter for classifying different astrophysical objects, the unit of comparison being one solar mass (denoted by $M_\odot$). We may then define three mass-categories, by decreasing mass:

- If $M > 0.08 M_\odot$ ($\sim 80 M_J$, where $M_J$ is the mass of Jupiter), the mass is sufficient for gravitational contraction to allow the object's core to reach the temperature at which hydrogen fusion reactions begin. The object is then described as a 'star', and its radius is proportional to its mass.
- If $0.013 M_\odot < M < 0.08 M_\odot$ ($13 M_J < M < 80 M_J$), the temperature at the core of the object does not start hydrogen fusion reactions, but does initiate deuterium fusion reactions. The object is called a 'brown dwarf', and its radius is inversely proportional to the cube root of its mass.
- If $M < 0.013 M_\odot$ ($M < 13 M_J$), the temperature at the core does not allow any nuclear fusion reactions to occur. The object is called a 'planet'. Within this category, a distinction is generally made between giant planets and terrestrial planets, where the mass of the latter is not sufficient for them to accrete any gas. The boundary between giant planets and terrestrial planets lies at about 10 Earth masses.

Notes:

- Unlike hydrogen fusion, deuterium fusion plays no role in determining the nature of the object. The limit of 13 $M_J$ between a planet and a brown dwarf is conventional (and also based on consensus).
- A planet is also (and always?) a body orbiting a star. There are brown dwarfs that are not bound to a central star.

## A.2 Gravitation and Kepler's Laws

Kepler's three laws, although predating Newton's theory, result from the law of universal gravitation, which postulates that 2 masses $m_1$ and $m_2$, separated by a distance $R$, exert on one another an attractive force $\vec{F}$, parallel to the radius vector $\vec{R}$ that joins their respective centres of mass (this force being described as 'central'). This force is expressed by the relationship:

$$\vec{F} = \left( \frac{G m_1 m_2}{R^2} \right) \left( \frac{\vec{R}}{R} \right)$$

where $G$ is the universal constant of gravitation ($= 6.67 \times 10^{-11} N.m^2.kg^{-2}$).

Kepler's laws are always valid for a system consisting of two bodies, but remain valid for multiple systems (several planets) when the approximation is made that the planets are of negligible mass relative to the central star. The laws were advanced as applying to the Solar System, but may be generalized to apply to any planetary system.

The first law, known as the law of orbits (1605) states that in the heliocentric reference frame, the orbit of each planet is an ellipse, of which one focus is occupied by the Sun.

The second law, known as the law of areas (1604) states that the motion of each planet is such that the section of the straight line joining the Sun and the planet (the radius vector) sweeps out equal areas in equal times.

The third law, known as the law of periods (1618) states that for all the planets, the ratio of the cube of the semi-major axis ($a$) of the orbit and the square of the orbital period ($T$) is constant, and is expressed as:

$$\frac{a^3}{T^2} = C^{te} = \frac{G(m_{star} + m_{planet})}{4\pi^2} \approx \frac{Gm_{star}}{4\pi^2}$$

where $m_{star}$ and $m_{planet}$ are the masses of the star and planet, respectively.

## A.3 Black-Body Emission – Planck's Radiation Law – Stefan's Law

By definition, a black body is an ideal physical body, isolated, consisting of a medium that is in thermodynamical equilibrium, and with a unique equilibrium temperature. It is a perfect absorber and an ideal emitter. A black body's radiation field is isotropic, and depends solely on the temperature. The spectral distribution of the radiation intensity is given by the Planck function, which gives the monochromatic brightness at a frequency $v$ of the black body as a function of its temperature $T$:

$$B_\nu = \frac{2h\nu^3}{c^2 e^{\left(\frac{h\nu}{kT}-1\right)}} \qquad W \cdot m^{-2} \cdot sr^{-1} \cdot Hz^{-1}$$

where $\nu$: frequency in Hz

$h$: Planck constant $= 6.62620 \times 10^{-34}$ J.s
$c$: velocity of light $= 2.9979 \times 10^8$ m.s$^{-1}$
$k$: Boltzmann constant $= 1.38 \times 10^{-23}$ J.K$^{-1}$

The Planck function may be drawn for various values of $T$ (Fig. A.1). This function reaches a maximum, which depends on the temperature. So each temperature of the black body may be associated with a colour (the wavelength at the peak emission) and, conversely, by determining the maximum emission of a black body, it is possible to determine what is known as the black-body temperature.

By integrating the Planck function over all frequencies and in all directions, it is possible to determine the total power (or flux) emitted by a black body at temperature $T$. This relationship is known as the Stefan-Boltzmann law:

$$F = \sigma T^4 \qquad W.m^{-2}$$

where $\sigma =$ Stefan's constant $= 5.66956 \times 10^{-8}$ W.m$^{-2}$.K$^{-4}$.

Conversely, for every source emitting a flux $F$ (as measured by a bolometer, for example), we may derive a temperature known as the 'effective temperature' $T_{eff}$, obtained from F by use of Stefan's law.

A direct application of Stefan's law is the calculation of the temperature $T_{eff}$ of a planet in radiative equilibrium, of radius $R_{pl}$, with a mean albedo (reflection coefficient) $A$, lying at a distance $D$ from a star that emits a flux $S$. The equality

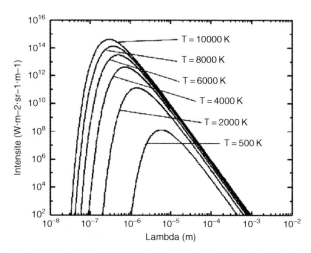

**Fig. A.1** Graphical representation of Planck's radiation law for different black-body temperatures. The higher the temperature of the body, the more the peak emission is shifted towards shorter wavelengths

between the flux received from the star and the flux emitted by the planet may be written in terms of the parameters mentioned.

$$S(1-A)\,\frac{\pi R_{pl}^2}{D^2} = 4\pi R_{pl}^2\,\sigma\,T_{eff}^4$$

It will be seen, in particular, that the equilibrium temperature ($T_{eff}$) does not depend on the size of the planet, but uniquely depends on the flux from the star, the albedo, and the planet's distance.

## A.4 The Hertzsprung-Russell Diagram and the Spectral Classification of Stars

The Hertzsprung-Russell diagram (known in what follows as the HR diagram) takes its name from the Danish astronomer E. Hertzsprung, who established it in 1911, and from the American astronomer H.N. Russell, who independently rediscovered it in 1913. This diagram shows the absolute luminosity of a star (that is, independent of its distance from Earth) as a function of the effective temperature, or of any other quantity linked to it (such as, for example, the difference in luminosity of the object measured through two coloured filters of different colours and therefore of different passbands). This diagram is shown in Fig. A.2. The specific feature of this diagram is that it is simultaneously a tool for visualizing both the morphological diversity of stars, and equally their evolution over the course of time. In fact, stars occupy a position on the HR diagram that evolves from their birth to their death.

Of the stars that we see, 90 per cent are at a stage of their lives that we may term 'adult' and occupy a portion of the HR diagram that is known as the 'Main Sequence' (the portion encircled in Fig. A.2). This is the case with our Sun, which was born about 5000 million years ago, and will die in about 5000 million years time. The special feature of stars on the Main Sequence is that there is an unequivocal relationship between their effective temperature, their absolute luminosity, their mass, and their lifetime. As effective temperature and colour are directly linked by Stefan's law, we can see that the HR diagram is therefore an extremely powerful tool, because, from the colour of an object, it allows us to deduce the other characteristics. For example, the average lifetime ($t_{life}$) of a star depends on its luminosity or its mass via the relationship:

$$t_{life} \approx 10^{10}\left(\frac{M_\odot}{M_\odot}\right)^2 \approx 10^{10}\left(\frac{L_\odot}{L_\odot}\right)^{2/3}$$

where $M_\odot$, $M_*$, $L_\odot$ and $L_*$ are the masses and absolute luminosities of the Sun and the star, respectively.

The HR diagram may be used to illustrate the concept of spectral classification, which consists of distinguishing stars from spectroscopic criteria. The best-known and most-used spectral classification is the 'Harvard' one, which dates from 1872.

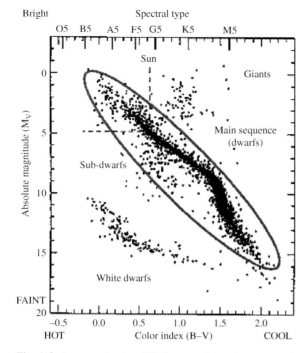

**Fig. A.2** An example of an HR diagram: the main sequence is circled

This classification, solely based on the temperature of stars, divides them into 9 categories, of which the first 7 (O, B, A, F, G, K, and M) cover about 99 per cent of the stars in the sky. Table A.1 describes the different categories in the Harvard classification. This classification is now extended to the coolest objects (stars of very low mass and sub-stellar objects) with the introduction of types L and T.

Each class is divided into 10 sub-classes, numbered 0–9, from the hottest stars to the coolest (a G0 star is hotter than a G9 star).

The same spectral classes are used to describe every star, whatever their type. However, to differentiate between objects (the effective temperature of a G0 giant is not the same at than of a star of the same G0 spectral class that is on the Main Sequence), the spectral type is followed by a number, in Roman numerals, describing the nature of the object, with the following convention:

- I: supergiants
- II: luminous giants
- III: giants
- IV: sub-giants
- V: dwarfs
- VI: sub-dwarfs
- VII: white dwarfs

**Table A.1** The Harvard spectral classification (after Léna, 1996)

| Spectral type | $T_{eff}$ of star | Colour of star | Spectroscopic signature |
|---|---|---|---|
| O | > 25 000 K | Blue | Ionized helium (He II) |
| B | 11 000–25 000 K | Blue-white | Neutral helium |
| A | 7500–11 000 K | White | Hydrogen |
| F | 6000–7500 K | Yellow-white | Hydrogen (weaker than in A stars), ionized calcium |
| G | 5000–6000 K | Yellow (∼ Sun) | Ionized calcium (Ca II), H, and K, CH bands |
| K | 3500–5000 K | Orange-Yellow | Metal lines |
| M | < 3500 K | Red | Molecular bands of titanium oxide TiO |
| C (rare) | < 3500 K | Carbon stars | Molecular bands of $C_2$, CN, CH, but no TiO |
| S (rare) | < 3500 K | Red | Zirconium oxide ZrO |

These different categories correspond to different groups of stars that are easily identifiable on the HR diagram.

The Sun, in this classification is a G2V star with $T_{eff} = 5777$ K.

## A.5 Resonances

When an oscillating system is subject to an excitation (a periodic perturbation), it may experience an increase in the energy of oscillation. This phenomenon, known as resonance, depends on the excitation frequencies and the natural frequencies of the system.

We speak of orbital resonances when the phenomenon is the motion of a planet around a star (or a satellite around a planet). The excitation is the gravitational perturbation by another planet (or by another satellite).

Several parameters in the elliptical orbit may give rise to resonances:

- The primary one is the motion of the planet along its orbit. The mean velocity is $n$ (known as the 'mean motion'). A mean-motion resonance occurs if the mean motions of two planets $n_1$ and $n_2$ form a rational ratio, i.e., if:

$p.n_1 + q.n_2$ is close to 0, p and q being integers.

This configuration is frequently found in the Solar System:

$$3.n_{(Neptune)} - 2.n_{(Pluto)} = 0$$

Such a relationship may involve several objects. The satellites of Jupiter, Io, Europa, and Ganymede are linked by the relationship:

$$1.n_{(Io)} - 3.n_{(Europa)} + 2.n_{(Ganymede)} = 0$$

Resonances may give rise to a paradoxical effect: despite the fact that gravitation is a force of attraction, the cumulative effect of resonance may lead to a force of repulsion. When a body orbits within a ring or disk, the particles of the ring are subject to resonances of higher and higher order as they approach the body. The combined effect of these resonances is a force of repulsion, which causes the edge of the ring to recede from the body's orbit.

- Resonances may also link the mean motion of a body and its rotation. A spectacular example is the synchronous rotation of many of the planetary satellites and, in particular, of the Moon. The mean motion of the Moon around the Earth is equal to its rotation velocity, which causes it to present the same face to the Earth at all times. Tidal forces have created frictional forces within the Moon, which have braked its rotation until it was in synchronous rotation, a configuration that causes the friction to disappear.
- Resonances may also link the motion of periapses or ascending nodes. These resonances are said to be 'secular' because these motions are slower than the mean motion.

# Index

# ASTRONOMY AND ASTROPHYSICS LIBRARY

**Series Editors:** G. Börner · A. Burkert · W. B. Burton · M. A. Dopita
A. Eckart · T. Encrenaz · E. K. Grebel · B. Leibundgut
A. Maeder · V. Trimble

**The Stars** By E. L. Schatzman and F. Praderie

**Modern Astrometry** 2nd Edition
By J. Kovalevsky

**The Physics and Dynamics of Planetary Nebulae** By G. A. Gurzadyan

**Galaxies and Cosmology** By F. Combes, P. Boissé, A. Mazure and A. Blanchard

**Observational Astrophysics** 2nd Edition
By P. Léna, F. Lebrun and F. Mignard

**Physics of Planetary Rings** Celestial Mechanics of Continuous Media
By A. M. Fridman and N. N. Gorkavyi

**Tools of Radio Astronomy** 4th Edition, Corr. 2nd printing
By K. Rohlfs and T. L. Wilson

**Tools of Radio Astronomy** Problems and Solutions 1st Edition, Corr. 2nd printing
By T. L. Wilson and S. Hüttemeister

**Astrophysical Formulae** 3rd Edition (2 volumes)
Volume I: Radiation, Gas Processes and High Energy Astrophysics
Volume II: Space, Time, Matter and Cosmology
By K. R. Lang

**Galaxy Formation** 2nd Edition
By M. S. Longair

**Astrophysical Concepts** 4th Edition
By M. Harwit

**Astrometry of Fundamental Catalogues**
The Evolution from Optical to Radio Reference Frames
By H. G. Walter and O. J. Sovers

**Compact Stars.** Nuclear Physics, Particle Physics and General Relativity 2nd Edition
By N. K. Glendenning

**The Sun from Space** By K. R. Lang

**Stellar Physics** (2 volumes)
Volume 1: Fundamental Concepts and Stellar Equilibrium
By G. S. Bisnovatyi-Kogan

**Stellar Physics** (2 volumes)
Volume 2: Stellar Evolution and Stability
By G. S. Bisnovatyi-Kogan

**Theory of Orbits** (2 volumes)
Volume 1: Integrable Systems and Non-perturbative Methods
Volume 2: Perturbative and Geometrical Methods
By D. Boccaletti and G. Pucacco

**Black Hole Gravitohydromagnetics**
By B. Punsly

**Stellar Structure and Evolution**
By R. Kippenhahn and A. Weigert

**Gravitational Lenses** By P. Schneider, J. Ehlers and E. E. Falco

**Reflecting Telescope Optics** (2 volumes)
Volume I: Basic Design Theory and its Historical Development. 2nd Edition
Volume II: Manufacture, Testing, Alignment, Modern Techniques
By R. N. Wilson

**Interplanetary Dust**
By E. Grün, B. Å. S. Gustafson, S. Dermott and H. Fechtig (Eds.)

**The Universe in Gamma Rays**
By V. Schönfelder

**Astrophysics.** A New Approach 2nd Edition
By W. Kundt

**Cosmic Ray Astrophysics**
By R. Schlickeiser

**Astrophysics of the Diffuse Universe**
By M. A. Dopita and R. S. Sutherland

**The Sun** An Introduction. 2nd Edition
By M. Stix

**Order and Chaos in Dynamical Astronomy**
By G. J. Contopoulos

**Astronomical Image and Data Analysis**
2nd Edition By J.-L. Starck and F. Murtagh

**The Early Universe** Facts and Fiction
4th Edition By G. Börner

# ASTRONOMY AND ASTROPHYSICS LIBRARY

**Series Editors:**  G. Börner · A. Burkert · W. B. Burton · M. A. Dopita
A. Eckart · T. Encrenaz · E. K. Grebel · B. Leibundgut
A. Maeder · V. Trimble